2016 International Conference on VLSI Systems, Architectures, Technology and Applications (VLSI-SATA 2016)

Bengaluru, India
10 – 12 January 2016

IEEE Catalog Number: CFP1684Z-POD
ISBN: 978-1-5090-0037-1

Copyright © 2016 by the Institute of Electrical and Electronics Engineers, Inc
All Rights Reserved

Copyright and Reprint Permissions: Abstracting is permitted with credit to the source. Libraries are permitted to photocopy beyond the limit of U.S. copyright law for private use of patrons those articles in this volume that carry a code at the bottom of the first page, provided the per-copy fee indicated in the code is paid through Copyright Clearance Center, 222 Rosewood Drive, Danvers, MA 01923.

For other copying, reprint or republication permission, write to IEEE Copyrights Manager, IEEE Service Center, 445 Hoes Lane, Piscataway, NJ 08854. All rights reserved.

***This publication is a representation of what appears in the IEEE Digital Libraries. Some format issues inherent in the e-media version may also appear in this print version.*

IEEE Catalog Number: CFP1684Z-POD
ISBN (Print-On-Demand):   978-1-5090-0037-1
ISBN (Online):   978-1-5090-0033-3

Additional Copies of This Publication Are Available From:

Curran Associates, Inc
57 Morehouse Lane
Red Hook, NY 12571 USA
Phone: (845) 758-0400
Fax: (845) 758-2633
E-mail: curran@proceedings.com
Web: www.proceedings.com

Contents

FPGA Implementation of Face Recognition System using Efficient 5/3 2D-Lifting Scheme   1
Satish Bhairannawar, Rajath Kumar, Varsha Mirji and Sindu P.S.

NEDA Based Hybrid Architecture for DCT - HWT   6
Vidhya Chandran, Mamatha I. and Shikha Tripathi

Security Situational Aware Intelligent Road Traffic Monitoring Using UAVs   12
Reshma R., Tirumale K. Ramesh and Sathish Kumar

Automatic Pressure Maintenance System for Tyres in Automobiles to Reduce Accidents   18
Rajesh Kannan Megalingam, Jayakrishnan C., Sriraj Nambiar, Rudit Mathews,
Vishnu Das and Pramesh Rao

Effect on Temperature and Time In Parallel Test Scheduling With Alterations In Layers
Arrangements of 3D Stacked SoCs   24
Indira Rawat, M.K. Gupta and Virendra Singh

Functional Verification of DSP based On-board VLSI Designs   30
Sourabh Kumar Jain, Parimal Govani and Kamal Poddar

An Arbitration on Cache Replacements Based on Frequency - Recency Product Values   34
Somak Das and Aikatan Banerjee

Clock skew optimized VLSI Architecture for Zero Frequency Filter   40
Radhakrishnan K.R. and Subha Rani S.

Effect of Split Manufacturing on Power Supply Requirements   46
Sharath K. Rangan, Shazia Afreen and Raghuram Srinivasan

Design of CMOS Programmable Output Binary and Fibonacci Switched Capacitor
Step-down DC-DC Converter   51
Mahesh Zanwar and Subhajit Sen

Design and Analysis of Novel Fuzzifer Circuits in CMOS Current Mode Approach   57
Deeksha M. and Abdullah Gubbi

Performance of Asymmetric Gate Oxide on Gate-Drain Overlap in Si and $Si1-xGex$ Double
Gate Tunnel FETs   63
Poorvasha S. and Lakshmi B.

CORDIC-based VLSI Architecture for implementing CI-OFDM and Its FPGA Prototype   67
Vikas Kumar, Kailash Chandra Ray and Preetam Kumar

A Hardware optimized Low power RNM Compensated three stage Operational amplifier with
Embedded Capacitance Multiplier Compensation   72
Karan Raj Singh and Anu Gupta

Design and Analysis of Different Low Noise Amplifiers in 2-3GHz   78
Prameela B. and Asha Daniel

Fully-Digital Time based ADC/TDC in 0.18um CMOS 84
Vineet Sharma, Nupur Jain and Biswajit Mishra

A Portable Platform to Estimate Power Consumption of Software Modules 90
Abhishek Bhardwaj and Saket Saurav

Towards Formal Verification of Adaptive Cruise Controller using SpaceEx 96
Ambuj Mishra and Subir K. Roy

Efficient Network on Chip (NoC) using heterogeneous circuit switched routers 102
Anuja Naik and Tirumale K.Ramesh

A circuit technique for leakage power reduction in CMOS VLSI circuits 108
Venkata Ramakrishna Nandyala and Kamala Kanta Mahapatra

A Gain Enhanced Low Voltage Bulk Driven Pseudo-Differential OTA design in CMOS 113
Antaryami Panigrahi and Abhipsa Parhi

Non-Intrusive FPGA based Profiler for Loop Execution Characterization 118
Pavan Kumar Nadimpalli and Subir K. Roy

A RISC-V Instruction Set Processor-Micro-architecture Design and Analysis 124
Aneesh R., Vinayak Patil, Sobha P.M., David Selvakumar and Vivian Desalphine

Reconfigurable Side Channel Attack resistant True Random Number Generator 131
Vijay Bahadur, Vijendran N., Sobha P.M. and David Selvakumar

RF Tracking Test System Design for Closed Loop Testing of Ku-Band Antenna 137
Rahul Mishra

Tunable Distributed Harmonic Voltage Controlled Oscillator for Generating Second and Third
Harmonic Microwave Signals in 180nm CMOS 141
Kalyan Bhattacharyya

Performance Enhancement of Slot Synchronization in W-CDMA 145
Mridula Korde

Design of 3C-SiC Symmetric and Asymmetric Double Gate MOSFET 150
Sudarshana Jilowa, Sandeep Singh Gill and Gurjot Kaur Walia

A hierarchical cluster - based model with runtime reconfigurable resource allocation on FPGAs 155
Amin Yoosefi and Hamid Reza Naji

An Efficient VLSI Architecture for Data Encryption Standard and its FPGA Implementation 160
Jai Gopal Pandey, A. Gurawa, Heena Nehra and Abhijit Karmakar

Implementation of RNS and LNS Based Addition and Subtraction Units for cryptography 165
Satish Kumar Ch, Prathiba A. and Kanchana Bhaskaran V.S.

Real Time Watermarking of Grayscale Images using Integer DWT Transform 170
Sakthivel S.M. and Ravi Sankar A.

Ultra Low Power Capacitive Power Management Unit in 0.18μm CMOS 176
Sanjay Kasodniya, Biswajit Mishra and Nilesh Desai

Switching based evaluation of substrate current in lightly and heavily doped CMOS at 45nm   181

Sanjay Sharma, R.P. Yadav and Vijay Janyani

OptMem: Dark-Silicon Aware Low Latency Hybrid Memory Design   184

Salman Onsori, Arghavan Asad, Kaamran Raahemifar and Mahmood Fathy

Ultra Low Power 12-Bit SAR ADC for Wireless Sensing Applications   189

Rajahari Gudlavalleti and Subash Chandra Bose

Distance Estimation and Direction Finding Using I2C Protocol for an Auto-navigation Platform   193

Rajesh Kannan Megalingam, Jeeba M. Varghese and Aarsha Anil

Mathematical Modeling and Analysis of New Modified Glitch Free Adiabatic Inverter Circuit
with Trapezoidal Power Supply   197

Alak Majumder and Rahul Kaushik

Design and Implementation of Reconfigurable Coders for Communication Systems   202

Manikandan J., Shruthi S., Mangala S. Joshi and Dr. V.K. Agrawal

2016 International Conference on

VLSI Systems, Architecture, Technology and Applications

VLSI SATA 2016
10 - 12 January 2016
Bengaluru, India

Technical Sponsors:

Organized by
Amrita Vishwa Vidyapeetham (University), School of Engineering, Bengaluru Campus

Preface

We are delighted to welcome you at Amrita Vishwa Vidyapeetham University, Bengaluru campus. We extend a very warm welcome to all the delegates participating in 2nd IEEE International Conference on VLSI Systems, Architecture, Technology and Applications (VLSI-SATA 2016). Bengaluru is known as Silicon Valley of India. The ecosystem of this city has attracted many key players in the VLSI and Embedded System domains to make their presence felt. It is a pleasure to organize a Conference in this hub of technology, innovation and entrepreneurship.

The 1st conference held in January 2015 was an astounding success. Our aim is to make this conference a flagship event. We have made all the efforts to maintain high standards in terms of quality of accepted papers, tutorials and keynote talks. This event will bring together several researchers and experts from academia, R&D organisations and industry professionals. We expect it to be a great opportunity and an inspiring occasion for learning, networking and collaborating.

The conference is organized in collaboration with IEEE Bangalore Section, VLSI Society of India, IETE and IESA. We are also indebted to our sponsors who have generously supported us in this endeavor. The success of the conference depends on the people who worked with us in planning and organizing the technical program and social logistics. We thank our Advisory Board Members and Program Chairs for providing valuable suggestions. We are highly indebted to the Technical Program Committee members for their thorough and timely reviewing of the papers and providing constructive feedbacks. We express our sincere thanks to the Tutorial Speakers, Keynote Speakers, Panel Members and Technical Session Chairs for being part of this conference.

Our sincere thanks to the Management and Administration of Amrita Vishwa Vidyapeetham (University) for all the organizational support to conduct this conference.

With immense pleasure and great pride we welcome all the delegates, speakers and invitees to VLSI-SATA 2016. We hope that the conference will provide stimulating discussions, networking and cooperation.

Steering Committee

VLSI-SATA 2016

About the Conference

Department of Electronics and Communication Engineering & Department of Computer Science & Engineering, Amrita School of Engineering, Amrita Vishwa Vidyapeetham Bengaluru Campus are jointly organizing the **Second IEEE International Conference on VLSI Systems, Architecture, Technology and Applications (VLSI SATA 2016)** from 10th - 12th January, 2016.

The conference will serve as an annual forum for researchers, academicians, and practitioners from around the world to present their current theoretical research efforts, system and design solutions and practical applications in VLSI.

The conference with a theme of **"System Solutions for Emerging Applications"**, aims to provide a common platform for discussing broader VLSI topics to better understand the integration of VLSI circuits and systems for today's emerging applications.

VLSI SATA 2016 will be a 3 day event that will include Day 1 devoted to tutorials to provide a broader technical platform to discuss recent topics in VLSI. Day 2 and Day 3 will include key note addresses and parallel technical sessions in three tracks: Systems and Architecture, Technology, and Applications with invited talks and contributed paper presentations. In addition, Day 2 and Day 3 will include industry exhibits from industrial sponsoring partners to the conference.

Topics to be discussed in this conference include (but are not limited to) the following:

- Multi / Many Core Architecture
- System-on-Chip and Network-on-Chip
- Heterogeneous Architectures and Models
- Embedded Systems
- Verification and Validation
- FPGA/GPU Hybrid Computing and FPGA Based System Design
- VLSI Testing and Reliability
- Novel and special purpose device structures
- Crypto VLSI Hardware and VLSI Hardware Security
- Video and Image Processing Applications
- Low Power VLSI Circuits
- Computer-Aided Design (CAD)
- 3D IC
- Hardware/Software Co-Design, Synthesis and Verification
- Energy-Efficient and Fault-Tolerance Design
- RF and analog and mixed signal circuit design

Organizing Committee

Patrons

Br. Abhayamrita Chaitanya, Pro Chancellor, Amrita University, India.

Dr. Venkat Rangan, Vice Chancellor, Amrita University, India.

Dr. Sasangan Ramanathan, Dean - Engineering, Amrita University, India.

Dr. Krishnashree, Dean - PGP, Amrita University, India.

Br. Dhanraj, Director, Amrita University, Bengaluru Campus.

Dr. Rakesh S.G., Associate Dean, Amrita University, Bengaluru Campus.

Steering Committee

T.S.B. Sudarshan, Amrita University, Bengaluru Campus, India

Shikha Tripathi, Amrita University, Bengaluru Campus, India

Conference Chair & Co-Chair

Chair - N.S. Murty, Amrita University, Bengaluru Campus, India

Co-Chair - Gopal Krishna, Maxim Integrated , Bengaluru, India

Technical Committee

Chair - Navin Kumar, Amrita University, Bengaluru Campus, India

Co-Chair - Madhura P., Amrita University, Bengaluru Campus, India

Co-Chair - Ananda Mohan, Chairman, IEEE CAS Bangalore Chapter, India

Publications Committee

Chair - Radhakrishnan G., Amrita University, Bengaluru Campus, India

Co-Chair - Vinodhini M., Amrita University, Bengaluru Campus, India

Co-Chair - Nandi Vardhan, Amrita University, Bengaluru Campus, India

Web Design Committee

Chair - Rajesh M., Amrita University, Bengaluru Campus, India

Co-Chair - C. Babu, Amrita University, Bengaluru Campus, India

Industry Liaison Committee

Chair - Sekhar Babu, Amrita University, Bengaluru Campus, India

Co-Chair - Shanmuga Rajan, Amrita University, Bengaluru Campus, India

R&D Organization Liaison

Chair - Dhanesh G. Kurup, Amrita University, Bengaluru Campus, India

Co-Chair - Sanjika Devi R.V. , Amrita University, Bengaluru Campus, India

Finance Committee

Chair - Rakesh N., Amrita University, Bengaluru Campus, India

Co-Chair - Kirti S. Pande, Amrita University, Bengaluru Campus, India

Publicity Committee

Chair - Karthikeyan R., Amrita University, Bengaluru Campus, India

Co-Chair - Nippun Kumaar A.A., Amrita University, Bengaluru Campus, India

Co-Chair - Sonali Agrawal, Amrita University, Bengaluru Campus, India

Local Arrangements Committee

Chair - Giriraja C.V., Amrita University, Bengaluru Campus, India

Co-Chair - Vignesh V., Amrita University, Bengaluru Campus, India

Co-Chair - Sathish Kumar P., Amrita University, Bengaluru Campus, India

Co-Chair - Ponni M., Amrita University, Bengaluru Campus, India

Tutorials Committee

Chair - Ramesh T.K., Amrita University, Bengaluru Campus, India

Co-Chair - Ganapathy Hegde, Amrita University, Bengaluru Campus, India

Student Research Forum Committee

Chair - Amudha J., Amrita University, Bengaluru Campus, India

Exhibition Committee

Chair - S. Ravishankar, Amrita University, Bengaluru Campus, India

International Advisory Committee

- Ahmad Taher Azar, Benha University, Egypt
- Alan George, University of Florida, USA
- Bill Grundmann, Xilinx, USA
- Bob Doering, Texas Instruments, USA
- Brian Bailey, Brian Bailey Consulting, USA
- Dennis Brophy, Mentor Graphics, USA
- Jimson Mathew, University of Bristol, UK
- Marius M. Balas, University "Aurel Vlaicu" of Arad, Romania
- Nagi Naganathan, Avago Tech., USA
- Paolo Ienne, EPFL, Switzerland
- Radhakrishnan M.K., Editor in Chief IEEE EDS Newsletter
- Raghuram Tupuri, AMD, USA
- San Murugesan, Cutter Consultancy, Australia
- Sandro Rigo, IC-UNICAMP, Campinas, Brazil
- Sartaj Sahni, University of Florida, USA
- Sridhar R., SUNY Baffalo, USA
- Steven Drager, Airforce Research Lab, USA
- Subhasish Mitra, Stanford University , USA
- Subra Ganesan, Okland University, USA
- Sundaram K.B., UCF, USA
- Tirumale Ramesh – Chair, Advanced Computing Consultant, USA
- Ulrich Rueckert, University of Bielefeld, Germany
- Valentina Balas, University "Aurel Vlaicu" of Arad, Romania
- Viktor Prasanna, University of Southern California, USA
- Zine El Abidine Alaoui Ismali, Mohamed V Souissi,Morocco

National Advisory Committee

- A. Ravi Kiran, Chair, IEEE Bangalore Section
- Ajitha Kumari, CVC, Bengaluru
- C.P. Ravikumar, Texas Instruments, Bengaluru
- Chandrasekhar, CEERI, Pilani
- Debabrata Das, IIIT, Bengaluru
- Guru Ganesan, ARM, Bengaluru
- Jaswinder Ahuja, Cadenc, Noida
- Jayakumar, Amrita University, Coimbatore Campus
- Kamakoti V., IIT, Madras
- Latha Parameswaran, Amrita University, Coimbatore Campus
- M.H. Kori, Chairman, TPCC
- Madhusudhan Atre, Vegashakthi Consultants
- Maneesha Ramesh, Amrita University, Amritapuri Campus
- Pamela Kumar, HP, Bengaluru
- Rajkumar P. Sreedharan, Amrita University, Bengaluru Campus
- Ramachandran Kaimal, Amrita University, Amritapuri Campus
- S.K. Nandy, IISc., Bengaluru
- Seetharam D.R., Synopsis, Bengaluru
- Shyam Vasudev, Forus Healthcare
- Smriti Dagur, President, IETE
- Sundar Gopalan, Amrita University, Amritapuri Campus
- Sundararajan Srinivasan, Intel, Bengaluru
- Vilas Bhade, Graphene Semiconductors

Technical Programme Committee

- M Achutha Kirankumar V, Intel
- Mahesh Panicker, GE
- Mario Porrmann, Bielefeld University
- Navin Kumar, Institute of Telecommunication, University of Aveiro
- Nirmala Devi, Amrita Vishwa Vidyapeetham, Coimbatore
- N S Murthy, NITW
- Pa Govindacharyulu, VCE, Hyderabad
- Paolo Ienne, EPFL-IC-LAP
- Pradeep Salla, Mentor Graphics
- Prasanna Kesavan, Broadcom Communications
- K Jayaram, Maxim Integrated
- Subir K Roy, IIIT-Bangalore
- Rajshekher Mitra, Synopsys
- Ramdas Mozhikunnath, Applied Micro
- Saif Abrar, IBM
- Sekhar G Yanamala, Intel
- Shashidhara PG
- Shivaling S Mahant-Shetti, Karnataka Miroelectronics Design Center Pvt Ltd., Manipal
- Shivraj Dharne, Intel
- Smitha Kavallur Pisharath Gopi, Nanyang Technological University
- Srikrishnan Venkataraman, Intel Corporation
- Srinath Naidu, IIIT Bangalore
- Subhajit Sen, International Institute of Information Technology
- Subhendu Kumar Sahoo, BITS Pilani Hyderabad campus
- Sumit Dara, IIT Delhi
- Swapna Banerjee, IIT Kharagpur
- Tirumale Ramesh, Consultant, Advanced Computing, USA
- Udaybhaskara Rao, IRAM TECHNOLOGIES PVT LTD
- Ulrich Rueckert, Bielefeld University
- Usha Mehta, Nirma University, Ahmedabad
- Vasanthanayaki C, Govt College of Technology, Coimbatore
- Vikas Mishra, Intel Technology India Pvt. Ltd
- Vinod Madan, MIET Jammu

FPGA Implementation of Face Recognition System using Efficient 5/3 2D-Lifting Scheme

Satish S Bhairannawar, Rajath Kumar,
Dept. of E&C, DSCE,
Bengaluru, India
rajatshetty999@gmail.com

Varsha Mirji, Sindhu P S
Dept. of Electronics and Communication,
DayanandaSagar College of Engineering,
Bengaluru, India

Abstract—**Face recognition is gaining more importance in today's real world for automated transactions. In this paper, we propose FPGA Implementation of Face Recognition System using Efficient 5/3 2D-Lifting scheme. The database image of FVC-2004 DB3_A is resized to 256x256 pixels. The resized image is convolved with 3x3 Gaussian mask kernelsto remove high frequency edges, which improves matching accuracy. The proposed 5/3 2D-Lift DWT is used to extract LL band features of 128x128 coefficients. Similarly, the test image LL band features are extracted and are compared with LL band features of database image using Euclidean distance classifier for accurate matching. The proposed face recognition architecture is implemented on Virtex5 xc5vlx110-2ff676 board. It is observed that the performance parameters such as area and speed are better compared to existing architectures.**

Keywords-Face Recognition; 2D-DWT; 5/3 Lift Scheme; FPGA; Euclidean Distance.

I. INTRODUCTION

Biometric is a unique, measurable characteristic of a human body that can be used to automatically recognize or verify an individual identity. The unique biometric features play an importantrole in accurate classification for big data analysis. The demand for real time biometric identification is increasing for the growing size of population for which recognition rate and response time becomes a critical criteria. Face is a unique and important characteristic, as it does not require human interaction for extraction. Human's ability to recognize another person looking at his face does not deteriorate with time. Hence, face recognition technique can be considered robust. Face recognition proves to be the more beneficial compared to other biometrics since the degree of security is improved. The social gestures like fever, happiness, sadness of a person can be recognized only by looking at the face which is responsible for the gestures, suggesting that face is an important feature for comparison in society. This comparison is the basic principle when it comes to security issues like fighting terrorism. The use of Field Programmable Gate Array (FPGA) for real time face recognition application is increasing due to its flexibility for parallel processing which enhances the speed of the system. The FPGA based face authentication interfacing to the cloud computing isrequired for future transactions suitable for its high-speed recognition and also to secure the confidential data of authorized cloud computing users.

The Efficient 5/3 2D-Lift architecture is designed and is used to extract LL band features of database face images. These features of database images are further compared with test image feature for high speed and accurate matching. In section II, some related work is discussed. In section III, the detailed proposed methodology offace recognition is described, section IV presents the results and discussions of our design and in section V conclusions are discussed.

II. RELATED WORKS

Dongil Han et al., [1] proposed architecture of face detection engine for mobile applications using Modified Census Transform and Adaboost learning technique. The proposed implementation has advantages like low cost, low power consumption and high performance. Chengjun Zhang et al., [2] proposed pipelined architecture for fast computation of the 2D-DWT implemented using low cost hardware. The task of computation is distributed among various decomposition levels, speeding up the process by maximizing the inter-stage and intra-stage computational parallelism.

Azadeh Safari et al., [3] proposed VLSI architecture of multiplier less Bi orthogonal DWT image processor which is capable of processing any image irrespective of its size with less storage space and high processing rate, in turn minimizing the chip overhead. Maria Angelopoulonet al., [4] proposed 5/3 lifting 2D-DWT computation schedules on FPGAs based on various input patterns. The lifting based filter-bank implementations optimize the throughput as well as the memory requirements.

John Wright et al., [5] proposed a classification algorithm for object recognition based on a sparse representation computed by l-minimization, which addresses two crucial issues in face recognition: face extraction and robustness to disturbances. This implementation is capable of handling error due to corruption achieving striking recognition performance with respect to error rates and recognition accuracy. Veera manikandan et al., [6] proposed a real-time face recognition system, where preprocessing like conversion of RGB to Gray scale image, followed by difference of Gaussian, masking and equalization normalization is performed and feature extraction is done using sub-space technique. The results obtained proved to be efficient in terms of speed and accuracy of recognition.

III. PROPOSED METHODOLOGY

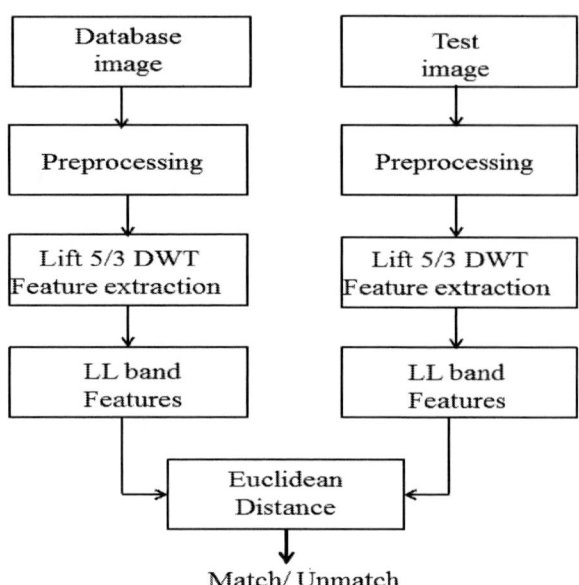

Fig 1. Proposed Methodology

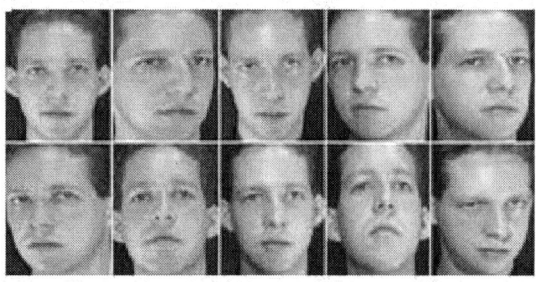

Fig 2. Samples of ORL face images of person

The proposed methodology for face recognition is shown in Fig 1.

A. Database

The face images are considered from the standard ORL Database. The Database consists of 40 persons with 10 samples per person. Each sample is of size 112x92 as shown in Fig 2.

B. Preprocessing

The preprocessing basically deals with Resizing, Blurring, Edge detection, Sharpening, Smoothing, etc. that results in increasing matching accuracy. The image is resized to 256x256 since it reduces the on chip memory of FPGA.The moving window architecture is used to obtain 3x3 pixel matrix of the input pixels which is then convolved with the Gaussian mask to remove high frequency edges in face resulting in smoothened image as shown in Fig. 3. For every clock cycle new

overlapping matrices are obtained through moving window to cover entire 256x256 images.The convolution of image pixel matrices with Gaussian filter mask for noise removal is given by equation 1.

$$y(n) = \frac{1}{16}\sum_{i=1}^{n} x(i) \times h(i) \qquad (1)$$

x(i) – Image pixels.

h(i) – Mask filter coefficients.

The convolution of 3x3 image pixels matrix with Gaussian filter mask coefficients of 3x3 with scaling factor of 1/16 is as shown in Fig. 4.

C. Feature Extraction

The features are extracted using transform domain technique. The Lift 5/3 DWT transform, which is used to extract the features, is as shown in Fig 5.

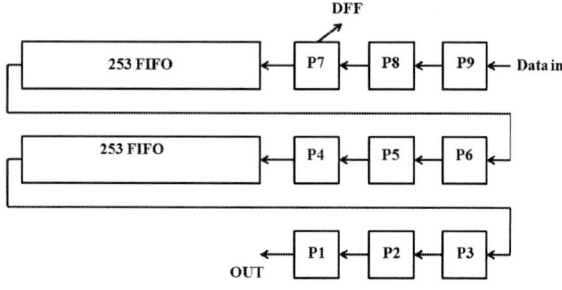

Fig 3. Moving window architecture

Image pixels				Mask filter		
P1	P2	P3		1	2	1
P4	P5	P6	* 1/16	2	4	2
P7	P8	P9		1	2	1

Fig 4. Gaussian architecture

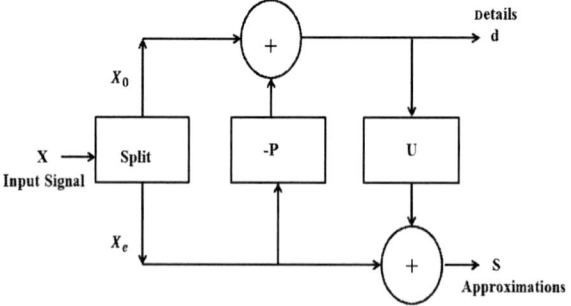

Fig 5. Lift based approach

The three stages of Lift 5/3 [4] involves

- *Split:*The input samples are split into even data samples and odd data samples.

  Odd data samples: $d_i^0 = x_{2i+1}$

  Even data samples:$s_i^0 = x_{2i}$ (2)

  Input data samples:

  i.e.,$x = \{x_1, x_2, \ldots \ldots \ldots, x_{2n}\}$

  Split into

  $x_e = \{x_2, x_4, \ldots \ldots, x_{2n}\}$ and $x_o = \{x_1, x_3, \ldots \ldots, x_{2n-1}\}$

- *Predict*: The odd samples are predicted using the even samples as given in equation 3

$$d_i^1 = d_i^0 + \left(\frac{1}{2}\right)(s_i^0 + s_{i+1}^0) \qquad (3)$$

- *Update*: The original odd and the predicted odd samples are subtracted and later updated using the even samples, giving the approximated coefficients using equation 4.

$$s_i^1 = s_i^0 - \left(\frac{1}{4}\right)(d_{i-1}^1 + d_i^1) \qquad (4)$$

1. Proposed 1D-Lift DWT Architecture

The filter coefficient values of predict and update stages given by equation 3 and 4 are modified to obtain low pass and high pass coefficients given by equations 5 and 6 respectively.

$$h_{LPF}(n) = [-0.125, 0.25, 0.75, 0.25, -0.125] \qquad (5)$$

$$h_{HPF}(n) = [-0.25, 0.5, -0.25] \qquad (6)$$

The Low pass filter coefficients of equation 5 is modified into powers of 2 ratios and multiplied with input samples given by equation 7, which leads to optimization of hardware using simple adders and shifters and is shown in Fig 6.The 1D-Lift DWT architecture consists of 5 shifters, 2 adder units and 4 D-flip-flops(Dff).The necessary shifting operations are performed in accordance with the filter coefficients considered. Later, the addition/subtraction operations are done and the required LPF features are obtained.

$$y_{LPF}(n) = -\frac{1}{8}x(n) + \frac{1}{4}x(n-1) + \frac{3}{4}x(n-2) + \frac{1}{4}x(n-3) - \frac{1}{8}x(n-4)$$

$$(7)$$

2. Proposed 2D-Lift DWT Architecture

The proposed 1D-Lift DWT architecture is extended to 2D-Lift DWT architecture using Mux, Demux, Memory unit and Control unit. The input data samples of 256x256 face image are fed through a multiplexer for which the select line (rd_wr) is set to 0. The low pass filtered output on the line LPF is fed to 1-D DWT block where the lift algorithm is applied yielding 128x256 features as row compression output and are stored in memory unit. The control unit is used to obtain transpose of 1D image to achieve column compression. The output of memory is again fed to 1D-DWT by setting rd_wr to 1. The 1D-DWT

compresses the features in column wise to obtain 2D features of 128x128.The clk_out signal is the clock divided signal used to synchronize the system architecture. The counter with output rst_out will be high for some counts until the garbage value is present. This rst_out signal is used for matching, which indicates that the $\overline{LL}$_band output will not be considered until the rst_out is low. The proposed 2D-DWT Lift architecture is as shown in Fig 7.

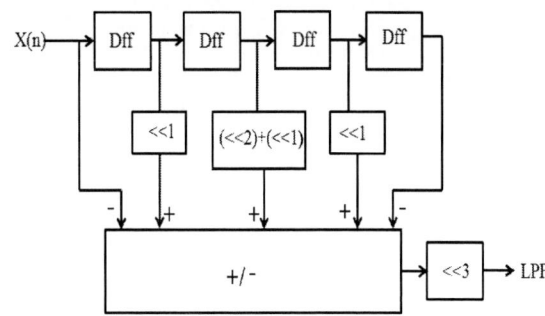

Fig 6. Proposed 1D-DWT Architecture

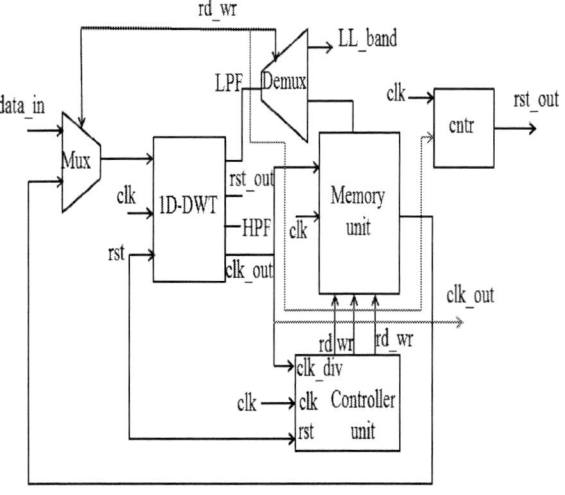

Fig 7. Proposed 2D-Lift DWT Architecture

TABLE I. COMPARISON OF PROPOSED 2D-DWT WITH EXISTING ARCHITECTURE

Architecture	Multiplier	Shifter	Adder
Wu et al., [7]	16	0	16
Shriram et al.,[8]	3	18	6
Proposed Work	0	18	12

978-1-5090-0037-1/16 $31.00 © 2016 IEEE   3

The performing of Lift 5/3 DWT involves complex operations such as multiplication of large number of filter coefficients by the filter input or its time shifted versions, which in turn consumes more hardware as well as time. It is observed that the proposed method is better in terms of speed and area compared to existing methods since it uses multiplier lessarchitecture, which are replaced by shifters and adders.

3. Controller Unit

The controller unit is used to generate address for the memory to store the features of Lift DWT in memory. The proposed controller unit consists of *three*counters, where the *first* counter counts to store 1D feature of 32768 pixels and the *second* and *third* counter is used to obtain transpose which stores 16384 2D features.

D. Matching

The Euclidean distance between database and test image features is computed for matching using equation 8.

$$\sqrt{\left(D_1(i) - D_1(j)\right)^2 + \cdots + \left(D_n(i) - D_n(j)\right)^2} \qquad (8)$$

D(i) – Database image pixel
D(j) – Test image pixel

IV. RESULTS AND SIMULATIONS

A. Simulation

The simulation was performed using MATLAB 2012a and the results of the Total Success Rate (TSR) obtained for best threshold of 0.16 is 93.33% for an ORL database of 15 persons taken randomly with 4 images per person. The comparison of percentage TSR of existing technique with proposed method is as shown in TABLE II. It can be observed from TABLE II that the proposed method has higher Total Success Rate (TSR) when compared to Ajay et al., [9].

B. Hardware Utilization

The VHDL language was used to design the hardware modules and is synthesized using Virtex5 xc5vlx110-2ff676 FPGA board. The hardware utilization of the proposed method is shown in Table III. It is observed from the above table that the Preprocessing unit uses 186 slice registers, 218 LUTs and 25 LUT-FF pairs. The 2D-Lift DWT uses 213 slice registers, 4899 LUTs and 186 LUT-FF pairs and the matching unit uses 47 slice registers, 64 LUTs and 30 LUT-FF pairs. The VHDL language was used to design the hardware modules and is synthesized using Virtex5 xc5vlx110-2ff676 FPGA board. The hardware utilization of the proposed method is shown in Table III. It is observed from the above table that the Preprocessing unit uses 186 slice registers, 218 LUTs and 25 LUT-FF pairs. The 2D-Lift DWT uses 213 slice registers,

4899 LUTs and 186 LUT-FF pairs and the matching unit uses 47 slice registers, 64 LUTs and 30 LUT-FF pairs.

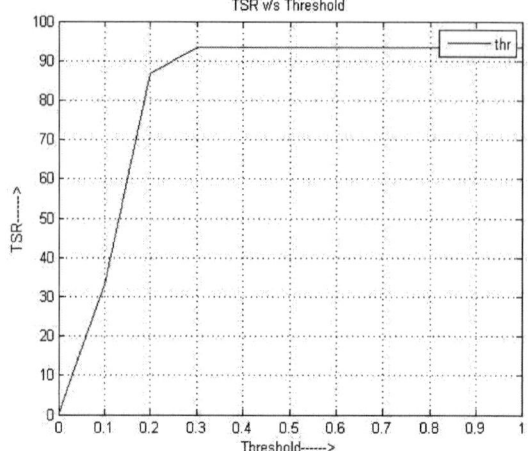

Fig 8. TSR v/s Threshold

TABLE II. COMPARISON % TSR OF EXISTING TECHNIQUES WITH PROPOSED METHOD

Authors	Techniques	%TSR
Ajay et al.,[9]	N-PCA	92.50
	PCA	90.00
Proposed method	LIFT 5/3 DWT	93.33

TABLE III. HARDWARE UTILIZATION OF PROPOSED METHOD

Hardware Module	No. of Slice Registers	No. of Slice LUTs	No. of LUT-FF Pairs
Preprocessing unit	186	218	25
2D-Lift DWT	213	4899	186
Matching unit	47	64	30

TABLE IV. DEVICE UTILIZATION AND FREQUENCY COMPARISON OF EXISTING TECHNIQUES WITH PROPOSED METHOD

Logic Utilization	Senthil Singh et al., [10]	Proposed Work
No. of slice registers	302	645
No. of slice LUTs	2643	5485
Speed in MHz	207.009	258.35

978-1-5090-0037-1/16 $31.00 © 2016 IEEE

The device utilization and frequency comparison between proposed and existing techniques is shown in Table IV. It is observed that the proposed methodology utilizes 645 slice registers and 5,485 slice LUT's with the maximum operating frequency of 258.35MHz, which is comparatively very less indicating efficient utilization of the hardware with a much higher frequency. The high performance of proposed method is due to the output of 1D-DWT is not stored on-chip memory. Hence it reduces the use of hardware. The speed of the proposed work when compared to Senthil Singh.C et al., [10] is high due to multiplier less architecture.

V. CONCLUSION

In this paper, we propose efficient FPGA based face recognition system. The preprocessing such as resizing and Gaussian smoothening is performed for both database and test images. The extracted LL band features using proposed multiplier less 5/3 2D-Lift DWT architecture are compared using Euclidean distance classifier for accurate matching. It is observed that the proposed system performs better compared to existing architectures.

REFERENCES

[1] Dongil Han, Jongho Choi, Jae-Il Cho and Dongsu Kwak, "Design and VLSI Implementation on of High-Performance Face-Detection Engine for Mobile Applications", IEEE International Conference on Consumer Electronics, pp. 705-706, 2011.

[2] Chengjun Zhang, Chunyan Wang, and M. Omair Ahmad, "Pipeline VLSI Architecture for Fast Computation of the 2-D Discrete Wavelet Transform", IEEE Transactions on Circuits and Systems – I: Regular papers, Vol. 59, pp. 1775-1785, August 2012.

[3] Azadeh Safari, Niras C V and Yinan Kong, "VLSI Architecture of Multiplierless DWT Image Processor", IEEE Tencon- spring,Vol. 1, No. 2, pp. 280–284, 2013.

[4] Maria E. Angelopoulou and Peter Y.K. Cheung, "Implementation and Comparison of the 5/3 Lifting 2D Discrete WaveletTransform Computation Schedules on FPGAs", Journal of VLSI Signal Processing, Vol. 51, Issue. 1, pp. 3-21,2007.

[5] John Wright, Allen Y. Yang, Arvind Ganesh, S. Shankar Sastry and Yi Ma, "Robust Face Recognition via Sparse Representation", IEEE Transaction on Pattern Analysis and Machine Intelligence, Vol. 31, No. 2, February 2009.

[6] K. Veeramanikandan, R. Ezhilarasi and R. Brinda, "An FPGA-Based Real-time Face Detection and Recognition System Across Illumination", International Journal of Emerging Science and Engineering, Vol. 1, Issue. 5, pp. 66-68, March 2013.

[7] P.C. Wu and L.G. Chen, "An efficient architecture for two-dimensional discrete wavelet transform",IEEE Transaction on Circuit and Systems and Systems for Video Technology, Vol. 11, No. 4, pp. 536-545, 2011.

[8] Shriram Hegde and S Ramachandran, "FPGA Implementation of CDF 5/3 Wavelet Transform", International Journal of Electrical, Electronics and Data Communication (IJEEDC), Vol. 2, Issue. 11, pp.36-38, November 2014.

[9] Ajay Kumar Bansal and Pankaj Chawla, "Performance Evaluation of Face Recognition using PCA and N-PCA", International Journal of Computer Applications, Vol. 76, No.8, pp. 14-20, August 2013.

[10] Senthilsingh. C and Manikandan. M, " Design and Implementation of an FPGA-Based Real-Time Very Low Resolution Face Recognition System", International Journal of Advanced Informmation Science and Technology, Vol. 7, No. 7, pp. 59-65, November 2012.

NEDA Based Hybrid Architecture for DCT - HWT

Vidhya Chandran, Mamatha I , Shikha Tripathi
Amrita School of Engineering, Bangalore Campus
Amrita Vishwa Vidyapeetham (University)
{vidhucnair, mamraj78, shikha.eee}@gmail.com

Abstract— Transforms are used in many signal processing applications. The VLSI implementation of a hybrid architecture to compute 8-point discrete cosine transform and Haar wavelet transform is proposed . The architecture is developed using NEw Distributed Arithmetic (NEDA) which is an efficient method for implementing inner products without using multipliers and ROM. The architecture developed is coded using Verilog HDL, simulated in ModelSim 6.4 and implemented using Xilinx ISE 14.7. Further, the hybrid architecture is implemented in 0.18μm CMOS technology using Cadence RTL compiler. Compared to standalone architectures, proposed architecture has 77.92% saving in register utilization, 41.80% savings in LUT utilization and 27.55% savings in number of adders used. The results show that the architecture is better in terms of power, hardware resources and complexity compared to earlier architectures.

Keywords—NEw Distributed Arithmetic, Discrete cosine transform, Haar Wavelet Transform.

I. INTRODUCTION

The modern communication filed demands a huge amount of image and video processing which takes large storage space if the data is stored in raw form. Also, such raw data requires large transmission bandwidth over the network. Data compression is the most efficient technique to reduce the storage space and to improve the speed of transmission. Network congestion can be avoided using data compression and reconstruction. Discrete Cosine Transform (DCT) and Wavelet Transform are the two transforms used for data compression.

In [1] authors presented a DCT architecture by replacing multiplications with a minimum number of additions and shifting operations which resulted in better frequency of operation, reduced resource usage and dynamic power consumption. Similar approach presented in [2] proven to be area and power efficient architecture for DCT (Discrete Cosine Transform) for a video conferencing application on mobile devices.A Canonical Signed Digit(CSD) representation for fixed point DCT coefficient is used for multiplier realization.. Liu et.al in [5] proposed a DA based 8x8 1-D architecture implemented in 0.18 μm CMOS technology operating at 20 MHz.

A convolution based 1-D DWT proposed by Ramsey Hourani in [4] achieved high throughput at low power dissipation with an acceptable design area. Data-interleaving technique used for reduced area and poly-phase structures for lower power dissipation, and a combination of pipelining and V_{DD} scaling to further reduce power and improve throughput. Zhang et.al in [5] proposed high speed and reduced 2-D DWT

using lifting scheme. Qing Sun in [6] proposed a reconfigurable DWT architecture adaptable to different kinds of filter banks and different set of inputs.

Chen et.al in [7] proposed a sparse matrix decomposition technique for designing architecture for 1-D DCT/DFT/Haar Wavelet transforms. Further, a hybrid architecture by exploiting common structures among the three transforms was developed. Another hybrid architecture for DCT/DFT/Wavelet transform using element-wise matrix factorization and row-permutation algorithm was proposed by Wahid et.al [8,9]. A NEw Distributed Arithmetic (NEDA) technique proposed in [10] computes inner products without using multipliers and ROMs. In [11], the implementation of 1-D DCT using NEDA, considering retiming technique was proposed.

A hybrid architecture using NEDA for 1-D DCT and 1-D HWT is proposed in this work. An 8 point Haar wavelet matrix and an 8 point DCT IV matrix is considered for developing the architecture. Hardware is optimized by using an array of shared adders between the transforms and few dedicated adders. Compared to the existing designs, the proposed hybrid scheme is low power and faster

Rest of the paper is organized as follows: Section II gives a brief introduction on NEDA. Section III explains the proposed architecture. Section IV discusses the results obtained and performance comparison with other similar approaches. Section V concludes the work.

II. A BRIEF INTRODUCTION ON NEDA

New Distributed Arithmetic (NEDA) is an efficient method for implementing inner products without using multipliers and ROM proposed in [10]. The basic building block in many digital signal processing systems is Multiply and Accumulate (MAC) unit. The general structure of a MAC unit consists of a multiplier, an adder and a shifter. The algorithms such as Distributed Arithmetic (DA) eliminate MAC multipliers. NEw Distributed Arithmetic (NEDA) technique eliminates the usage of ROM unit. Thus NEDA eliminates the use of multipliers and ROM to carry out many computations such as Fast Fourier Transform (FFT), Discrete Cosine Transform, etc. Mathematical derivation of NEDA technique is described below. The sum of products can be expressed as:

$$Y = \sum_{i=1}^{L} A_k \times X_k \qquad (1)$$

where A_k are fixed coefficients and X_k are input data words.

978-1-5090-0037-1/16 $31.00 © 2016 IEEE

$$A_k = -A_k^M \, 2^M + \sum_{i=N}^{M-1} A_k^i \, 2^i \qquad (2)$$

Substituting (2) in (1), Y can be written as

$$Y = \sum_{k=1}^{L}(A_k^M \, 2^M + \sum_{i=N}^{M-1} A_k^i \, 2^i)X_k$$

$$= \left[\sum_{k=1}^{L}(A_k^M \, X_k)\right] 2^M + \sum_{i=N}^{M-1}\sum_{k=1}^{L}(-A_k^i \, X_k)2^i$$

$$= -(Y^M 2^M) + \sum_{i=N}^{M-1} Y^i \, 2^i$$

Where i=N,N+1,....., M

In Fig. 1 the architecture implementing NEDA is shown. The sign extended ($\log_2 L$ bits) inputs are fed to the adder matrix, the outputs of which is determined by DA precision. A multiplexer is used to chose one of the outputs of the adder butterfly and fed to the 'shift and add' module, which produces the final output Y. Only addition and shift operations are involved in this architecture. Shifting is done at final stage for achieving significant savings in hardware. Further, unlike conventional DA based architectures, NEDA based architectures are scalable. Extra DA bit outputs can be added if higher precision is required.

Consider matrix [A] as a sparse matrix. Adder arrays can be compressed by realizing matrix [A] which is a critical step in NEDA. The following algorithm is used to implement the sparse matrix [A] which can be used to reduce the number of addition operations:

The initial step in this algorithm is to calculate the coefficient DA matrix [A] considering the DA precision. Each and every element in matrix [A] will be binary digits of 0's and 1's. Next step in this algorithm is to detect rows of same content so that only one row can be considered in the generation of the adder array. Next we search for a pair of rows with maximum number of 1's at the same position which can be represented using an adder. Similarly the number of adders can be reduced. After this all 1's which are already computed are replaced by 0's. This row matching operation is continued until there remains a single entry of "1" is left out. The single entry of "1" indicates that it is the output line. For all rows with "0" entry indicates that no processing is required. Finally a butterfly structure can be made using this compression scheme.

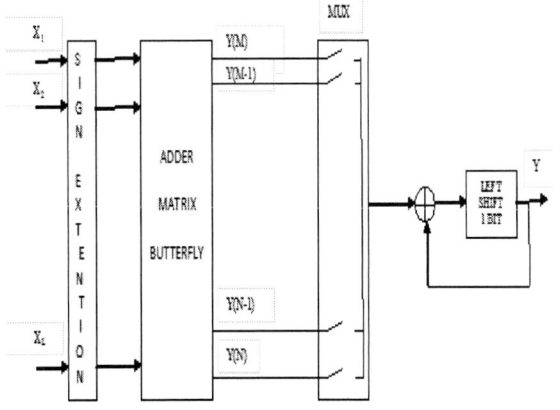

Fig. 1. Block diagram of NEDA architecture[10]

III. PROPOSED ARCHITECTURE

Using the NEDA technique, an architecture is developed for 1-D DCT and 1-D HWT. Further, a hybrid architecture is designed by exploiting the commonality between the two transform structures leading to resource sharing.

A. 8x8 DCT Implementation using NEDA

8x1 DCT is defined as:

$$F(u) = \frac{c(u)}{2} \sum_{x=0}^{7} f(x)\cos\frac{(2x+1)u}{16} \qquad (3)$$

$$\text{where } c(u) = \begin{cases} \frac{1}{\sqrt{2}}, & for \; u = 0 \\ 1, & otherwise \end{cases}$$

Assume that

$$A_u = [A_u(x)] \qquad \text{where x=0,1,2,.....7}$$

and

$$A_u(x) = \cos\frac{(2x+1)u}{16}$$

$$F(u) = \frac{c(u)}{2} \sum_{x=0}^{7} [A_u(x)]f(x)$$

$$= \sum_{x=0}^{7} A_u \, f(x). \qquad (4)$$

Each and every coefficient is converted to binary. DA precision is chosen to be 13 and $F_u(0)$ indicates the sign bit and $F_u(-12)$ represents the LSB.DA matrix for coefficient A_0 is shown below :-

$$A_0 = [A_0(0) \; A_0(1) \; A_0(2) \; A_0(3) \; A_0(4) \; A_0(5) \; A_0(6) \; A_0(7)]$$

$$= \begin{bmatrix} 0 & 0 & 0 & 0 & 0 & 0 & 0 & 0 \\ 0 & 0 & 0 & 0 & 0 & 0 & 0 & 0 \\ 0 & 0 & 0 & 0 & 0 & 0 & 0 & 0 \\ 1 & 1 & 1 & 1 & 1 & 1 & 1 & 1 \\ 0 & 0 & 0 & 0 & 0 & 0 & 0 & 0 \\ 1 & 1 & 1 & 1 & 1 & 1 & 1 & 1 \\ 0 & 0 & 0 & 0 & 0 & 0 & 0 & 0 \\ 1 & 1 & 1 & 1 & 1 & 1 & 1 & 1 \\ 1 & 1 & 1 & 1 & 1 & 1 & 1 & 1 \\ 0 & 0 & 0 & 0 & 0 & 0 & 0 & 0 \\ 1 & 1 & 1 & 1 & 1 & 1 & 1 & 1 \\ 0 & 0 & 0 & 0 & 0 & 0 & 0 & 0 \\ 0 & 0 & 0 & 0 & 0 & 0 & 0 & 0 \end{bmatrix}$$

The top row of A_0 represents the LSB's of $A_0(x)$, for x = 0,1, ... , 7 and the bottom row represents the sign bits of $A_0(x)$. By examining A_0, we can observe that there are 5 rows with eight 1's. So it needs 35 adders since 7 adders are required for each row. Since each row performs the same operation only 7 adders are sufficient and this result can be shared among all other rows also. This is how compression is done. The architecture for $F_0 = A_0 * x[n]$ is shown below. As shown in Fig. 2(a), only 7 adders are required and the result is shared with $F_0(-2)$, $F_0(-4)$, $F_0(-5)$, $F_0(-7)$. and $F_0(-9)$, allowing for a compression ratio of 5:1 in number of addition operations.

2016 International Conference on VLSI Systems, Architectures, Technology and Applications (VLSI-SATA)

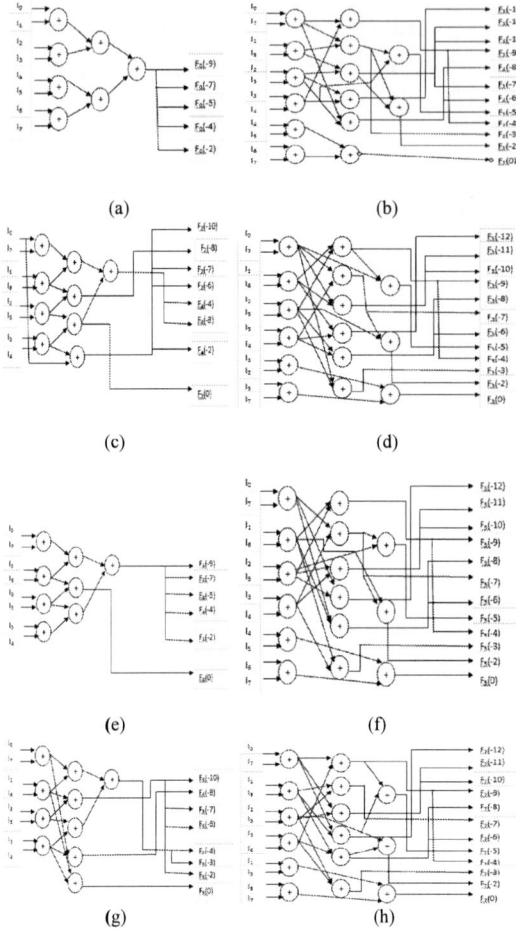

(a)   (b)

(c)   (d)

(e)   (f)

(g)   (h)

Fig. 2. Butterfly structures for computing DCT

Similar way, F_1 to F_7 are represented as a matrix showing bitwise representation and an adder butterfly is obtained as shown in Fig. 2(b) - Fig.2(h).

B. 8x8 HWT Implementation using NEDA

An un-normalized 8-point HWT matrix can be represented as:

$$H_8 = \begin{bmatrix} 1 & 1 & 1 & 1 & 1 & 1 & 1 & 1 \\ 1 & 1 & 1 & 1 & -1 & -1 & -1 & -1 \\ 1 & 1 & -1 & -1 & 0 & 0 & 0 & 0 \\ 0 & 0 & 0 & 0 & 1 & 1 & -1 & -1 \\ 1 & -1 & 0 & 0 & 0 & 0 & 0 & 0 \\ 0 & 0 & 1 & -1 & 0 & 0 & 0 & 0 \\ 0 & 0 & 0 & 0 & 1 & -1 & 0 & 0 \\ 0 & 0 & 0 & 0 & 0 & 0 & 1 & -1 \end{bmatrix}$$

8x8 HWT is defined as:

$$Y[K]=$$

$$\begin{bmatrix} r^3 & r^3 & r^3 & r^3 & r^3 & r^3 & r^3 & r^3 \\ r^3 & r^3 & r^3 & r^3 & -r^3 & r^3 & r^3 & r^3 \\ r^2 & r^2 & -r^2 & -r^2 & 0 & 0 & 0 & 0 \\ 0 & 0 & 0 & 0 & r^2 & r^2 & -r^2 & -r^2 \\ r & -r & 0 & 0 & 0 & 0 & 0 & 0 \\ 0 & 0 & r & -r & 0 & 0 & 0 & 0 \\ 0 & 0 & 0 & 0 & r & -r & 0 & 0 \\ 0 & 0 & 0 & 0 & 0 & 0 & r & -r \end{bmatrix} \times x[n],$$

where $x[n]$ is input data and $r = 1/\sqrt{2}$. DA matrix for coefficient H_0 is shown below where $H_0(0) = H_0(1) = H_0(2) = H_0(3) = H_0(4) = H_0(5) = H_0(6) = H_0(7) = 0.3535$.
$H_0 = [H_0(0)\ H_0(1)\ H_0(2)\ H_0(3)\ H_0(4)\ H_0(5)\ H_0(6)\ H_0(7)]$

$$= \begin{bmatrix} 0 & 0 & 0 & 0 & 0 & 0 & 0 & 0 \\ 0 & 0 & 0 & 0 & 0 & 0 & 0 & 0 \\ 0 & 0 & 0 & 0 & 0 & 0 & 0 & 0 \\ 1 & 1 & 1 & 1 & 1 & 1 & 1 & 1 \\ 0 & 0 & 0 & 0 & 0 & 0 & 0 & 0 \\ 1 & 1 & 1 & 1 & 1 & 1 & 1 & 1 \\ 0 & 0 & 0 & 0 & 0 & 0 & 0 & 0 \\ 1 & 1 & 1 & 1 & 1 & 1 & 1 & 1 \\ 1 & 1 & 1 & 1 & 1 & 1 & 1 & 1 \\ 0 & 0 & 0 & 0 & 0 & 0 & 0 & 0 \\ 1 & 1 & 1 & 1 & 1 & 1 & 1 & 1 \\ 0 & 0 & 0 & 0 & 0 & 0 & 0 & 0 \\ 0 & 0 & 0 & 0 & 0 & 0 & 0 & 0 \end{bmatrix}$$

The matrix obtained for F_0 and H_0 are same, thereby, adder butterfly structure is shared among them. Likewise H_1 to H_7 is calculated and adder butterflies are made as shown in Fig.3

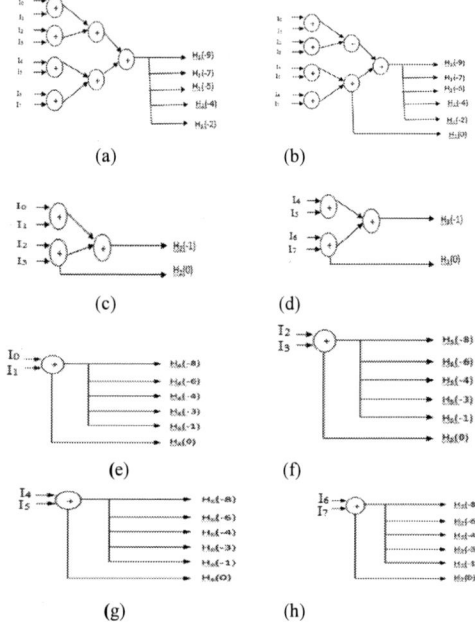

(a)   (b)

(c)   (d)

(e)   (f)

(g)   (h)

Fig. 3. Butterfly structures for computing HWT

IV. RESULTS AND DISCUSSION

The basic block diagram for the proposed hybrid architecture is shown in Fig. 4.

A recorded voice signal of 8kB is considered as input signal for this work. Initially the voice signal is converted to hex file by using Matlab. This hex file is then used for entire processing. For this proposed work, samples are represented using 4 bits and 8 samples are fed to the architecture simultaneously. The samples are stored in a hex file which is then loaded to RAM. The RAM gives 8 samples at a time. The remaining samples are loaded according to a control signal generated by the controller module. These 8 samples are then fed to the adder matrix butterfly

978-1-5090-0037-1/16 $31.00 © 2016 IEEE

2016 International Conference on VLSI Systems, Architectures, Technology and Applications (VLSI-SATA)

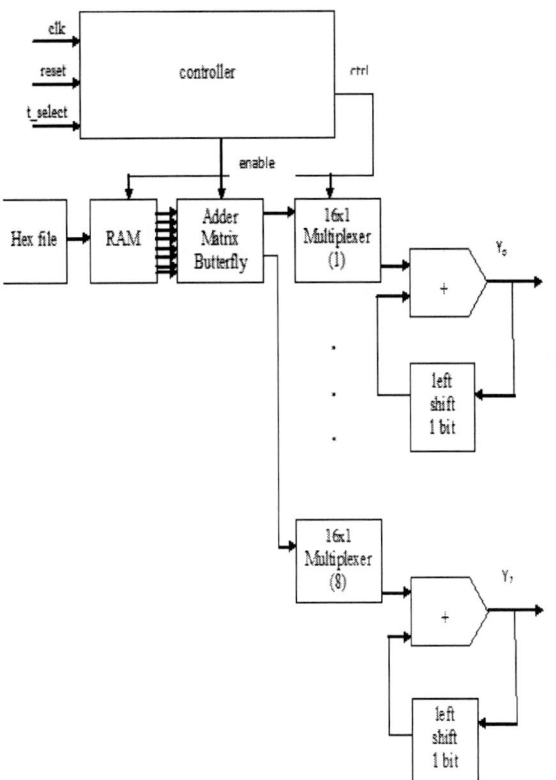

Fig. 4. Block diagram for proposed work

During simulation it takes 13 clock cycles to obtain the transform output after loading 8 samples. So the latency is 13 clock cycles. For better accuracy the output bit precision is chosen to be 17bits.

The architecture is simulated in ModelSim 6.4 and implemented using Xilinx ISE 14.7. Thus obtained results are compared with results from Matlab. The output obtained for t_select=0, which is for DCT using ModelSim 6.4 is shown below in Fig. 5. The output obtained when t_select = 1 which is for HWT in ModelSim 6.4 is shown in Fig. 6.

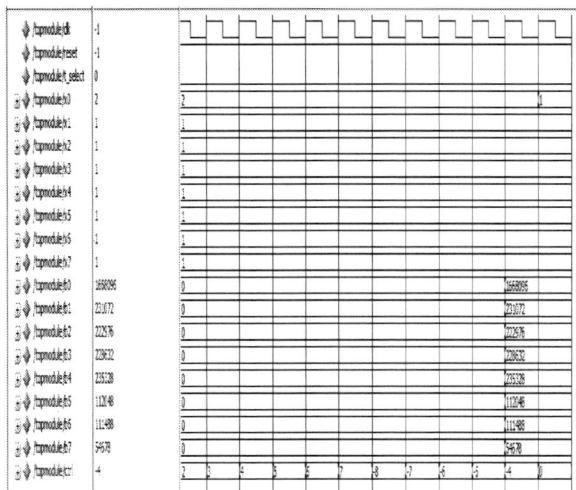

Fig. 5. Output obtained for DCT

The intermediate enable signal is generated by controller based on t_select pin. If the t_select = 0, then enable =0 which will give DCT coefficients and if t_select =1 ,then enable = 1 which will give HWT coefficients. Since the DA precision considered is 13 bits, the adder matrix butterfly will provide 13 intermediate outputs for each transform coefficient which is then fed to a 16x1 multiplexer. The intermediate output is fed to eight 16x1 multiplexer since 8 samples are processed in parallel. The select signal to the multiplexer is generated by the controller. The multiplexer provides 1 output at a time which is fed to adder. The first input to the adder is initialized by 0 and the other input is fed from the multiplexer. It is then added and the output is left shifted by 1 bit and fed back to the adder. Other input then changed according to the control signal of the multiplexer. The add and shift operation is continued for 13 clock cycles. After 13 clock cycles the output for first 8 samples are obtained. Likewise another 8 samples are loaded and after 13 clock cycles the transform output for these 8 samples can be obtained.

The complete architecture is controlled by the controller. The inputs to the controller are clock signal, reset and t_select pin. The control signals generated by controller are enable and ctrl. The controller starts generating controlling signals only when reset = 1. The enable signal will decide which transform to be obtained and it is fed to the adder butterfly. The ctrl signal is used in RAM as well as in 16x1 multiplexer. In RAM , this ctrl signal is provided to give a delay of 13 clock cycles between successive 8 samples whereas in 16x1 multiplexer it will act as a select signal to opt for an intermediate output from adder butterfly to be loaded to the adder.

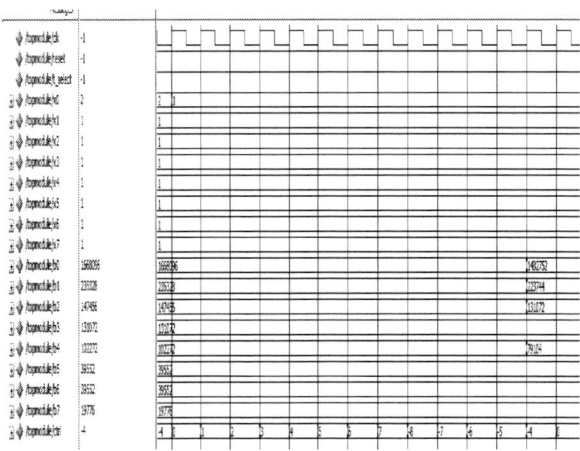

Fig. 6. Output obtained for HWT

The standalone and hybrid architectures are then synthesized in Xilinx ISE 14.7 and post synthesis place and route(PAR) report is summarized in Table I. Table II shows the power consumption details of the hybrid architecture on the target devices Spartan 6 at a clock frequency of 50 MHz.

Table II summarizes the resource utilization for the standalone architectures for DCT and HWT and the proposed hybrid architecture while synthesizing using Spartan 6 device. If the three transforms are implemented individually and the total hardware is compared against the hybrid architecture, the proposed architecture uses about 27.5% less adders, 41.8% less LUTs and 77.9% less registers compared to standalone units.

978-1-5090-0037-1/16 $31.00 © 2016 IEEE   9

TABLE I.   POST PLACE AND ROUTE(PAR) AND POWER REPORT

Logic Utilization	Spartan 6	
	Used	Available
No. of Slice registers	217	126576
No. of Slice LUT's	1158	63288
No. of LUT-FF pairs	163	1208
No. of Bonded IOB's	175	296
Static Power (W)	0.081	
Dynamic Power(W)	0.010	
Total Power (W)	0.091	

TABLE II.   COMPARISON BETWEEN STANDALONE AND HYBRID ARCHITECTURE (PAR REPORT FROM XILINX)

Topology	Transform	Adders	LUT	Registers
Standalone	DCT	84	1176	263
	HWT	43	814	231
	Total	127	1990	494
Hybrid	DCT-HWT	92	1158	213
Overall savings(%)		27.55	41.8	77.92

Table III shows comparison between various architectures in literature and the proposed architecture implemented on a FPGA target devices. Hybrid architecture by Wahid et.al is to implement 1-D DFT,1-D DCT and 1-D HWT. Proposed architecture although is for 1-D DCT and 1-D HWT, has almost 60-70% less hardware and can run at a maximum frequency of 2.5 times than that of [7]. Architectures by Sun et al [6] and Mankar et.al [11] are standalone architectures dedicated to single transform. Compared to these architectures, proposed hybrid architecture has almost half of the resource usage at a reduced power consumption of about 10%. It shows that the proposed architecture is better in terms of hardware complexity, frequency and power.

TABLE III.   COMPARISON BETWEEN PREVIOUS AND HYBRID ARCHITECTURE IN XILINX ISE 14.1

	Wahid et al.[2009] Hybrid DCT-DFT-DWT 2009	Qing Sun et al. [5] Convolution based DWT 2013	Mankar et al. [10] NEDA BASED DCT 2013	Proposed Hybrid DCT-HWT
Device used	Xilinx Virtex4 (XCV4LX 15SF363)	Xilinx Virtex6 (XC6VLX 240T)	Xilinx Virtex2 (XC2VP30 7FF896)	Xilinx Virtex4 (XCV4LX 15SF363)
No: of slice LUT's	2017	29875	1012	787
No: of slice registers	1479	23825	238	161
Maximum Frequency (MHz)	95	200	311.944	213.563
Power(w)	NA	NA	2.7599	0.206

Further, the hybrid architecture is implemented in 0.18μm CMOS technology using Cadence RTL compiler. Table IV shows the comparison between the various architectures in 0.18 CMOS technology in Cadence RTL Compiler operating at 1.8 V.

TABLE IV.   COMPARISON BETWEEN PREVIOUS AND HYBRID ARCHITECTURE IN CADENCE RTL COMPILER

	Liu [2007] DA based DCT	Wahid et al. [2010] Hybrid DCT- DFT-DWT	Proposed hybrid DCT - HWT
Technology used	0.18	0.18	0.18
Area (mm²)	0.601	0.20368	0.36703
Frequency (MHz)	20	100	200
Power (mW)	15.2	15.38	5.573

Proposed architecture has almost 50% of the area compared to standalone unit proposed in Liu et.al[3] and has almost 30% less power consumption. Proposed structure can operate at almost twice the frequency compared to the other architectures.

V. CONCLUSION AND FUTURE WORK

In this work, a hybrid architecture for 1-D Haar Wavelet Transform (HWT) and 1-D DCT using NEDA technique is designed and implemented. Compared with standalone architectures, 77.92% saving in register utilization, 41.80% savings in LUT utilization and 27.55% savings in number of adders is achieved by the proposed design. Proposed structure is efficient in terms of frequency of operation and power consumption compared to standalone or other similar hybrid architectures. Simulation is carried out using ModelSim SE 6.4 and the architecture is implemented using Xilinx 14.7 ISE using Virtex 4 and Spartan 6 as the target device. Two devices

978-1-5090-0037-1/16 $31.00 © 2016 IEEE

are chosen to enable reasonable comparison with existing architectures. The hybrid architecture is implemented in 0.18μm CMOS technology using Cadence RTL compiler. In order to process an 8-sample input sequence, the architecture requires 13 clock cycles.. The proposed architecture is for one dimensional transform and can be extended to 2-D or 3-D transforms as well.

REFERENCES

[1] El Aakif, M., S. Belkouch, N. Chabini, and M. M. Hassani. "Low power and fast DCT architecture using multiplier-less method." In IEEE Faible Tension Faible Consommation (FTFC), pp. 63-66, 2011

[2] Yuan-Ho Chen and Tsin-Yuan Chang,"A High Performance Video Transform Engine byUsing Space-Time Scheduling Strategy", IEEE transactions on very large scale integration (VLSI) systems, vol. 20, no. 4, pp: 655-664, April 2012

[3] Z. Liu, T. Arslan, T. Erdogan, A novel reconfigurable low power distributed arithmetic architecture for multimedia applications, in Proc. of the IEEE Design Automation Conference , pp. 908–913, 2007.

[4] Hourani, Ramsey, Ishita Dalal, W. Rhett Davis, Christopher Doss, and Winser Alexander. "An efficient VLSI implementation for the 1D convolutional discrete wavelet transform." In 51st Midwest Symposium on Circuits and Systems, pp. 870-873. 2008.

[5] Wei Zhang, Member, IEEE, Zhe Jiang, Zhiyu Gao, and Yanyan Liu, "An Efficient VLSI Architecture for Lifting-Based Discrete Wavelet Transform",IEEE Transactions on Circuits and Systems—II: Express briefs, vol. 59, no. 3, March 2012

[6] Sun, Qing, Jiang Jiang, Yongxin Zhu, and Yuzhuo Fu. "A Reconfigurable Architecture for 1-D and 2-D Discrete Wavelet Transform." In 2013 IEEE 21st Annual International Symposium on Field-Programmable Custom Computing Machines (FCCM), , pp. 81-84, 2013.

[7] Winser AlexanderZ. Chen, M.H. Lee," On fast hybrid source coding design", in Proc. of the International Symposium on Information Technology Convergence, pp. 143–147, 2007

[8] K. Wahid, S. Shimu, M. Islam, D. Teng, M. Lee, S.-B. Ko," Efficient hardware implementation of hybrid cosine-Fourier-wavelet transforms on a single FPGA", in Proc. of the IEEE International Symposium on Circuits and Systems, pp. 2325–2328, 2009

[9] Khan A. Wahid , M.A. Islam , Samia S. Shimu ,Moon Ho Lee , Seok-Bum Ko," Hybrid Architecture and VLSI Implementation of the Cosine–Fourier–Haar Transforms", Springer Circuits Syst Signal Process,vol. 29 , pp:1193-1205, 2010

[10] Pan, Wendi, Ahmed Shams, and Magdy Bayoumi. "NEDA: a new distributed arithmetic architecture and its application to one dimensional discrete cosine transform." In 1999 IEEE Workshop on Signal Processing Systems, (SiPS 9)., pp. 159-168,1999.

[11] Mankar, Abhishek, Narayan Prasad, and Ansuman Diptisankar Das. "FPGA implementation of retimed low power and high throughput DCT core using NEDA." In 2013 IEEE Students Conference on Engineering and Systems (SCES), , pp. 1-4, 2013.

Security Situational Aware Intelligent Road Traffic Monitoring Using UAVs

Reshma R*
M.Tech Scholar
Department of CSE
rshmr01@gmail.com

Dr. Tirumale Ramesh*
Prof Senior Member IEEE
Department of CSE
tiru_ramesh@blr.amrita.edu

P.Sathishkumar*
Assistant Professor
Department of ECE
p_sathishkumar@blr.amrita.edu

*Amrita Vishwa Vidhyapeetham,Bangalore,India

Abstract—Roadway networks span large distances and can be difficult to monitor for managing road traffic. Most efforts to collect usage data from roadways either require a large fixed infrastructure or are labour intensive. Numerous number of road traffic monitoring techniques exists today. Technological advances in electronics and communication have recently enabled an alternative, such as use of Unmanned Aerial Vehicles (UAVs). UAVs are very flexible and fast compared to the normal road traffic monitoring techniques. In addition, security situations can also impact the road traffic management. This paper presents a security situational aware intelligent traffic monitoring using UAV. Such security situational aware road traffic management can also support research and development of intelligent transportation systems (ITS). In this paper, various real time traffic as well as security issues are analysed and mitigations are presented as re-routing of the traffic. The decision commands for re-routing or shortest traffic paths are given by the UAV and these informations are sent to the traffic management control centre. The model was validated using ARM microcontroller simulation using Keil µvision and Proteus tools.

Keywords-Unmanned Aerial Vehicles (UAVs), Intelligent Transportation Systems, ARM microcontroller.

I. INTRODUCTION

An unmanned aerial vehicle [1] is an aircraft that flies without a human crew on board the aircraft. It uses state-of-the art microprocessors and microsensors to navigate and track targets. There are wide variety of UAVs and Micro UAV (MUAVs) in shapes, sizes, configurations, and characteristics. MUAVs come in two varieties: some are controlled from a remote location, and others fly autonomously based on pre-programmed flight plans using more complex dynamic automation systems. UAVs can be used in wide variety of applications like search and rescue operations [2], pollution monitoring [3], oil or gas pipeline monitoring etc. The road traffic monitoring using UAVs is an area of significant research these days. There has been voluminous research which proved that unmanned aerial vehicles (UAVs) is a viable and less time consuming alternative to real-time distributed traffic monitoring and management, providing the eye-in-the-sky solution to the problem. Some of the most commonly used road traffic monitoring techniques according to [4] are: (i) Airborne road traffic monitoring using radar. (ii) Statistical profile

generation for traffic monitoring using real-time UAV based video data. (iii)Vehicle detection from aerial imagery (iv) Visual MTI for UAV systems (v) A vision algorithm for dynamic detection of moving vehicles. Also, various types of image processing algorithms like the one according to [5], uses a histogram based adaptive threshold algorithm to detect possible road regions in an image. According to the authors in [6], UAVs are used to monitor roadway traffic and thereby develop and demonstrate several applications using data collected from a UAV flying in an urban environment.

The main emphasis of our work is to analyse the role of a UAV in a road traffic management system and mitigate various situational security issues involved in traffic management. In our approach, the UAV intelligence analyses various real time traffic as well as security issues and accordingly provide the required mitigation commands to the traffic management control centre for re-routing. Typically, UAVs use image processing algorithms for detecting the road ways and vehicles. But in this paper, we present an alternate simple approach to road traffic monitoring using UAVs by combining the concepts of sensor networks and graph theory representing the road network. Various situational security scenarios are formulated to validate the road traffic management.

The rest of the paper is organized as follows: Section II describes the overall system design that is used in the approach. Section III discusses the infrastructure graph models used in the system. These graph models are actually an abstracted view of the environment which is under the UAV surveillance. Section IV describes the system implementation and the results. Finally, conclusions and future research directions are given in section V.

II. SYSTEM DESIGN

There is an increased demand for intelligent traffic information for real time traffic monitoring and management. This work basically aims at developing a security situational aware intelligent traffic monitoring using unmanned air vehicle. The system analyses various real time traffic as well as security issues and accordingly provide the required mitigation or re-routing. The decision commands regarding mitigations and re-routed paths

978-1-5090-0037-1/16 $31.00 © 2016 IEEE

are given by the UAV and these informations are sent to the security control centre.

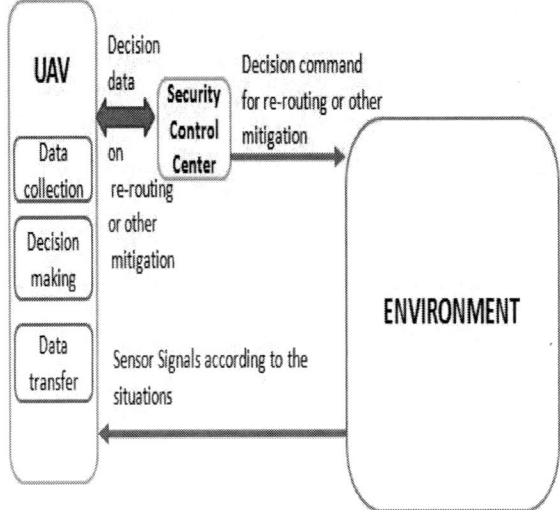

Figure 1: Block Diagram for a traffic monitoring system using unmanned air vehicles

With reference to Figure 1, there are three primary steps that UAV performs:

1. Data aggregation/collection

Different data from different nodes of the layout are collected and the resulting situations and scenarios are analysed.

2. Decision making

Various decisions regarding shortest path selection and mitigations for various security related situations are provided with the help of an intelligent smart UAV processor.

3. Decision Transfer

After collecting the data from various sensors, decisions regarding re-routing and mitigations for security related issues are send to the central control centre for further activities.

The following are the reasons for UAVs preferred over the traditional technologies:
➢ Mobility
➢ Low cost of operation.
➢ Higher speeds than ground vehicles.
➢ Do not have any restrictions compared to ground traffic monitoring techniques.

➢ With autonomous flight capabilities UAVs can potentially free personnel from time-consuming travel to remote field locations.

This paper focuses on decision making aspects of the UAV regarding different situations in a secure road traffic environment.

III. INFRASTRUCTURE GRAPH MODEL

A. Introduction

Networks can be easily represented by graphs. The network nodes are vertices and the communication links are edges. Routing protocols often use shortest path algorithms.

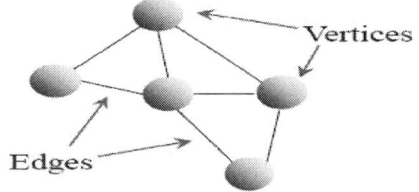

Figure 2: General form of a graph

A graph $G(V,E)$ is two sets of object:
• Vertices (or nodes), set V
• Edges, set E

A graph is represented(Figure 2) with dots or circles (vertices) joined by lines (edges).The magnitude of graph G is characterized by the number of vertices $|V|$ (called the order of G) and number of edges $|E|$ (size of G).The running time of algorithms are measured in terms of the order and size.

B. Situational graph

Using the above concepts of graph theory, we construct the environment or the road network and the associated infrastructures like buildings which are under the surveillance of the UAV in the form of a graph. We call the resulting graph model as situational graphs as shown in Figure 3. We consider an arbitrary graph as the environment under surveillance of the UAV.

From the situational graph given in Figure 3 we can infer that all the nodes A,B.C,D,E,F,G,I represents different traffic junctions with each of the junctions having traffic lights. The edges of the graph connecting the nodes are the roadways connecting different junctions. The distance from one junction to the other junction is represented in the form of weights as assigned to the edges of the graph. The infrastructures associated with road network are also shown in the graph. Here we have included many critical infrastructures which either directly creates security awareness or indirectly creates a security situation such as school, government building, bank, hospital, defence area etc. We formulate our problem by having different types of sensors to detect the different situations that arise. For example if a car without an RFID tag is entering the defence area or government building area, it is prompted as a security situation. The sensors deployed for purpose of tag identification identifies if the vehicle is having an RFID tag or not and correspondingly send signals to the UAV. In this case, the UAV provides its mitigation for re-routing to the security aware traffic control station. Another example where a theft in a bank can be promoted as an indirect security situation that can eventually lead to a security attack. We assume that the bank lockers are equipped with sensors that can detect a theft and

communicate it to the UAV and thereby UAV sends the mitigation to the security control station. So, in general using our model, multiple situations can be extracted and the UAV can provide a comprehensive mitigation to protect against security situation.

Some of the security situations that can arise are listed in Table I along with its respective mitigations. For this work, it has been considered that for seven distinct security situations resulting in 2^7 security scenarios. Each scenario can be formed into a scenario graph as shown in Figure 4.

One or more of the above mentioned situations (Table I) can occur concurrently thereby leading to different combinations of security situations leading to multiple security scenarios. These multiple scenarios can be represented in the form of scenario graphs as stated before.

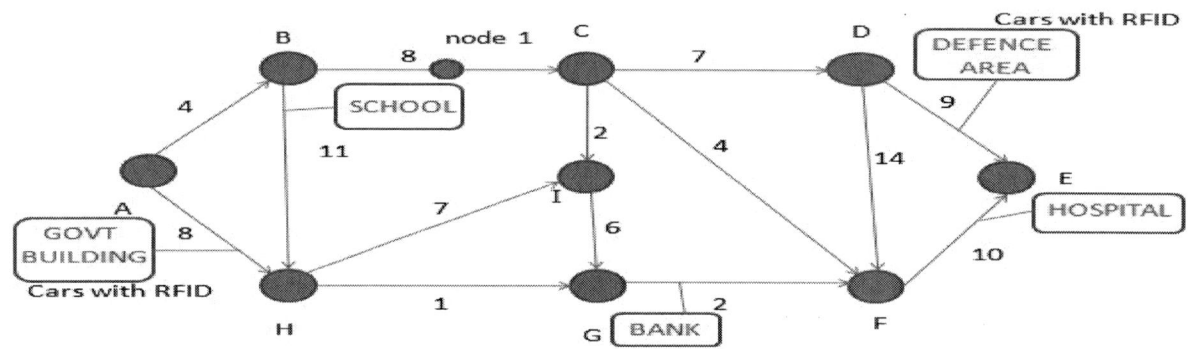

Figure 3: An illustrative situational graph

Sl.no	Security situation	Mitigation
1.	VIP car (eg : in A to I)	Shortest path for VIP car (eg: ABCI) .
2.	Theft in the bank	Give signal from bank lockers to UAV.
3.	Incident-attack(in node-1)	Normal traffic re-routed (from B to C because its a blocked path and gives signal to school via UAV).
4.	Accident (eg: in H to I path	Normal traffic re-routed (from H to I because its a blocked path and gives signal to school via UAV).
5.	Overspeed	Speed detector detects speed of the vehicles if value exceeds beyond a certain threshold.
6.	Cars with RFID	Cars without RFID tags not permitted to government and defence area.
7.	Ambulance (eg:in H to E)	Shortest path for ambulance (eg: HGFE) .

Table I: Security situations and its mitigations

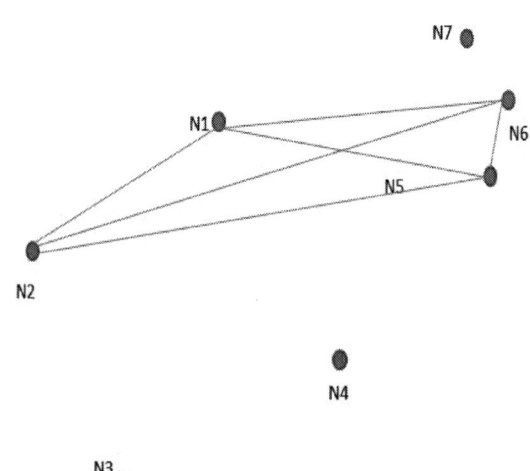

Figure 4: A Sample Scenario Graph

The scenario graph given in Figure 4 depicts a combination of four situations occurring concurrently. Here the nodes N1, N2, N3, N4, N5, N6 and N7 represents the seven situations as mentioned in the table. Now the graph shows the interconnection of N1, N2, N5, N6 which indicates that all these four situations occur concurrently.

The UAV decision capability includes identification of a particular scenario and provides the respective mitigation and sent them to the security control station. The system can be abstracted as shown in Figure 5 and Figure 6.

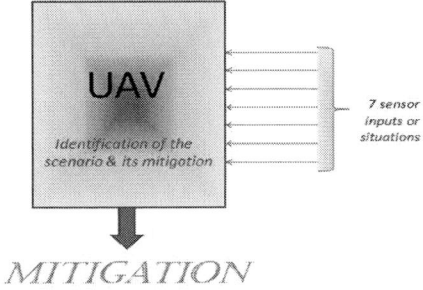

Figure 5: Identification of the scenario

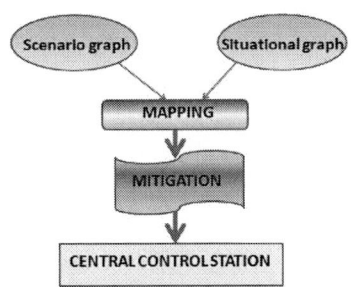

Figure 6: Entire system flowchart

The decision making capability of the UAV basically concentrates on the graph mapping between the scenario graph and the situational graph as shown in Figure 6. The intelligence of the UAV after mapping between scenario graph and the situational graph provides the required mitigation and sends it to the central traffic control station to control the actual traffic flow.

Algorithms
a) Top level algorithm
1) Start
2) Enter the infrastructure graph model or the situational graph in the form of a matrix.

Here it can be written as a distance matrix,

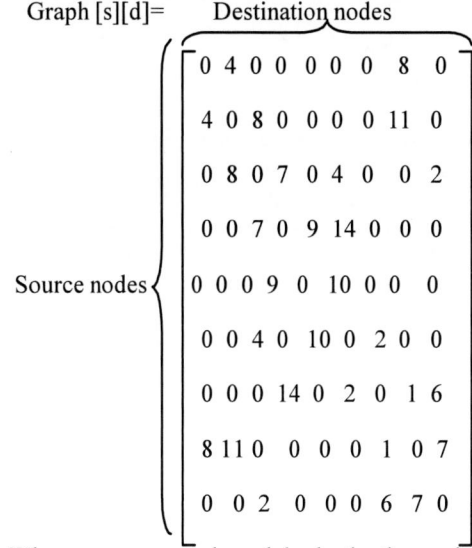

Where, s=source node and d= destination node

3) There are seven situations and the occurrence of each situation will produce the corresponding mitigation that is required.

For eg: (With reference to Table I, let us consider two situations)
If (Situation1)
{
VIP car detected
Find shortest for the VIP car (Use Dijktras algorithm: A)
}
If (Situation 2)
{
 Theft in the bank detected
 Red LED glows
}
A: Dijktras algorithm
1) Start
2) Enter source(s) and destination (d)
3) While (graph[s][d]= =0) // (sub paths required)
  {
      Find the smallest value in the row „s‟
      other than zero.
      Find its corresponding column value.
      That will be the nearest neighbor to
      source(s)
       s<= current position
  }
If graph[s][d]= =0 (no subpaths found)
  {
      Direct connection path between source and
      destination.
      }

IV. SYSTEM IMPLEMENTATION AND RESULTS

A. Implementation

The system is implemented using LPC2138 ARM microprocessor, associated switches and LCD display to show the occurrence of a particular situation and the source and destination. The LPC2138 microcontrollers are based on a 16/32 bit ARM7TDMI-S CPU with real-time emulation and embedded trace support that combines the microcontroller with embedded high speed Flash memory ranging from 32 kB to 512 kB. A 128-bit memory interface and unique accelerator architecture enable 32-bit code execution at maximum clock rate.

In case of critical code size applications, the 16-bit Thumb Mode reduces code by more than 30 % with minimal performance penalty. We use seven switches to represent the seven situations and four switches for the source and 4 switches for the destination.

For example, the switching combination 0001 represents source(s)=1 or destination(d)=1. Similarly, 0111 represents source(s) =7 or destination (d)=7. The algorithm is implemented using embedded C language and was simulated using Keil µVision4 tool [7]. The system implementation is done using Proteus simulator [8] as shown in figure 7. Table II shows some scenarios and its mitigations.

B. Results

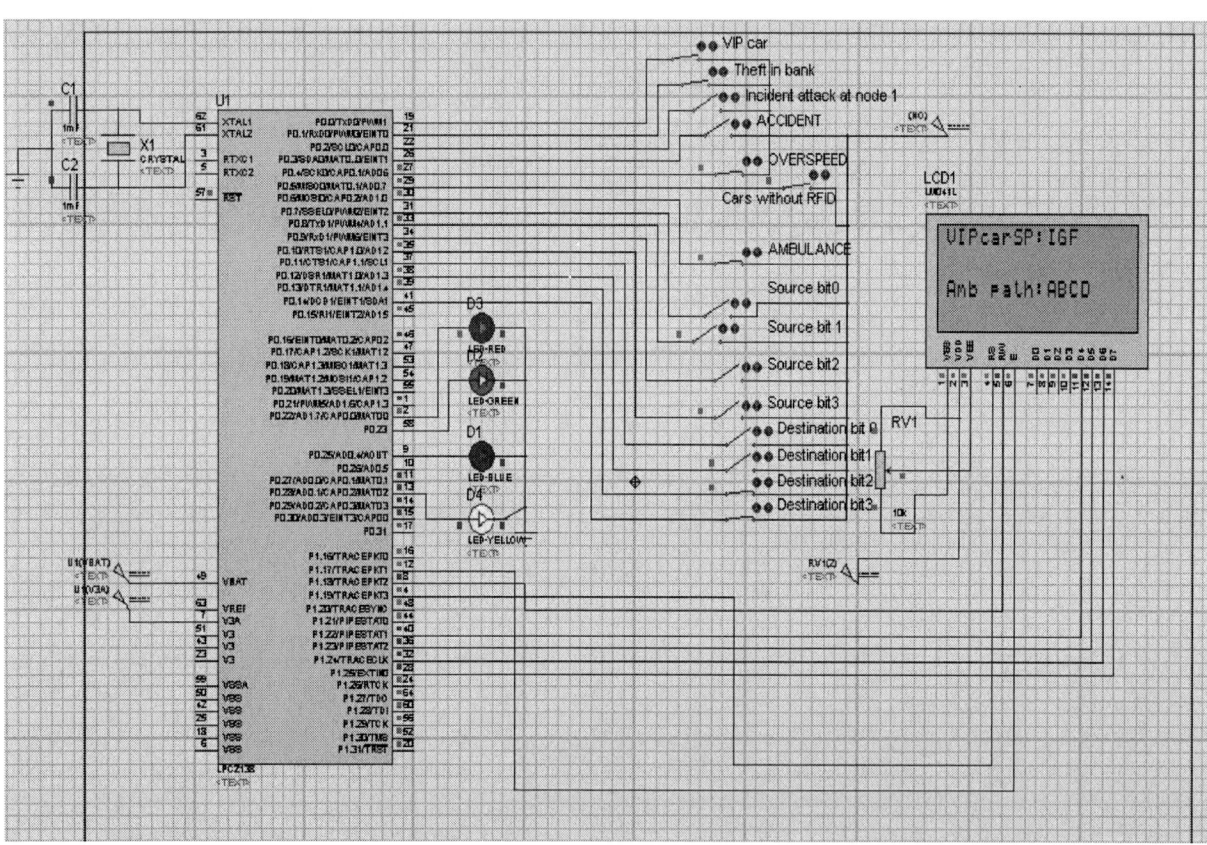

Figure 7: Here,
Switch 1=on => VIP car detected and shortest path displayed(IGF=> I=source ;   G=sub paths and F= destination.)
Switch 7=on=>Ambulance detected and shortest path displayed(ABCD=>A=source; B,C=sub paths and   D=destination.)(Also, yellow LED glows to indicate the presence of the ambulance)
(Eg. Source bit combination = 0000(A) and Destination bit combination = 0011(D))
Switch 2=on => Theft in bank detected so, RED LED glows
Switch 3, Switch 4 =off => No incident attack and no accident
Switch 5=on => Overspeed of the vehicles detected so, GREEN LED glows
Switch 6=on => Cars without RFID detected so, BLUE LED glows.

Scenario combinations							Description of the scenario (Multiple situations occurring concurrently)	Mitigation
S7	S6	S5	S4	S3	S2	S1		
1	1	1	0	0	1	1	VIP car, theft in the bank, overspeed vehicle, cars without RFID, Ambulance detected. (Simulation result in figure 7)	Display shortest paths for VIP car and Ambulance (yellow LED), and red, green, blue LED glows.
1	1	0	0	0	1	0	Theft in the bank, cars without RFID, Ambulance detected.	Display shortest path for ambulance (yellow LED), and red , blue LED glows.
0	1	1	0	0	1	0	Theft in the bank, overspeed vehicle, cars without RFIDdetected	Red, Green, Blue LED glows.
1	0	0	0	0	0	1	VIP car, Ambulance detected.	Display shortest paths for VIP car and Ambulance (yellow LED), and no LED glows.
0	0	0	0	0	1	1	VIP car, theft in the bank detected	Display shortest paths for VIP car and red LED glows.

Table II: Different scenarios and its mitigations

V. CONCLUSION AND FUTURE WORKS

An UAV based security situational aware intelligent road traffic management system has been implemented. UAVs enable real time and quick responding to various security scenarios and situations because it need not face any obstacles to wireless signal transmission compared to ground traffic monitoring systems. Using shortest distance algorithm for the graph model, a re-routing is provided to protect against security situation. The system is very flexible and efficient. The algorithm is highly scalable as any additional nodes into the infrastructure can be easily added which provides the mitigation per generic algorithm developed.

The future work is to integrate the system into an intelligent vehicular transportation system (ITS) [9] by interfacing video cameras to capture visuals of the different real time security situations and scenarios in both road and train network for an optimized coordination and comprehensive transportation system security.

REFERENCES

[1] M. Kontitsis, K. P. Valavanis, R. Garcia "A simple low cost vision system for small unmanned VTOL vehicles" IEEE 2005.

[2] RenSuzuki"Teleoperation of a tail-sitter VTOL UAV" IEEE/conference on intelligent robots and systems, pp.1618-1623, 18-22 Oct 2010.

[3] Lang.Wenqian,Lin.Jiayuan,Wang.Yangchu,Tao Heping "Investigating small-scale water pollution with UAV remote sensing technology." IEEE conference on world automation congress (WAC), pp.1-4, 24-28 June 2012

[4] Konstantinos Kanistras ,Goncalo Martins "A survey of unmanned aerial vehicles(UAVs) for traffic monitoring" IEEE conference on unmanned aircraft systems 2013.

[5] Yucong Lin , Srikanth Saripalli "Road detection from aerial imagery" IEEE conference on robotics and automation 2012.

[6] B. Coifman, M. McCord "Roadway traffic monitoring from an unmanned aerial vehicle". IEEE conference on intelligent transport systems 2006.

[7] www.keil.com/uvision/ ide_prj.asp

[8] www.labcenter.com/index.cfm

[9] Padmadas M "A deployable architecture of intelligent transport system-a developing country perspective" IEEE international conference on computational intelligence and computing research (ICCIC),pp.1-6,28-29Dec2010.

Automatic Pressure Maintenance System for Tyres in Automobiles to Reduce Accidents

Rajesh KannanMegalingam, Jayakrishnan C, SrirajNambiar, Rudit Mathews, Vishnu Das, Pramesh Rao,
rajeshkannan@ieee.org, c.jayakrishnan4@gmail.com, srirajnambiar2@gmail.com, rudit1994@gmail.com
vishnudas11292@gmail.com, pramesh.r123@gmail.com
Department of Electronics &Communication
Amrita VishwaVidyapeetham, Amritapuri, Kollam

Abstract—One of the important factors that play key role in increasing automobile efficiency and reducing number of road accidents, is optimum tyre pressure maintenance. Variation of tyre pressure from the optimum value can result in reduction of fuel mileage, tyre life, and safety and often in tyre bursts causing accidents. We propose the design of an automatic system that can ensure and maintain tyre pressure to its optimum safe value. The design consists of real time tyre pressure monitoring system for taking periodic pressure readings, a pressure control system that makes adjustments to achieve the desired pressure settings and a wireless communication link to integrate both these systems together. The design in addition allows adjustments to predefined terrain based tyre pressure settings and to manually entered pressure settings. The implementation, testing,and analysis of the system is provided in this research work.

IndexTerms: fuel mileage,pressure monitoring system, tyre bursts

I. INTRODUCTION

According to the World Health Organization, one person is killed in road accidents for every 25 seconds. Approximately 124 million deaths occurred due to road accidents in 2010. Researches tell us that major causes of road accidents are over speed (78%), driving under the influence of drugs (17%), not using helmets and seat belts (1.5%), and tyre bursts (3.5%). It is also taken into note that if tyre burst occurs in extremely high speed the death rate is nearly 100%. Tyre bursts are mainly caused due to low quality tyre material and irregular amount of air inside the tyre. One of the main reasons for the above problems are driver's lack of knowledge about tyre pressure and indeed the absence of tyre pressure measuring device in demand.

Studies show that if tyre pressure is maintained to its optimum value and pressure changes are discovered and adjusted within time tyre burst can be avoided. This facts show that a system is required to maintain optimum tyre pressure for effective automobile performance in a variety of conditions. The system designed here makes periodic adjustments to ensure that optimum tyre pressure is maintained. The design mainly comprises of a tyre pressure monitoring system that includes a differential pressure sensor MPX5500DP interfaced using an ATmega328 microcontroller, a pressure control system that has a solenoid valve electronically controlled by another ATmega328 microcontroller and a Bluetooth transceiver module that provides the communication link between two systems. Throughout 2012, Bridgestone carried tyre checks at over 28,000 vehicles which showed a massive 78% of those tyres under recommended pressure level and 25% tyres worn out below the legal limit. Due to under inflation and over inflation can cause its failure and eventually tyre burst. Hence optimized tyre pressure is needed to experience better tyre performance and fuel economy.

Conventionally, drivers measure the tyre pressure manually at fuel stops. But this is not done regularly. High end vehicles have electronic displays which display the tyre pressure. Even in this case, if the display indicates that the tyre pressure is low, the driver has to stop the vehicle and fix the tyre pressure at the next fuel stop. If the vehicle is driven on uneven terrain and the tyre pressure has to be increased or decreased, then it is not possible at a remote place. Our proposed design not only relieves driver being alert about the tyre pressure, but also can adjust the tyre pressure, i.e. increase or decrease the tyre pressure automatically without the need for the vehicle to be stopped.

The paper is organized as follows: Section two describes the current research going on in this field followed by the system architecture and design methods in section three. The implementation of the system is explained in the fourth section followed by system integration and testing in the fifth and sixth section respectively. The observations are presented in the last section.

II. RELATED WORKS

Many researchers and engineers on the aim to tackle the above problems are extensively working on tyre pressure monitoring systems (TPMS). TPMS can be divided into two groups: The first one is called indirect TPMS and is based on wheel speed. It monitors speed of different tyres and then compares through antilock brake system (ABS) wheel speed sensor for monitoring the tyre pressure. The main disadvantages include that it doesn't operate when two or more tyres are underinflated and when the vehicle is moving at speed above 100 km per hour. The second one is direct TPMS which uses pressure sensor fitted on each tyre to display and monitor the tyre pressure. With the development of integrated circuits and microcontrollers, the cost of direct TPMS can be reduced.

Due to the concerns in unpredictable durability of underinflated tyres, present need should not be just of monitoring tyre pressure but also maintaining it to the desired value. The monitoring system will just alert the driver of the presence of any under inflated tyre and the driver must

978-1-5090-0037-1/16 $31.00 © 2016 IEEE

manually fill the tyre to the recommended optimum level using an external source. This is highly undesirable in long trips and in cases of unavailability of any external source for refilling before eventual wear out. Hence TPMS concepts must be extended to maintaining tyre optimum tyre pressure.

Commercial and military trucks have been using systems that can automatically inflate tyres using on board air source. Also as early as in 1984, GM offered central tyre inflation systems (CTIS) on commercially utility cargo vehicles (CUCV) blazers and pickups. These types of trucks have been extensively used for military purposes [3] but the concept was never developed to present day automobiles. Ref [2] discusses about an automatic inflation system for tyres but uses a pressure gauge to measure the pressure from the tyre before inflating them. This less cost effective and accurate. Ref [4] comes up with an efficient mechanical design to pump air to the moving tyre using a centralized compressor system but is silent about the efficiency of pressure monitoring system and pressure sensor. Ref [5] comes out with ways of improving the efficiency of tyre pressure monitoring system and pressure sensors but does not discuss the needs of self-inflating the underinflated tyres. Ref [6] deals with improving the efficiency of direct TPMS sensors and proposes a fuel leakage detection system but do not discuss about self-inflating the underinflated tyres. Ref [7] deals with warning the driver of an underinflated tyre based on its proposed tyre pressure monitoring system and do not address the needs of maintaining tyre pressure. Ref [8] deals with improving the design concepts in existing TPMS technology. Ref [9] deal with mechanical design of self-inflating truck tyres based on the values received from the pressure sensor fitted in tyres. However it does not deal in detail with working of tyre pressure monitoring system and its efficiency and is absolutely designed to heavy vehicles like trucks

III. SYSTEM DESIGN AND ARCHITECTURE

A. *System Architecture*

The general system architecture of the design is shown in the figure 2: MPX5500DP sensor, Bluetooth transceiver, ATmega328 microcontroller, TIP 120 transistor and solenoid control circuit and the solenoid valve. From the architecture it can be seen that the entire electronic phase of the project is divided into two parts: first one is the pressure monitoring system and second one is the pressure control system. In the pressure monitoring system the sensor control circuit and the Bluetooth transmitter are connected to ATmega328

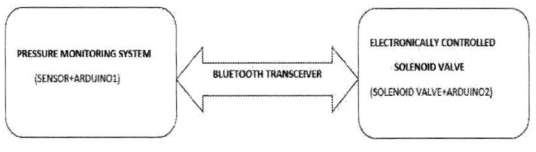

Fig. 1. Simplified block diagram

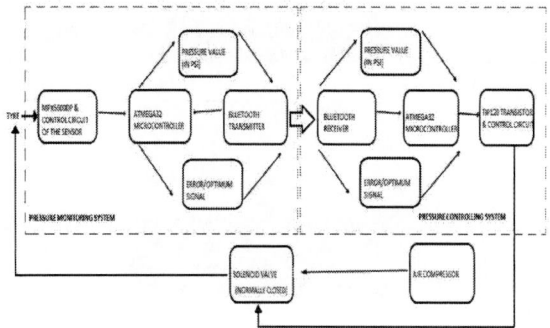

Fig2. General system architecture of the design

microcontroller. In the pressure control system the solenoid control circuit and the Bluetooth receiver are connected to the ATmega328 microcontroller. The normally closed solenoid valve is connected to the air compressor.

B. *System Design*

The design is divided into its electronical and mechanical phases. The electronic phase is broadly subdivided into following sub phases:

- Pressure monitoring system: Used for obtaining and monitoring periodic real time pressure value from the tyre using the sensor MPX5500DP, comparing it with the optimum terrain specific pressure value subsequently finding out the error and transmitting the pressure value and error through a Bluetooth transmitter interfaced with the microcontroller ATmega328.

- Pressure control system: The Bluetooth receiver gets the transmitted data and depending on the error value the solenoid valve is made open by setting TIP120 transistor base high for sufficient amount of time.

- Wireless communication link between the pressure monitoring system and pressure control system.

- LCD displaying the real time pressure value and warning on under inflation.

The mechanical phase of the design include quick joint coupling between the air outlet of the compressor and hollow axle of the tyre, tyre – hollow axle assembling and the belt drive assembly to run the tyre in demonstration set up.

IV. DESIGN IMPLEMENTATION

A. *Pressure monitoring system*

Pressure monitoring system consists of the differential pressure sensor MPX5500DP, and a Bluetooth transmitter connected to an ATmega328 microcontroller. The sensor fixed

978-1-5090-0037-1/16 $31.00 © 2016 IEEE   19

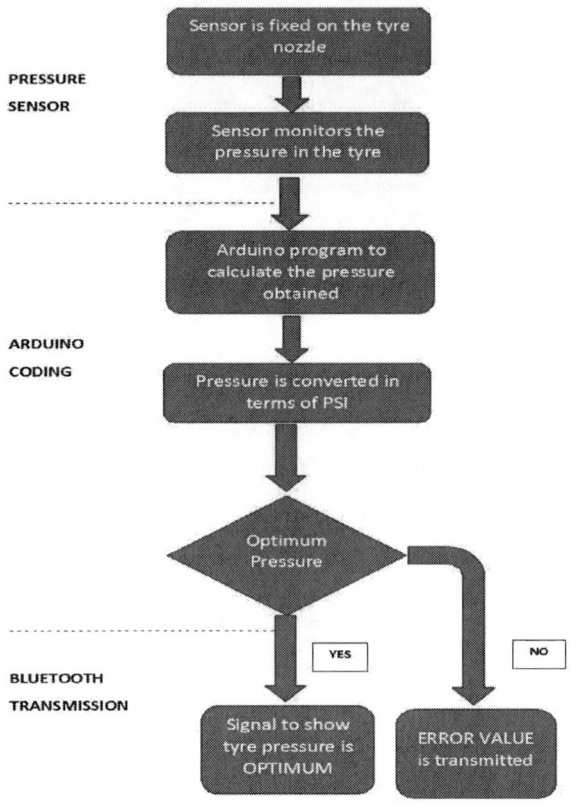

PRESSURE
SENSOR

ARDUINO
CODING

BLUETOOTH
TRANSMISSION

Fig.3. Block diagram of pressure monitoring system

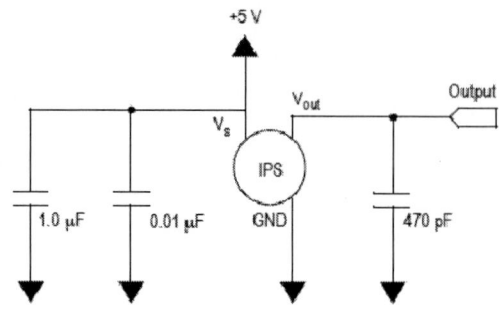

Fig. 4. Output filtering of sensor

2) ATmega 328 microcontroller

The design uses ATmega328 microcontroller embedded in Arduino Uno board for programming. Two microcontrollers of the same kind is used in the design one each in pressure monitoring system and pressure control system. It's also interfaced with Bluetooth HC-05 transceiver modules in each system. The microcontroller has 14 digital I/O pins of which 6 can be used as PWM outputs, 6 analog inputs, a flash memory of 32 KB, SRAM 2KB, clock speed of 16MHz and an operating voltage of about 5V. ATmega328 microcontroller in the Arduino board is programmed by the Arduino software

B. Pressure control system

Pressure control system consists of a control circuit of solenoid valve, Bluetooth receiver and a normally closed solenoid valve. The transmitted error value signal along with the real time pressure value measured is received by a Bluetooth receiver which is interfaced to a microcontroller ATmega328. If there is no error and the pressure value is optimum the valve remains closed. Depending upon the error signal, the amount of time the solenoid valve should be kept opened is determined. Subsequently sufficient amount of air start flowing from the compressor to the hollow axle through the solenoid valve.

The hollow axle is coupled with the air pipe from the compressor using a quick joint coupler. The air flow stops after the tyre pressure retains its optimum value and the valve goes back to the previous closed stage. The solenoid valve can't be interfaced directly with the microcontroller since the operating voltage of valve is 24 V which is much higher compared to that of microcontroller. To tackle this, a control circuit is used to interface the solenoid valve with the microcontroller. The control circuit consists of a TIP120 transistor, a 2.2k resistor and an IN4007 diode. The solenoid valve gets opened as long as the transistor base pin is kept high by the microcontroller digital pin. The block diagram of the pressure control system is given in figure5.

on the tyre periodically monitors the tyre pressure and output a voltage signal. The corresponding pressure is obtained from the microcontroller from the sensor sensitivity and offset voltage. The error value is obtained as the difference between pressure measured and the optimum pressure and then transmitted along with pressure measured using a Bluetooth transmitter. If the pressure in the tyre is optimum and safe, an error value of 0 is transmitted by the transmitter.

1) Sensor

The de1sign uses a MPX5500 series piezoresistive transducer provided by FREESCALE for its operation. The pressure sensor MPX5500DP combines advanced micro matching techniques, thin film metallization and bipolar processing to provide an accurate, high level analog output signal that is proportional to the applied pressure. Figure 4b shows the circuit for interfacing the output of the integrated circuit to A/D input of the microcontroller ATmega328. Table 1 shows the operating characteristics of the sensor (supply voltage= 5.0 V DC, pressure at the first port should be greater than that at the second port P1>P2).

2016 International Conference on VLSI Systems, Architectures, Technology and Applications (VLSI-SATA)

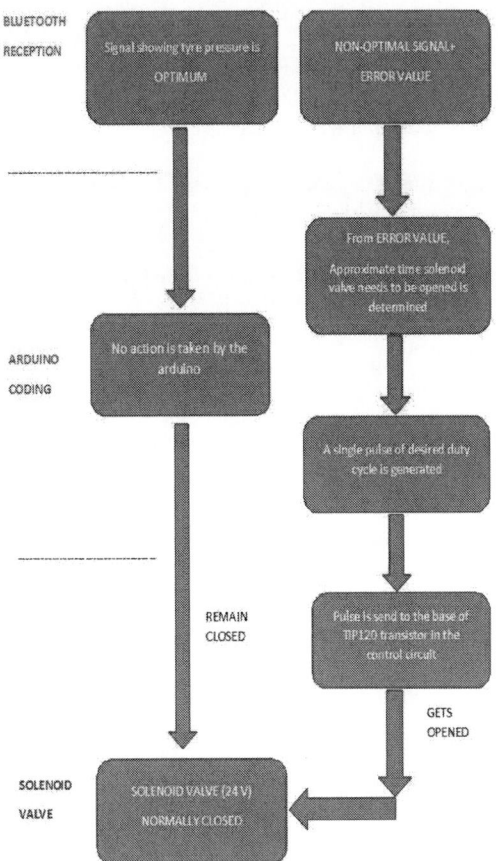

Fig5. Pressure control system block diagram

1) Solenoid valve and control circuit

The design uses a 2 way ¼" normally closed solenoid valve operating at 24 V DC voltages. A solenoid valve is electromechanically operated valve which is controlled by an electric current through solenoid. The valve is closed when the solenoids are not energised and hence called normally closed. The valve is opened only when the two ports are connected, the solenoids are energised and there is air flow between two ports. The control circuit consists of TIP120 transistor, a 2.2 K diode and an IN4007 diode. This is used to interface solenoid valve with the microcontroller.

The TIP120 transistor allows a smaller voltage from the microcontroller to switch a larger voltage to the solenoid. When a small voltage from the microcontroller is applied to the base of the transistor, allows current to flow from its collector to emitter so that 24V supply gets connected to solenoid valve power input. When the solenoids are stopped being powered a reverse voltage spikes back. In order to prevent this voltage from killing the transistor a diode is used which allows current flow only in one direction. The microcontroller depending on the values error value received on it set the transistor pin high for sufficient amount of time. This energizes the solenoid with voltage supplied (24V) and opens the valve. The control circuit is shown in figure6.

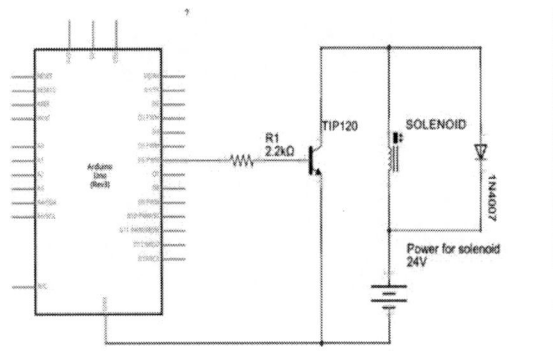

Fig6.. Solenoid valve and control circuit

C. Wireless Communication link between systems

Wireless communication link between pressure monitoring system and pressure control system is established using two Bluetooth HC-05 transceiver modules. Bluetooth transceivers was preferred over normal RF transceivers and zigbee modules considering the availability, affordability and reliable data transfer range of the same compared to others. The Bluetooth module transmits two data values

- Error value which is measured as the difference between pressures measured and optimum predefined pressure of the specified terrain.
- The real time pressure value measured from the tyre by MPX5500DP sensor.

HC-05 is chosen because it's better than all other Bluetooth modules and leaves lots of space for the user.HC-05 is an easy to use Bluetooth serial port protocol module designed for transparent wireless serial connection setup. Serial port Bluetooth is fully qualified Bluetooth V2.0 + enhanced data rate 3Mbps modulation with complete 2.4 GHz radio transceiver and baseband. It has a low power 1.8V operation and 3.3 to 5V I/O. The Bluetooth HC-05 module in pressure monitoring system acts as the transmitter and the one in pressure control system acts as the receiver. Both the Bluetooth modules are interfaced to the ATmega328 microcontroller

V. MECHANICAL DESIGN AND OVERALL INTEGRATION

After giving an electronic design to the concept, its mechanical design and integration is taken into concern. The main problem in the mechanical design is in transferring air from static vehicle chassis to the moving tyre. A mechanism that could reliably transfer air between these two components need to be designed. In order to avoid the complexities involved with the concept of solid axle, a demonstration model of the concept is designed. The main difference between the demonstration model and present day automobiles is in the concept of hollow axle which is mainly solid in automobiles. The design incorporates a hollow axle so that it acts as a passage for the air flow from the compressor to the tyre.

978-1-5090-0037-1/16 $31.00 © 2016 IEEE 21

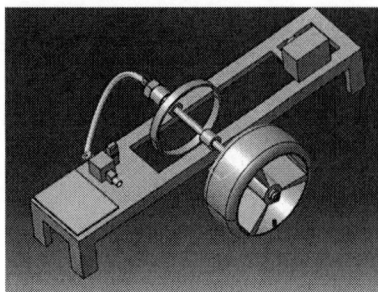

Fig.7. Mechanical demonstration setup of design

The demonstration setup as shown in the figure 7 have a one tyre- axle assembly run by a motor connected belt drive. The air flow from the compressor to the axle inlet is regulated by the electronically controlled solenoid valve. The hollow axle inlet is coupled with the air pressure line outlet from the solenoid valve by using a quick joint coupler. The coupler has two components: a static component to which air pressure line is connected and a dynamic component to which the hollow axle inlet is connected. These function similar to a rotary seal assembly. The motion of the dynamic arm never affects the static arm which helps in reliable air transfer through the axle to the tyre. The hollow axle on the outlet side is attached firmly to the tyre and is fixed at the end to a nozzle. This nozzle protrudes outside the tyre hub and is connected to the wall tube using a small pipe. The pressure monitoring system is integrated with the wall tube to form a single mechanical unit to which the air pressure lines coming out of the axle is attached. The pressure monitoring system package and the air pressure line from the axle is integrated with tyre wall tube using a foot pump adapter. The mechanical demonstration setup of design is shown in figure7.

VI. TESTING AND EVALUATION

B. *Measurements and Calculations*

The pressure sensor output after A/D conversion needs to be monitored and converted into pressure values in terms of PSI before transmission. Figure8 shows the sensor output signal relative to the pressure input. Typical, minimum and maximum output curves are shown on operation over temperature range of 0- 85 C.

- The output voltage of the sensor depends upon the offset voltage, pressure applied and sensitivity of the sensor.

$$Vout = Voff (mV) + Sensitivity (mV/KPa) * P (KPa) \quad (1)$$

Where Vout is the voltage output from the sensor (in mV), Voff is the offset voltage of the sensor (in mV) and P is the pressure applied in KPa.Offset voltage is the voltage output of the sensor when the pressure applied is 0 KPa. The typical value of the offset voltage of MPX5500DP sensor is 0.2 V. Sensitivity is the ratio of change in output voltage to that of the pressure applied and is measured in mV/KPa. Sensitivity of MPX5500DP sensor is 9mv/KPa. Typically sensor output

produces voltage at range of 0.2-4.7V for an input pressure range of 0-500 KPa. Since the maximum acceptable voltage of the A/D converter in the microcontroller is 5V, extra op-amps for voltage matching were unnecessary.

The sensor output in voltage for different pressure values of the tyre is shown in the figure8.

- The error value is obtained from the measured pressure value as follows:

Error value= Mod (Pressure measured- Optimum predefined pressure of the terrain).If the pressure is optimum error value of 0 is transmitted by the Bluetooth module in pressure monitoring system

- The design could control the pressure of the tyre in a range from 20-40 PSI. The design warns the user about an under inflation whenever the pressure drops below 1 PSI tolerance level that is, on or below 31 PSI of tyre pressure. The tyre system in an automobile is said to be flat if its pressure is less than 10 PSI and is said to be leaked when any tyre is 20% below the pressure of any other tyre.

A total of 50 measurements were taken by the sensor in 50s, which is averaged to evaluate the mean pressure of the tyre. This mean is found whether to be within the tolerance level of allowable pressure range, and is sent by the Bluetooth transmitter to the pressure control system. The pressure sensor take readings for every 50s and then stops monitoring for the next 70s for providing the time span for necessary adjustments from the pressure control system. The Bluetooth module in the control system receives the value, and makes adjustments within the next 70s so that it could receive the next set of values. This time span is largely enough as the control system is said to inflate tyre for an error value of 15 PSI within 1min.

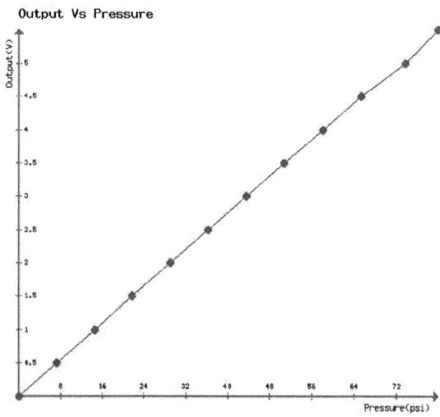

Fig 8. Sensor output for different pressure values from tyre

C. *Testing*

To ensure that the design satisfied all its technical requirements, test plans were devised. Initially the system was turned off for some days and the pressure loss in a stationary tyre is noticed by using a pressure gauge. Then same readings are observed in a tyre kept in constant motion for 5 hours. Tyres while in motion tend to lose more pressure. Then tyre pressure losses are noted in different set of terrains. Tyres tend to lose more pressure in rough terrains. The same set of readings was noted with the system turned on. Pressure readings are verified to be within the tolerance level of allowable pressure range. As next step inflation times were noted for different set of error values and are noted to meet the inflation time specified in the technical requirements of 70 sec.

VII. CONCLUSION

The demonstration set up is designed for hollow axle based wheel set up, which should be expanded to tyres using solid axle in present day automobiles. The electronics package of the system could be easily integrated with any of the automobile configuration. Wireless transmission is done using Bluetooth modules, which could be done by zig bee modules in a better budget. The demonstration set up is devised for one tyre assembly which could be expanded to 4 tyre automobile systems by having an individual tyre monitoring system for each of the tyre. Multiple solenoid valves or 3 way solenoid valves could be used to make inflation and deflation simpler. A touch LCD display can be maintained not only for the user to be informed on underinflated tyres but also to set desired pressure settings which can be easily incorporated in the present design. Design also gives space to multiple microcontrollers to be used in pressure control system for heavy load vehicles with increased chances of under inflation.

The self-inflating system is designed to address the consumer needs of maintaining appropriate tyre pressure conditions for reduced tyre wear, increased fuel economy, improved vehicle performance, reduced accident risk factor and reduced maintenance cost of the automobile. This automated system design regularly keeps the tyre at optimum pressure. The system could be expanded to incorporate many predefined tyre pressure settings allowing user to select them and also to manually enter the desired pressure settings. The design when compared to its related concepts has extra stability in having minimum increased load on vehicle. Also the electronics package is designed to work as accurate as possible and can be easily incorporated in any automobile.

REFERENCES

[1] Doran Manufacturing LLC," Keep your tyres at proper inflation".

[2] Hemant Soni, Pratik Golar, AshwinKherde, "Design of automatic tyre inflation system", Industrial Science,Vol 1,Issue 4,April 2014

[3] Lee Ann Obringer."How self inflating tyres work"

[4] Jennifer Drain, Rodney Hall, Christopher Pentland, Michael Snedeker, Aaron Thurber, " The Pressure management system", University of Arizona, April 2008.

[5] Avinash D. Kale, Shubhada S. Thakare, Dr. D. S. Chaudhari, " Wireless tyre pressure monitoring system for vehicles using SPI protocol, International journal of advanced research in computer engineering and technology, Vol 1, Issue 4, June 2012.

[6] Loya Chandreshkumar, Joshi Pranav, Chaudhari Hemraj, Prof. Gayatri Bokade ," Tyre pressure monitoring system and fuel leak detection", International journal of engineering research and applications, Vol. 3, Issue 3, May-June 2013.

[7] Jiaming Zhang, Quan Liu, Yi Zhong, " A tyre pressure monitoring system based on Wireless sensor Networks Technology", 2008 International conference on Multimedia information technologies.

[8] Tianli Li, Hong Hu, Gang Xu, Kemin Zhu and Licun Fang," Pressure and temperature micro sensor based on surface acoustic wave in TPMS" Shenzhen University.

[9] Ammer Khan,Amirdharaj. S, "Automatic inflation of truck tyres", IEFT college of engineering.

[10] Federal Motor Vehicle Safety Standards; Tire Pressure Monitoring Systems; Controls and Displays," Department of Transportation and National Highway Traffic Safety Administration.

Effect on Temperature and Time in Parallel Test Scheduling with Alterations in Layers Arrangements of 3D Stacked SoCs

Indira Rawat
Department of Electrical Engineering,
Govt. Engineering College, Ajmer,
Rajasthan , India.
indirarawat@gmail.com

M.K. Gupta
Deptt. of Electronics & Commn.
Engg., MANIT,
Bhopal, India.
gupta4@yahoo.com

Virendra Singh
Deptt. of Electrical Engineering,
IIT, Bombay
Mumbai, India
virendra@computer.org

Abstract— In modern electronic complexity, testing of products has become an important area of concern. Low cost and good defect coverage are the basic goals of testing, which are again determined by fault models, test volume and time. Time depends on efficiency of test scheduling scheme. Test scheduling has therefore become an important area of research. Authors have done work on test scheduling and have proposed Test Scheduling Algorithm in previous works done on 3D SoCs. 3D technology fulfils the demand of faster and compact design but there is a sharp rise in power density in such arrangement and therefore there is a sharp rise in temperature especially for the layers far from heat sink. Consequently formation of hotspots may occur which may lead to device failure. Testing dissipates more power than the functional power because of the high switching activity that takes place during testing. All this requires efficient test scheduling so that temperature does not rise above limits. Here temperature and time performance in 3D stacked SoCs with alterations in the arrangement of layers is presented. The modeling of 3D structure is done on HotSpot which is a validated tool widely used in VLSI Testing for temperature determination. The results so obtained are compared with the earlier works of the authors in the same field and an effort is made to draw an inference on temperature rise of cores on account of the variation in placement in 3D SoCs.

Keywords—3D SoCs, sequential testing, Adjacency exclusion, RHDF, HHDF, VHDF.

I. INTRODUCTION

In VLSI testing, test schedule generation is an important aspect. In present days, testing of VLSI circuits has become a mandatory task. Each manufactured product needs to undergo testing for its correct function. This warrants a test to be cheap, good and fast. Cheap means it has to be economical, as testing is cutting into the finances of manufacturing a chip or IC. It constitutes a major portion of the cost of the chip. Good test means it should have an efficient defect coverage and time of testing should be minimum possible. Multiple ICs are integrated on a single IC which can be a CPU, a PCI , SRAM, DSP or any other as per the application, hence called a System on chip. When these functional blocks are stacked vertically

they are called 3D SoCs. The advantages of system chips are many, like they can be used in complex applications, power dissipation is less in such systems, performance is very good, design time is very short and also have small volume and weight. At the same time they are characterized by a very large transistor count on a single IC, mixed technology on the same IC, have multiple clock frequencies and also the testing strategies can be different. In this work the testing methodologies of these 3D stacked SoCs are being referred.

Test scheduling refers to application of test vectors in a manner or sequence which reduces time as well as fulfils certain constraints which may be hardware related or conditions like temperature rise, voltage or frequency requirements. The test scheduling of SoCs which is the integration of a complete system onto a single IC has been taken up by authors in their earlier works also.

The challenges faced in SoC testing are many out of which power consumption during test is a main issue. The switching activity which takes place in a circuit under test are very fast and due to this the average power consumption is 3 to 10 times higher than the functional power. It is more in scan chain architectures due to generation of a large number of scan register shift-in and shift-out operations. This increases the overall chip temperature and also creates spots of localised heating called hotspots. Hotspots can cause permanent damage to silicon, high cooling costs and reliability failure. The heat has to be removed from the surface of the microprocessor die at the same rate as generation and these cooling solutions have become expensive. A remarkable work in this matter has been reported in [1]. The problems become more aggravated in 3D SoCs.

Authors have proposed a test scheduling algorithm for testing of cores concurrently which is different from traditional testing of layers in sequential manner. The work presented here is an extension of the earlier work where effort is made to explore the effect on maximum temperature rise of the cores reached with alterations in the arrangement of layers in the 3D stack. Here, temperature of each core is calculated after each test schedule and displayed in graphical form.

In Section II, brief background in the field of testing is discussed. In Section III, problem formulation is elaborated. In Section IV the implementation scheme is discussed. Section V details out the implementation of generated test schedules on HotSpot-5.02 along with the results generated on HotSpot-5.02 and graphical representation of the results. In Section VI conclusion and inference is briefed based on this work.

II. BRIEF BACKGROUND

The research in SoC testing can be roughly divided into the following heads: TAM optimization,test scheduling and a standard has also been developed called IEEE P1500. A lot of work has been done under the head of test scheduling which deals with the proper application of test vectors to the circuit under test so that the testing time is minimised. Yao, Saluja and Ramanathan have provided a collection of good work on test scheduling [2]. Research was also carried out by IBM, NEC, Siemens and Fraunhofer in 1990's. There has been an enormous advancement in technology over the past decades resulting in improved performance and productivity. 3D integration is considered as a key player in technology improvement in the years to come. The transformation to 3D offers many advantages over 2D including reduced interconnect lengths, better performance and heterogeneous system integration. Many difficulties are also faced in the adoption of 3D, the major being testing issues, thermal constraints and the EDA tools. Test scheduling is one of the major area to be considered in 3D. The most important challenge for 3D includes the thermal management. A 3D circuit will have multiple layers of devices and high density of interconnects, consequently several heat generating surfaces or sources. Excessive temperature gradients can occur which can cause permanent damage and cause major setback for implementation of this technology.

Test issues also assume importance in these circuits. Testing generates heat which is to be dissipated at a rate equal or faster than the heat generation. If the heat is not removed, permanent damage to the chip may be caused. The other problems of testing are encountered here also, mainly the test access, testing time and combined with the thermal problem the issue gets aggravated. Through Silicon Vias or TSVs are important components of these circuits providing power, clock signals and test access. Issues related to TSVs are also to be addressed. They have to be tested for defects and problems arising due to thermal effects. Test time minimization problems are discussed by Z. He, Z. Peng et al in [3] for 2D circuits. Very recent works are done by Millican and Saluja [4,5] in 2012 and 2015 in 3D test scheduling.

Various materials, technologies and functional components can be stacked together as shown in the Fig. 1. It shows that the stack can consist of memory, processor, ADC/DAC sensors etc. stacked vertically. The Fig. 2 also shows a 2D SoC which consist of many ICs . If the same are split in two and stacked vertically, a 3D SoC is obtained. For looking into the heat/thermal problem, take a look at the analysis available in the literature [6,7]. In this paper the research by the authors is devoted to test scheduling of 3D SoCs which is an emerging technology that forms vertically stacked integrated systems.

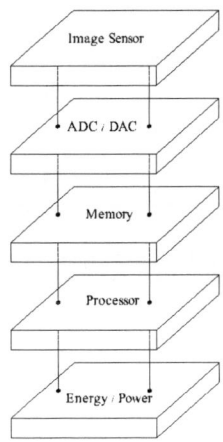

Fig. 1. Typical 3D SoC comprising different components stacked vertically

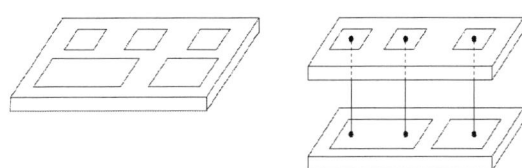

Fig. 2. Typical 2D SoC and a vertically stacked 3D SoC

The difference with previous works is that the thermal aspects are also being covered, implies that the temperature rise of various cores which constitute the SoC is also taken into account. Temperature affects the performance, reliability and life of a chip hence it is an important aspect which requires attention right from the design, to scheduling and testing. This way, the avoidance to chip damage can be done by preventing Hotspot formation. The life of an device is reduced to half by an increase in temperature by 10-15°C. The circuit performance also gets deteriorated and corresponding increase in gate delays is also introduced. With rise in temperature the leakage power of the circuit also increases. Possibility of timing errors also increases in overheated system. In author's paper [8], thermal aware test scheduling has been dealt with, which has nowadays become a very challenging job. Development of a simple and efficient thermal model has been suggested by authors. The test schedules have been implemented by authors on a well known linear model which is also a very popular RC model. This is the model which has been used in the HotSpot [1,9] tool.

III. PROBLEM FORMULATION

Given are benchmark circuits out of which 2 and 3 layered stacked structures are formed. Area, test length and placement of cores of each circuit is given. It is required to find the best arrangement of the circuits in a 3D stack so that the temperature rise of the cores is minimum when test scheduling is done as per the Test Scheduling Algorithm.

TABLE I. PARAMETERS OF SIMULATION MODEL

Parameters	Values
Chip thickness	0.00015 m
Heat spreader thickness	0.001 m
Heat spreader size	0.03 m
Heat sink thickness	0.0069 m
Heat sink size	0.06 m
Ambient temperature	300 K
Silicon thermal conductivity	100.0 W/m.K
Silicon specific heat	1.76 e6 J/m^3-K
Thermal interface material (TIM) thickness	2.0 e^-05
TIM thermal conductivity	4.0W/m.K
HotSpot calling interval	10 K cycles at 3 GHz

The test schedules are generated as per Test Scheduling algorithm employing Relative Heat Dissipation Factor (RHDF) and Adjacency Exclusion Principle [8]. These test schedules so generated are implemented on HotSpot to get the temperatures of cores after each test schedule. Temperatures of all cores after completion of each schedule and after complete testing are recorded. Highest temperature reached in every schedule and in complete test, mean temperature of all layers and complete chip after the test are noted and compared to infer the relationship between temperature rise and placement of layers in the stack.

The benchmark circuits used in this paper are d695 (10 cores), d281 (8 cores), f2126 (4 cores) [10]. The size of all the dies in the stack are same. Size of the chip has been kept as 4mm x 4mm. The parameters considered in this simulation model are shown in Table 1 which are mentioned in HotSpot-5.02 tool [1].

IV. THE IMPLEMENTATION SCHEME

In earlier paper [8] scheme of generating test schedules for parallel testing of cores in 3D stacked SoCs has been discussed. On account of their placement in stack, each core has different heat dissipation capability in horizontal and vertical direction. These capabilities have been addressed by the authors in [8]. The Horizontal Heat Dissipation Factor (HHDF) determines heat dissipation factor of any core in planar direction and is calculated based on the fact that heat dissipation takes place exponentially with resistive path as a criteria in time taken for dissipating heat. Similarly Vertical Heat Dissipation Factor (VHDF) of all layers is calculated which is based on the distance of layer from Heat Sink. More the distance of layer from Heat Sink more will be the VHDF thereby indicating that the layer will take more time in dissipating the heat generated during testing. One more factor which determines the heat dissipation factor is the P_Trace value applied to the core during testing. P_Trace value of any core is the averaging of 400:1 test cycles at 1.2 GHz test frequency. Here, the test frequency is 3 GHz and thus P_Trace is averaging of 1000:1 test cycles. Combination of HHDF, VHDF gives RHDF which is combined with P_Trace to get RHDF P_Trace product which is an indicator of heat dissipation capability of any core in the stack. More the RHDF P_Trace product, more time the core will take to dissipate heat. In the algorithm these cores were arranged in descending order of RHDF P_Trace product values and then test schedules were prepared from this sorted list considering the fact that no more

than two adjacent cores in horizontal or vertical direction are selected in a particular test schedule. This is Adjacency Exclusion Principle employed in generating test schedules.

In Fig. 3, Fig. 4 and Fig. 5 are shown the benchmark circuits used for implementing stacks. Fig. 6 and Fig. 7 shows the 3D Stacked SoCs used by authors with the arrangements of benchmark circuits as shown there upon. As seen from the Fig. 6 and Fig. 7, the benchmark circuit d695, which has 10 cores, is closest to heat sink.

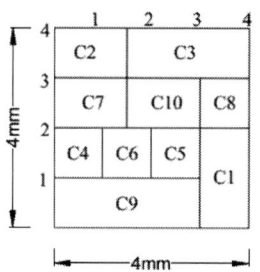

Fig. 3. The d695 benchmark circuit

Fig. 4. The d281 benchmark circuit

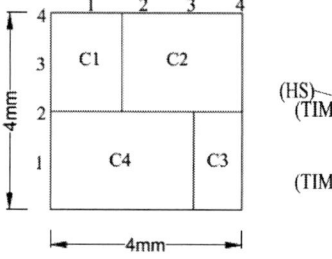

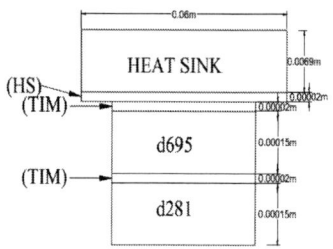

Fig. 5. The f2126 benchmark circuit

Fig. 6. The 2 layered stack

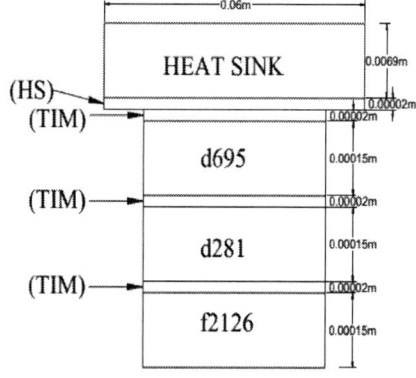

Fig. 7. The 3 layered stack

TABLE II. THE 2 LAYERED STACK TEST SCHEDULE (D695 NEAR HEAT SINK)

Schedule No.	Selected Cores (Layer – Core No)
Sch 1	2-3, 2-6; 0-2, 0-7, 0-8:
Sch 2	2-2, 2-4, 2-8; 0-3, 0-5, 0-6:
Sch 3	2-5, 2-7; 0-1, 0-4:
Sch 4 & Sch 5	2-9, 2-10; 2-1:

TABLE III. THE 3 LAYERED STACK TEST SCHEDULE (D695 NEAR HEAT SINK)

Schedule No.	Selected Cores (Layer – Core No)
Sch 1	4-6, 4-8; 2-2, 2-5, 0-2:
Sch 2	4-3, 4-4, 4-5; 2-7, 2-8; 0-1, 0-3:
Sch 3	4-2, 4-9, 4-10; 2-1, 2-3, 2-4:
Sch 4 & Sch 5	4-1, 4-7; 2-6; 0-4:

The test schedules generated with this arrangement (with d695 near heat sink) in 2 and 3 layered stacked structures are shown in Table II and Table III.

In this paper, the layer arrangement are altered such that every layer gets a chance to be closest to heat sink and then the test schedules are generate accordingly. The temperature profile after implementing these schedules on HotSpot-5.02 is studied.

In 2 layered stacked SoC, when d281 is put closer to heat sink, the test schedules, as shown in Table IV, are generated for parallel testing of cores. Here since d695 is away from heat sink it gets numbered as layer 0 whereas d281 gets numbered 2. Results are displayed with this convention only. Core no. 0-2 means core 2 of d 695 circuit. Alternate numbering takes place because of the reason that in between two circuit, Thermal Interface Material is placed (Fig. 6 and Fig. 7) which gets numbered as odd layers during test schedules implementation. Fig.8 shows in pictorial form these test schedules generated for 2 layered stacked SoC with d281 near heat sink.

Similarly, test schedules for 3 layers with d281 closest to heat sink and f2126 closest to heat sink are generated. The results of test schedules with d281 near heat sink are shown in Fig. 9 (in pictorial form) and Table V. Here also the core farthest from heat sink i.e. f2126 is numbered layer 0, d695 gets numbered 2 and d281 is numbered 4, which is closest to heat sink. The test schedules generated for 3 layered stacks with f2126 closest to heat sink are shown in Fig. 10 (in pictorial form) and Table VI.

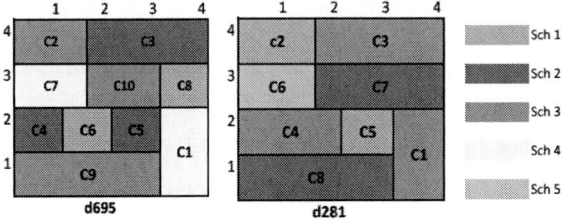

Fig. 8. The 2 Layer test schedules generation with d 281 closer to heat sink.

TABLE IV. THE 2 LAYERED STACK TEST SCHEDULE (D281 NEAR HEAT SINK)

Schedule No.	Selected Cores (Layer – Core No)
Sch 1	2-2, 2-5; 0-6, 0-8:
Sch 2	2-7, 2-8; 0-3, 0-4, 0-5:
Sch 3	2-1, 2-3, 2-4; 0-2, 0-9, 0-10:
Sch 4 & Sch 5	2-6, 0-1; 0-7:

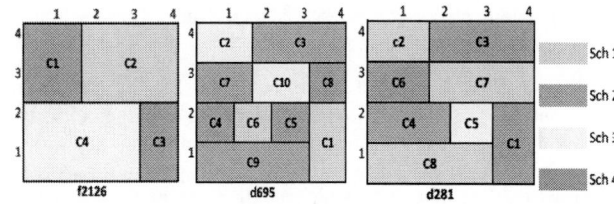

Fig. 9. The 3 Layer test schedules generation with d 281 closest to heat sink.

TABLE V. THE 3 LAYERED STACK TEST SCHEDULE (D281 NEAR HEAT SINK)

Schedule No.	Selected Cores (Layer – Core No)
Sch 1	4-2, 4-7, 4-8; 2-6, 2-1; 0-2:
Sch 2	4-1, 4-3, 4-6; 2-4, 2-5, 2-8; 0-1, 0-3:
Sch 3	4-5; 2-2, 2-10; 0-4:
Sch 4	4-4; 2-3, 2-7, 2-9:

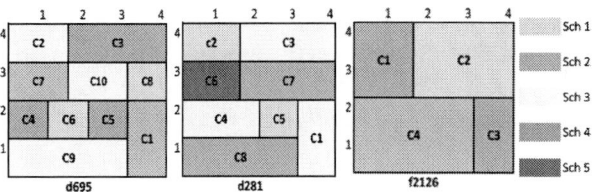

Fig. 10. The 3Layer test schedules generation with f 2126 closest to heat sink.

TABLE VI. THE 3 LAYERED STACK TEST SCHEDULE (F2126 NEAR HEAT SINK)

Schedule No.	Selected Cores (Layer – Core No)
Sch 1	4-2; 2-2, 2-5; 0-6, 0-8:
Sch 2	4-1, 4-3; 2-7, 2-8; 0-3, 0-4, 0-5:
Sch 3	2-1, 2-3, 2-4; 0-2, 0-9, 0-10:
Sch 4 & Sch 5	4-4; 2-6; 0-1, 0-7:

V. IMPLEMENTATION OF TEST SCHEDULES ON HOTSPOT

The HotSpot tool is an accurate and fast model based on an equivalent circuit of thermal resistances and capacitances that make up the micro architecture blocks and essential aspects of the thermal package. Validation of this model has been performed using finite element simulation [1,8]. Test schedules have been implemented by applying power traces on HotSpot-5.02. The chips today comprise the die placed on a spreader plate, made of aluminium, copper, or some other highly conductive material, which is in turn placed against a heat sink of aluminium or copper. This is the configuration modelled by HotSpot. Our stacks have the same configuration, i.e. heat spreader, heat sink and with interfacing material designed in between layers. HotSpot dynamically generates the RC circuit when provided with an input consisting of the blocks' layout and their areas. It is also provided with a power input values (these are the values for the current sources) over any time step and the present temperature of each block. HotSpot generates the temperatures at the centre of each core. Details of stacks,

978-1-5090-0037-1/16 $31.00 © 2016 IEEE

like floorplan, power trace files, area and initial temperatures are given as input in HotSpot.

Test schedules as generated above for different arrangements of layers in 2 and 3 layered stacks are implemented on HotSpot and the results so obtained are shown in graphical form in following Fig. 11 to Fig. 15. Where Fig. 11 and Fig. 12 show the results of all cores of the 2 layered stack after each test schedule. In Fig. 11 the results are of 2 layered stack where d695 is closest to heat sink. Fig. 12 is also the result of 2 layered stack but here the d281 circuit is closest to heat sink. Similarly, in Fig 13, 14 and 15 results of 3 layered stacks are shown with d695 closest to heat sink in Fig. 13, d281 closest to heat sink in Fig. 14 and f2126 closest to heat sink in Fig. 15.

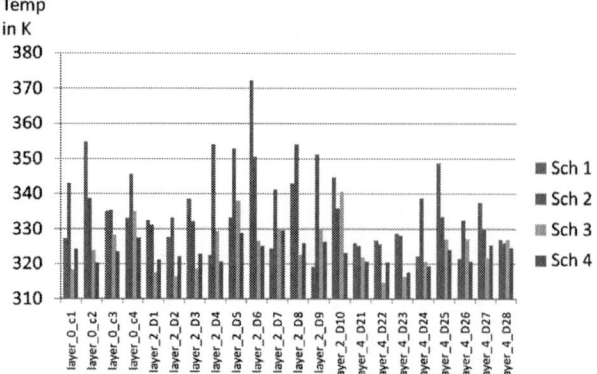

Fig. 14. Temperature in 3 layers scheduled testing (d281 near heat sink)

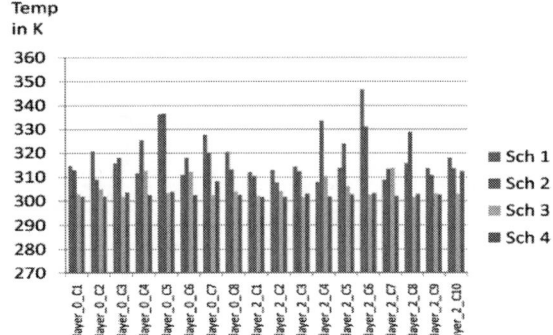

Fig. 11. Temperature in 2 layers scheduled testing (d695 near heat sink)

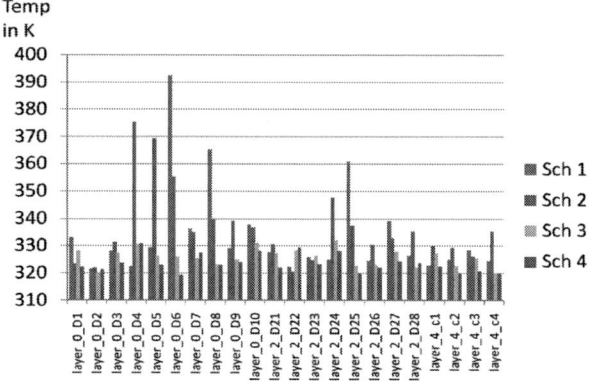

Fig. 15. Temperature in 3 layers scheduled testing (f 2126 near heat sink)

TABLE VII. COMPARISON RESULTS OF 2 LAYERS

Test Sch	D 695 near HS			D 281 near HS		
	Core	T_{max}	Time	Core	T_{max}	Time (ms)
1	C 2,6	346.71	18.5	C 0,6	375.14	18.5
2	C 0,5	336.65	13.3	C 0,4	357.72	13.3
3	C 2,7	313.76	12.32	C 0,10	325.12	12.32
4	C 2,10	312.53	12.74	C 0,7	321.96	11.09
Total Time			56.86	Total Time		55.21
Test Max Temp = 346.71 K				Test Max Temp = 375.14 K		
Layer	Max. Temp K			Layer	Max. Temp K	
D281	336.65			D281	341.91	
D695	346.71			D695	375.14	
Chip mean Temp	311.35 K			314.61 K		

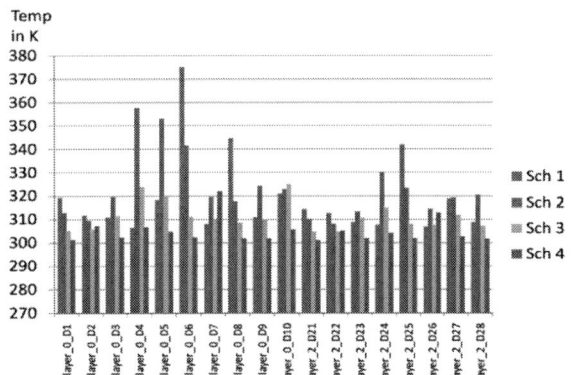

Fig. 12. Temperature in 2 layers scheduled testing (d281 near heat sink)

TABLE VIII. COMPARISON RESULTS OF 3 LAYERS

Test Sch	D 695 near HS			D 281 near HS			F 2126 near HS		
	Core	T_{max}	Time	Core	T_{max}	Time	Core	T_{max}	Time
1	C 4,6	367.5	23.9	C 2,6	372.4	23.9	C 0,6	392.6	23.9
2	C 2,5	351.6	13.4	C 2,4	354.1	13.4	C 0,4	375.5	13.4
3	C 4,5	344.2	17.01	C2,10	340.6	15.02	C0,10	332.1	15.02
4	C6,10	331.5	12.8	C 2,7	329.7	12.5	C 0,7	330.9	12.5
Total Time			67.11	Total Time		64.82	Total Time		64.82
Test Tmax = 367.5 K				Test Tmax = 372.4 K			Test Tmax = 392.6 K		
Layer	Tmax K			Layer	Tmax K		Layer	Tmax K	
F 2126	364.41			F2126	354.66		F2126	335.37	
D281	358.98			D281	348.74		D281	360.97	
D695	367.51			D695	372.38		D695	392.6	
Chip mean Temp	333.23 K			330.15 K			330.05 K		

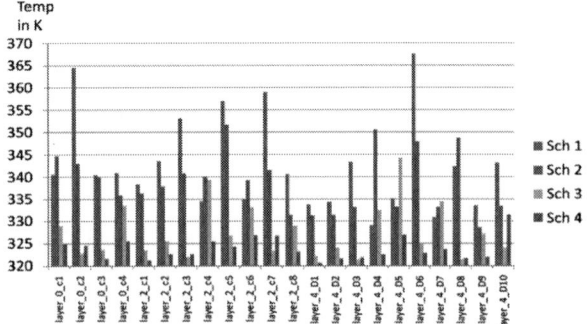

Fig. 13. Temperature in 3 layers scheduled testing (d695 near heat sink)

978-1-5090-0037-1/16 $31.00 © 2016 IEEE

These results are tabulated for 2 and 3 layers in Tables VII and VIII respectively.

From Tables VII and VIII it is very much clear that the temperature rise of cores vary with the placement of layers in the stack. In 2 layered stack, the variation in maximum temperature is 28.4 K whereas in 3 stacked structure the variation in maximum temperature in all 3 types of arrangements is 25.1 K. The variation in time taken in complete test is small in different arrangements of layers as in 2 layered stack the variation in time is only 1.6 ms and in 3 stacked structure the variation in total test time is 2.29 ms in different layers arrangements. It is seen that in 2 and 3 stacked structures, the maximum temperature is low when the d 695 is placed near the heat sink.

VI. CONCLUSION AND INFERENCE

It is known that all cores have different P_Trace values. The maximum temperature reached in all layers is seen to vary with the placement of layers in the stacks. When d695 is placed near heat sink the maximum temperature reached during the test in 2 and 3 layered stack remains low. Other than P_Trace values, all cores have different area also. On the basis of P_Trace value and area of cores, P_Trace density of all cores is calculated and tabulated below in Table IX. If P_Trace value of a core is high but its area is large then P_Trace density of the core will be low but in case of a small area core, even relatively small value of P_Trace value of the core will cause it to attain a relatively high P_Trace density. High P_Trace density will have a great impact on the temperature rise of the core.

Here it can be seen that P_Trace density of core 6 of d695 is highest among all cores. This causes maximum heat generation and therefore placement of d695 closest to heat sink keeps a check on the temperature rise during testing of cores. In all three arrangements, core no. 6 of d695 reaches the highest temperature. When d695 is placed nearest to heat sink, the maximum temperature is 367.5 K, when d695 is in the middle, the maximum temperature of core 6 of d695 is 372.4 K. When d695 is placed farthest from heat sink, core 6 of this circuit has maximum temperature of 392.6 K.

It is further proposed to extend this test to 4, 5, 6, 7 and 8 layers SoCs having highly varying P_Trace values and P_Trace density to draw inferences on higher layered circuits. Authors also propose to derive test partitioning and test schedules interlacing schemes in addition to test schedules based on Adjacency Exclusion Principle to further control the temperature rise during testing of cores.

TABLE IX. P_TRACE DENSITY OF ALL CIRCUITS

Core No.	d 695	d 281	f 2126
Core 1	0.01	0.14	0.50
Core 2	0.05	1.47	1.44
Core 3	1.00	0.86	0.45
Core 4	5.83	0.87	0.15
Core 5	5.11	2.62	
Core 6	9.87	0.67	
Core 7	2.24	1.61	
Core 8	4.61	1.08	
Core 9	0.24		
Core 10	2.58		

REFERENCES

[1] W. Huang, S. Ghosh, S. Velusamy, K. Sankaranarayanan, K. Skadron and M. R. Stan, "HotSpot: A compact thermal modeling methodology for early-stage VLSI design," in IEEE Transactions on Very Large Scale Integration (VLSI) Systems, 14(5), pp. 501–513, 2005.

[2] C. Yao, K. K. Saluja and P. Ramanathan, "Test scheduling for deep submicron technologies," in International Conference on VLSI Design, 2011.

[3] Z. He, Z. Peng and P. Eles, "A heuristic for thermal safe SoC test scheduling," in IEEE International Test Conf, pp. 1-10, 2007.

[4] S. Millican and K. K. Saluja, "Linear programming formulations for thermal aware test scheduling of 3D stacked integrated circuits," in 21st IEEE Asian Test Symposium, pp. 37-42, 2012.

[5] S. Millican and K. K. Saluja, "Optimal test scheduling of stacked circuits under various hardware and power constraints," in 28th International Conference on VLSI Design and 14th International Conference on Embedded Systems, pp. 487-492, 2015.

[6] K. Banerjee, S. L. Souri, P. Kapur and K. C. Saraswat, "3D ICs: A novel chip design for improving deep submicrometer interconnect performance and system on chip integration," in Proc. IEEE, 89(5), pp. 602-633, 2001.

[7] S. L. Souri, K. Banerjee, A. Mehrotra and K. C. Saraswat, "Multiple Si layer ICs: Motivation, performance analysis and design implications," in 37th ACM Design Automation Conference, pp. 873-880, 2000.

[8] I. Rawat, M. K. Gupta, V. Singh, "Temperature and time efficient parallel test scheduling for 3D stacked SoCs," in Proc. IEEE International Conference on Research in Computational Intelligence and Communication Networks, Kolkatta, pp. 306-311, Nov. 20-22, 2015.

[9] K. Skadron et.al, "Temperature aware microarchitecture," in Proc. ISCA-30, pp. 2-13, June 2003.

[10] E. J. Marinisen, V. Iyenger and K. Chakrabarty, "A set of benchmarks for modular Testing of SoCs," ITC, 2002.

Functional Verification of DSP based On-board VLSI Designs

Sourabh Jain[1], Parimal Govani[2], Kamal B Poddar[3], A K Lal[4], R M Parmar[5]

Space Applications Centre

Indian Space Research Organization

Ahmedabad, India

[1]sourabhjain@sac.isro.gov.in, [2]parimal@sac.isro.gov.in, [3]kamal_p@sac.isro.gov.in,
[4]aklal@sac.isro.gov.in, [5]rmparmar@sac.isro.gov.in

Abstract— The Usage of Field Programmable Gate Arrays (FPGA) and Application Specific Integrated Circuits (ASICs) with complex functionalities such as Digital Signal Processing (DSP) is increasing in onboard space applications. Verification of these complex designs within limited schedule and resources is challenging. In order to ensure reliable functioning of these designs in all possible run time conditions, functional verification is required to be carried out thoroughly. Development of an automated self-checking verification environment or test benches, including generation of bit-accurate golden reference values, is complex and time consuming task even with the use of state-of-the-art Hardware Verification Languages (HVLs) and methodology such as System-Verilog (SV) and Universal Verification Methodology (UVM) respectively.

This paper discusses a method for functional verification of DSP based VLSI design using SV and Matlab. The architecture of verification environment and technique for coupling of Matlab with SV based verification environment and generation of bit-accurate golden references, in real time is also discussed in detail, along with two case studies.

Keywords— DSP, VLSI, UVM, predictor, coverage driven verification, DPI

I. INTRODUCTION

To meet ever increasing functional requirements, digital VLSI designs are getting complex. Design teams are packing more & more logic gates onto a single chip to achieve desired functionality and performance within the specified footprint. Functional verification of such designs with tradition approach of using directed test benches does not provide sufficient confidence within given time schedule. Test benches written in SV gives advantages in terms enabling constrained random stimulus generation, self-checking and assertion based verification along with defining functional coverage matrix. Random testing improves productivity over manual testing, in terms of number of test vectors produced and generates test cases not explicitly thought by verification engineer. Binding assertions to a design during simulation phase identifies design flaws in real-time and reduces debugging time significantly over non-assertion based design. Simulated design's output is compared with golden reference values generated using HVL & verified automatically during run-time. Functional simulation is considered complete when the goal of 100% functional coverage is achieved.

In this SV based test bench, control signals output of the designs are validated using assertions and data processing functionalities of the designs are validated with predictor or checker. These checkers are generally hand-coded using higher level of abstraction by the verification engineer.

Development of checker for complex functionalities such as DSP is complex. This becomes even more challenging for an onboard design implementing multiple DSP IP cores having functionalities such as sine–cosine lookup table, fixed to floating point conversion & vice versa, FFT, FIR filter, etc.

DSP algorithms are available as standard functions in MATLAB. If these functions can be used in the test bench as golden reference models/checker, the test bench can be simplified and overall efficiency of the verification can be significantly improved. This paper discusses the verification environment development using SV and methodology to couple Matlab with SV.

II. PREDICTOR IN VERIFICATION ENVIRONMENT

Verification of a VLSI design consists of two major steps.
1. Stimulus generation
2. Analysis of the design response

In stimulus generation step, the design is configured in a particular mode and stimulus is applied. In analysis part, the actual verification is performed. A sample test bench architecture performing both these operations automatically is illustrated in Figure 1.

The test bench (verification environment) developed in SV using UVM is composed of reusable verification environment called verification components. Each component is encapsulated, ready-to-use, and configurable which can be used for verification of any interface protocol, design sub-module or a full system. The verification components along with device under test (DUT) is used to verify implementation of the protocol or design architecture.

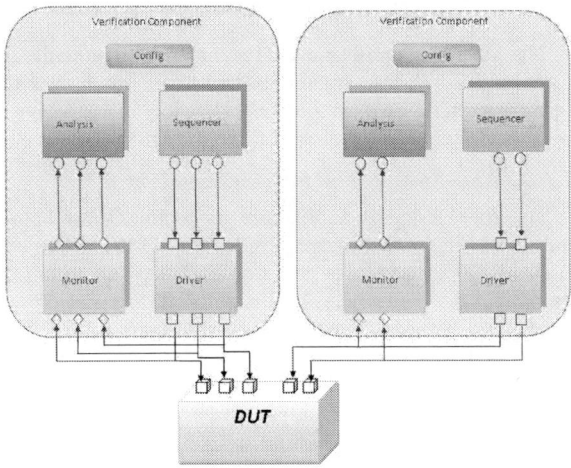

Figure 1: Verification Environment

The analysis component shown in Figure 2, consists of components that observe behavior of the DUT. The major parts for analysis component are coverage collection and scoreboard. Scoreboard determines whether the design is functioning properly or not. The scoreboard architecture separates its tasks into two areas of concern viz. prediction and evaluation. A predictor model, sometimes referred to as a 'Golden Reference Model', receives the same stimulus stream as the DUT and produces known response transaction streams. The predictor implements the DUT functionality at a higher level of abstraction written in C, C++, SV or System-C. After the correct functionality is predicted, the scoreboard can evaluate the actual results observed on the DUT with the predicted results.

III. USE OF MATLAB IN PREDICTOR

Matlab is proven industry standard for implementing DSP algorithms. The DSP functions available or algorithms developed using Matlab can be directly used to evaluate the performance of the HDL designs. Such DSP functions will have the same functionalities as the HDL designs. Figure 3 shows the usage of Matlab DSP function inside the predictor component to generate the golden reference values. It significantly simplifies the verification of complex design. However, integrating Matlab DSP functions directly in predictor is challenging task as it does not support HVL or HDL constructs directly.

Following methods are explored to use Matlab functions in verification environments -

A. Direct Programming Interface (DPI) and MATLAB Engine :- Matlab can be interfaced with SV test bench using DPI. A program written in 'C' language containing Matlab engine routines makes the bridge between them.

B. HDL Verifier:- Matlab tool box named HDL verifier, provides EDA link to get connection for supporting simulators like Cadence IEV and Questasim

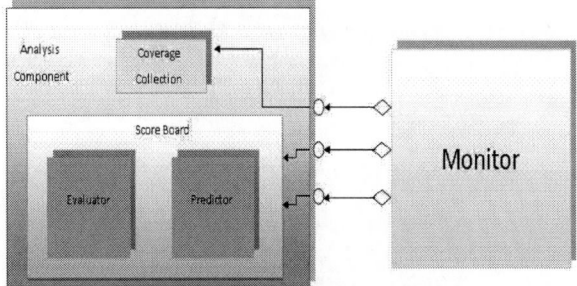

Figure 2: Analysis Component

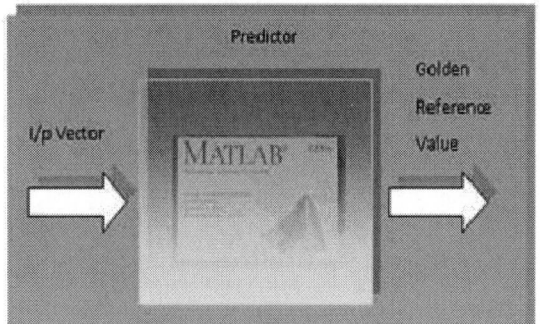

Figure 3: Matlab in Predictor

C. MATLAB communication with TLM2 transactions:- Both, UVM library and Matlab support TLM2 transactions with System C environment. Utilizing this method, an interface between *UVM* to *System C TLM2 and System C TLM2* to *Matlab* can be established.

Method A is independent of EDA simulator, configurable and does not require any specific tool kit. Hence, it was chosen for implementation.

DPI is an interface between SV and 'C', which allows direct inter-language function calls on either side of the interface. Matlab provides engine library functions, containing routines that allow calling Matlab from programs written in 'C' and FORTRAN. The engine library contains nine routines for controlling Matlab computation engine from a 'C' program. Table 1 summarizes these routines.

'C' program employing these routines can be used to establish the link between SV and Matlab [1] as shown in Figure 4.

Figure 5 shows a sample C program written for interface between a SV test bench and a Matlab algorithm.

Table 1: Matlab engine routines

C Routines	Matlab Functionality
eng{open|close}	Used to Start/Close its engine
engEvalString	Used for execution of its command
eng{get|put}	Used to get/put its variable array value from/to the engine
engOutputBuffer	Used for reading its buffer text content
eng{get|set}Visible	Used for making its engine visibility on /off
engOpenSingleUse	Start for its engine session

978-1-5090-0037-1/16 $31.00 © 2016 IEEE

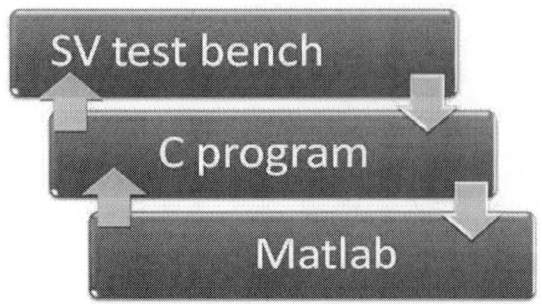

Figure 4: SV- Matlab Interface

```c
#include "engine.h"
#include "matrix.h"
#define BUFSIZE 80000

Engine *ep;
mxArray *T = NULL, *result = NULL;
char buffer[BUFSIZE+1];
int  buffer_x1 ;
double  result_1[6];
mxArray *result ;

int matlab_engine_start(char *cmd) // starting matlab engine
{
        if (!ep && !(ep = engOpen(cmd))) {
                fprintf(stderr, "\nMATLAB engine can not be started \n");
                return 0;
        }
        engSetVisible(ep ,1 );
        engOutputBuffer(ep, buffer, BUFSIZE);
        return 1;
}

int matlab_cmd( const char** cmd)  // execution of matlab command/functions
{    printf("\ncmd is\n %s" , cmd);
        return engEvalString(ep, cmd);
}

void matlab_variable( char** cmd,double z[] ,int k) //reading of matlab buffer
{
double *y;
int i;
mxArray *result = NULL;
```

Figure 5: Interface C program

System Verilog and Matlab have different data types. Matlab's default data type is a matrix double, whereas SV test bench sends and receives data in binary format. Hence, additional routines were developed in 'C' program to perform the required conversion on the received data for compatibility with target environment.

Here, Matlab files (.m) and C files are compiled into a shared library which creates a shared library file (.so) and a header file (.h). To communicate with the shared library, Matlab requires usage of its existing matrix functions. The compilation of 'C' file is done using gcc compiler and Matlab API header files. The compiled unit is also stored in a shared library.

IV. DSP CASE STUDY

The functional verification of two onboard IP core designs are considered as case studies. In these case studies, Matlab algorithms were used as golden reference in real-time. A coverage driven functional simulation for both the designs.

A. Case-Study 1: Fast Fourier Transformation

The first design is an IP core of complex Fast Fourier Transform (FFT) having transform length of 8192. It provides unscaled output and uses the Cooley-Tukey algorithm for computing the FFT. The Verification of this design was carried out by using Matlab FFT algorithm of the same transform size as a golden reference. Constrained random test vectors generated using SV were applied on both DUT and Matlab model. The DUT outputs were compared bit-by-bit against Matlab output in real time.

Further, signal of a standard frequency with random noise was generated in Matlab and applied to both reference model as well as DUT for shape evaluation. The outputs from both were plotted in Matlab in real time. Plotting of the DUT response over the Matlab output revealed that the design outputs were not exactly matching and there were minor differences due to DUT introducing undesired frequency components as shown in Figure 6.

Further debugging of DUT shown that these differences are introduced by compression of the processed data due to limited register length at every stage of processing and is confirmed by the design team.

B. Case Study 2: Sine-cosine LUT

The another design is an IP core of sine cosine look up table generator. In this design, the input angle widths is 16 bits which provides resolution (step size) better than 0.005 and output width is 10 bits long for better precision.

It was verified for all possible input angle values using constrained random test vector by running Matlab algorithm in a loop. Design output (sine /cosine values) were compared with Matlab generated values. Angle values (16 bits) generated by SV test bench, act as addresses for sine or cosine LUT in the design. To get the angle value from the Matlab algorithm for same input angle value, additional routines for angle conversion operation were required to be developed in 'C'. To make Matlab algorithm output (double - 64 bits long) as DUT equivalent (10 bits), fixed point conversion of these values was carried out in Matlab. Final output of DUT was successfully compared and matched with Matlab output (i.e. golden reference values).

V. ADVANTAGES

Use of MATLAB algorithm as a golden reference has resulted in reduction of test bench code development time and complexity. Comparison of outputs on MATLAB plots has made debugging of complex DSP functions easier. Hence, the overall efficiency of the functional verification was considerably improved.

978-1-5090-0037-1/16 $31.00 © 2016 IEEE

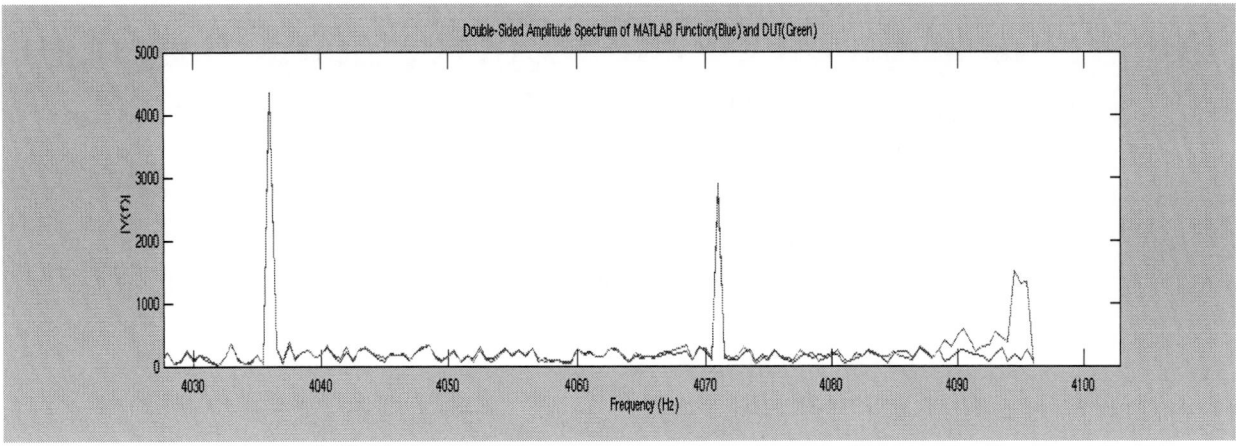

Figure 6: FFT responses of DUT (green) and Matlab function (blue)

VI. CONCLUSION

An innovative automatic self-checking functional verification method has been employed to ensure the functional correctness of complex DSP based designs. Usage of this approach enables verification of DSP based functionalities in lesser time with higher coverage. Here, Matlab is used to generate some specific input stimulus signals, which are difficult to generate in SV. It is also feasible to carry out extended analysis of the DUT output using spectrum analysis, filtering at RTL level.

ACKNOWLEDGMENT

The work presented in this paper reflects the interdependent efforts of many individuals. The authors wish to acknowledge the contributions of entire design and verification team, who provided their guidance in successful implementation of this method. We would also like to express our immense gratitude to Director SAC for providing the opportunity to carry out this task.

REFERENCES

[1] Dhaval Modi, Integrating MATLAB with verification HDLs for Functional Verification of Image and Video Processing ASIC" International Journal of Computer Science & Emerging Technologies" Volume 2, Issue 2, April 2011"

[2] Cookbook–UVM by Verification Academy Mentor graphics.

[3] Writing Testbenches using System Verilog by Janick Bergeron Synopsys, Inc.

[4] SystemVerilog for Verification -A Guide to Learning the Testbench Language Features   by Chris Spear

An Arbitration on Cache Replacements Based on Frequency - Recency Product Values

Somak Das and Aikatan Banerjee
Dept. of Computer Science & Engineering
University Institute of Technology, The University of Burdwan
Burdwan, India
somakdas2@gmail.com

Abstract— Evolving an efficient cache replacement policy has been a challenge since the last few decades. LRU (Least Recently Used) and LFU (Least Frequently Used) cache replacement techniques and a variety of their combinations were the most sought after. This paper proposes a new combination of the LRU and LFU in such a style that the time and complexity to replace moves below the current benchmarks. Here, a frequency-recency product value is computed which dictates the cache replacement arbitration. It out performs the existing methods by a significant reduction in computational overhead.

Keywords—LRU; LFU; frequency; recency; LFRU; set-associative mapping; minima; replacement policy;

I. INTRODUCTION

Processor accesses the cache memory before the main memory in search of data. If data is not present in cache, then the block of main memory with the desired data content is loaded into the cache and delivered to the processor. Because of the phenomenon of locality of reference, when a block of data is fetched into the cache to satisfy a single memory reference, it is likely that there will be future references to that same memory location or to other words in the block [1].

There are three basic cache organizations - direct, fully-associative and set-associative. In direct mapping, any main memory data block has a one specific line in cache where the block of data can reside. It has a very high miss rate. For fully-associative cache mapping, any data block from the main memory can be stored in any cache line. But it is very expensive in implementation and has also hardware complex [2]. The third organization is set-associative, where a mixture of direct and fully associative mapping is exploited. It divides the cache into a number of sets and allows a block of data from the main memory to be loaded at any empty cache line within the designated set. If there are no empty cache lines, a replacement policy will be effectively needed within the set. Set-associability provides optimal solution in mapping of cache memories and reduces the chances of cache miss [3]. Also it is relatively less hardware complex with respect to the fully-associative one.

All cache replacement algorithms work in a unified goal, which is reduction of miss rate. When a desired block of data

is not found in cache, processor loads the data block from main memory into the cache. This extra delay is called miss penalty. The hit rate of a cache is defined by the probability of a data block being found within the cache memory [2] [4]. This results in a higher bandwidth of data transfer. Selection of a cache replacement algorithm in set-associative cache memory plays an important role in improving the cache hit rate [2].

Random cache replacement is a policy where, the desired block of data if not available in the cache, is placed in any of the cache lines randomly replacing the content of the cache line. FIFO (first-in-first-out) is another replacement technique where such blocks of data within a cache line is emptied which had been occupied the cache first in sequence [4].

There are two main cache replacement parameters – recency and frequency. Recency spans from the current access time to the last access time of a certain data block. This concept is used in Least Recently Used (LRU) policy [4] [5] [6]. The second one is frequency, which is used by the Least Frequently Used (LFU) policy [4] [7]. In this policy, each cache line holds a counter that gets incremented each time the block of data placed in a cache line within a set is accessed. Whenever the referred block of data is not available in the cache, the cache line with block of data with minimum counter value i.e. minimum frequency is replaced [7]. LRU ignores the frequency of use of the cache lines within a set, so the most accessed cache line may be the victim of replacement. LFU ignores the latest referred cache line and thus chances of cache miss increases. For these reasons, a combination of frequency and recency became popular uplifting the hit ratio [4] [7].

LRU/LFU cache replacement policy adjusts a CRF (Combined Recency and Frequency) value [2] which quantifies the probability that a block may be referred in near future.

ARC (Adaptive Replacement Cache) is a cache replacement policy [8] which is self-tuned and dynamically adaptive. It balances the recency and frequency parameters in a continuous fashion. ARC displays constant complexity values per request and it allows sequential request once to pass through without any cache pollution. SF-LRU (Second Chance – Frequency – Least Recently Used) [6] is a cache replacement policy where block of data having minimum

RFCV (Recency-Frequency Control Value) is replaced. RFCV takes into account the number of times a block of memory has been referred and the second chance of the maximum referred block of memory before it is being replaced. Combination of LRU and LFU (CRFP) [9] cache replacement policies were also suggested where a switching of any one of these policies took place based on the memory access pattern. Designing efficient switching method and optimizing the cache directory overhead remain shortfall of this policy. Least Recently Plus Five Least Frequently Replacement Policy (LR+5LF) [7] replaces such cache line with data blocks having minimum weighting least recently frequently used (WLRFU) values.

A new replacement policy proposed to be named as Least Frequently and Least Recently Used (LFRU), which combines both LRU and LFU, is presented in this paper. It uses both of the access recency and frequency of the each cache line and is simple to implement and maintain.

Each cache line uses two access parameters, the Recency (R) and the Frequency (F) for all set index and for all line index in the cache. According to this proposed policy, the data content of a cache line in a set that contains minimum least recently and least frequently access parameter will be replaced by new block of data within the same set. The analysis of results clearly shows that this proposed policy significantly reduces the number of cache misses than the already existing ones.

II. POPULAR CACHE REPLACEMENT POLICIES

A. FIFO (First In First Out)

FIFO is a very simple replacement policy; here the block of data in a cache line that that resides in the cache memory for the longest amount of time [4] is simply replaced by the new block of data from main memory and the next replacement will be the second cache line in turn and so on. The operating system designates the oldest block of data present at the cache memory in a round-robin style. It is successful only in assumptions that processor needs newer blocks of data more frequently.

B. LRU (Least Recently Used)

This policy replaces such block of data from any cache line which has not been referred by the processor since the longest time interval [4] [5]. A counter may be associated with each cache line which gets incremented each time any data block residing in another cache line is invoked. A higher value of this counter gives the time interval a particular cache line not being referred.

C. LFU (Least Frequently Used)

This policy replaces such data block of the cache memory which has the lowest number of memory reference [4] [5] [7]. Associating a counter with each cache line and incrementing the counter as and when the particular cache line is referred by the processor provides the value of frequency.

D. LRFU (Least Recently / Least Frequently Used)

The decision to replace with the help of LRU or LFU is taken by weighting of a parameter λ. Here the time complexity ranges between that of LRU and LFU individually [2]. But, the limitation is that λ could not be dynamically adjusted as the workload varies.

E. LR+5LF (Least Recently Plus Five Least Frequently)

It [7] was supposed to be the policy with the lowest rate of cache miss compared to the above ones. Two priority constants are attached each with LRU and LFU respectively and WLRFU values are calculated for each cache lines. The cache line with minimum WLRFU value is replaced.

III. PROPOSED LFRU CACHE REPLACEMENT POLICY

This newly proposed replacement policy is a modification of LRFU policy [2] that clubs up both LRU and LFU together.

A. LFRU Algorithm

```
Set G: =0;      //General Counter G as Recency Counter
Set min: = FR [0] [0];

If Memory_Content exist in ith set index and jth line index;
    /* checking whether desired data block is present in the
                    Cache Memory */
begin
                        R[i] [j] <=G;
 /* If it is found in the Cache Memory then the Recency
Counter for ith Set Index and jth Line Index is set to current
value of G */
                        G<=G+1;
// Then incremented the General Counter G by 1
                        F[i] [j] <=F[i] [j] +1;
// Rate is incremented by 1 for same
end
        for all set index i
                for all line index j
/* If desired data block is not present in the Cache Memory
then calculate the minimum FR and The Cache Line
having minimum FR is replaced for all ith set index and for
all jth line index */
                    FR[i] [j]<=F[i][j]*R[i][j] ;

                    end for
        end for

                                        Continued next
```

```
Else
        begin
                for all set index i
                        for all line index j

                        if(min>FR[i][j])

                                begin
                                min<=FR[i][j];
                                end
                        end for
                end for
        for all set index i
                for all line index j

                        If(min==FR[i][j])
                        begin
                                Cache [i][j]<=Memory_Content;
                                // Replacement
                                R[i][j]<=G;
/* If it is found in the Cache Memory then the Recency
Counter for i^th Set Index and j^th Line Index is set to current
value of G */
                                G<=G+1;
// Then incremented the General Counter G by 1
                                F[i][j] <=0;
/*Rate is set to 0 because the block of data which is placed
in the specific cache line will become new block of data in
the Cache */
                                end
                        end for
                end for
end
```

B. Details of the Proposed LFRU Technique

In this method if there are i number of sets and in each set j number of lines, then the cache line to be replaced for loading the desired data block from main memory can be settled by the minimum value of FR[i][j] where F is the frequency of accessing the cache line and R is the recency of accessing a cache line. When a required data is present in cache, both F and R are increased.

If desired data is not present in cache then the proposed replacement policy comes into picture.

Here, | FR[i][j] = F[i][j]*R[i][j] |

An example of cache replacement based on the proposed policy is being discussed here. Fig. 1 depicts loading of blocks of data within a cache-set (with say, 0th index i.e. i=0) which has 4-cache lines. At first, a data block X is loaded into the 0th line (j=0) of the set. The frequency (F) and the recency (R) values in this case are 1 and 1 respectively. Similarly, data blocks Y, Z and A are also loaded in rest of the 3-cache lines. The F, R values for each of the cache lines becomes (1,2), (1,3) and (1,4) respectively. Once the cache-set is filled with the data blocks, the processor may access data block Z, Y, X and Y in sequence. Now, the F, R values corresponding to the

cache lines which stores the accesses data blocks changes to (2,5), (2,6), (2,7) and (3,8) respectively. Data block B may now be referred which is designated to reside in the same cache-set. But, B is not available in any of the cache lines of the said set. There comes the requirement of cache replacement. As per the proposed LFRU policy, the F-R product values are determined.

FR[0][0] = 2*7=14
FR[0][1] = 3*8=24
FR[0][2] = 2*5=10
FR[0][3] = 1*4=4

The F-R product value of the 3rd cache line is found to be the minimum and thus fit to be replaced by the data block B arriving from the main memory. The F,R value of this cache line now changes to (1,9) and FR[0][3] = 1*9=9. Fig. 2 highlights some more LFRU replacements under way for loading of D and A data blocks. So, after the sequence of data access shown in fig. 2, the F-R product values become

FR[0][0] = 2*13=26
FR[0][1] = 3*8=24
FR[0][2] = 3*11=33
FR[0][3] = 2*12=24

Further referral of data block A and X is shown in fig. 3. As data block A is accessed, the F-R product value corresponding to the cache line 0 changes to FR[0][0] = 3*14=42. The next referral of data block X poses a tricky situation for LFRU as the required replacement has to deal with multiple F-R minima values.

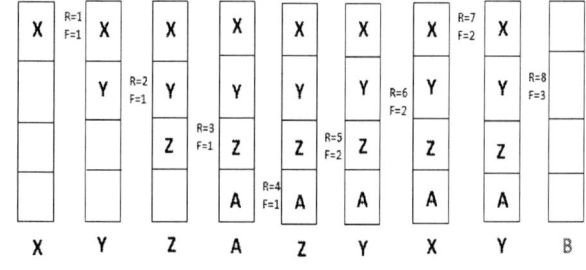

Fig. 1. Loading of the Cache lines

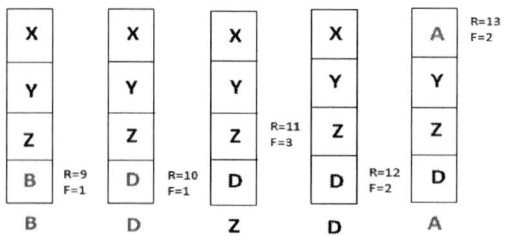

Fig. 2. Replacement by Proposed LFRU Policy

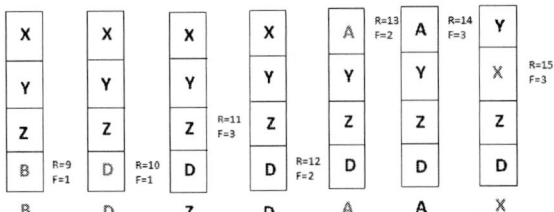

Fig. 3. When the 2 F-R Product Values are same

C. Multiple Minima of F-R product value: a Special Consideration

In general the cache line with minimum Frequency Recency (FR) product value is supposed to be replaced, but in any case, if two or more cache lines have the same minimum Frequency Recency product values as shown in fig. 3 then, special consideration is needed. Replacement of the cache line with minimum FR product value that is encountered first in sequence is to be replaced.

In fig. 3, after block of data A is loaded in the cache set, if it is again accessed by the processor, then the FR product values against each cache line will be

FR[0][0] = 3*14=42
FR[0][1] = 3*8=24
FR[0][2] = 3*11=33
FR[0][3] = 2*12=24

Therefore, it is found that FR [0][1] and FR [0][3] values are minimum within a set. As FR [0][1] is encountered first in sequence, the corresponding cache line which currently stores Y will be replaced by the new block of data X.

IV. SIMULATION AND IMPLEMENTATION

Custom 2-way, 4-way and 8-way set-associative cache memory structures were designed using Verilog HDL on Xilinx ISE 8.2i platform and simulated for replacements based on the proposed LFRU method using ISE simulator. Fig. 4, fig. 5 and fig. 6 are these simulation views. The 2-way set-associative cache with proposed replacement facility was also implemented on Xilinx Spartan-3E FPGA board. The placement of this design is shown in fig. 7.

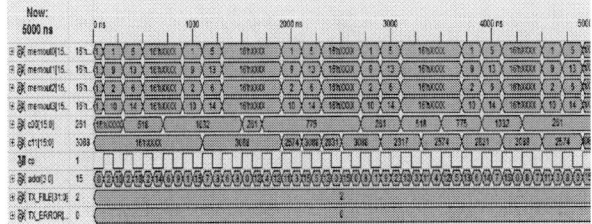

Fig. 4   Simulation View of 2-way Set-Associative Cache

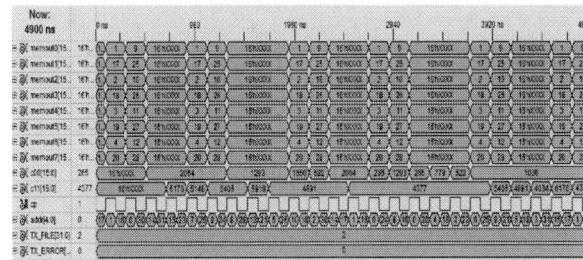

Fig. 5   Simulation View of 4-way Set-Associative Cache

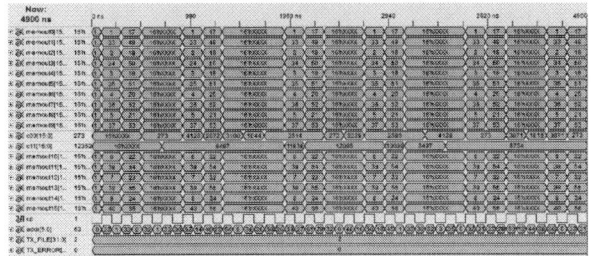

Fig. 6   Simulation View of 8-way Set-Associative Cache

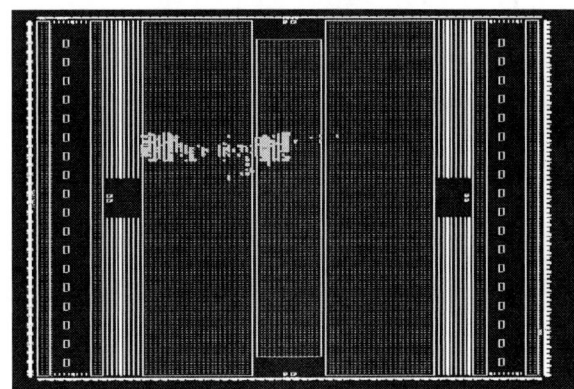

Fig. 7   Placement of the Design of 2-way Set-Associative Cache

V. COMPARATIVE CASE STUDY ON PRESUMED SEQUENCE OF DATA BLOCKS REFERRED

For a sequence of data blocks being referred by the main memory which has to be located inside a designated set within a 4-way set-associative cache memory, the access and replacement results were analyzed for the newly proposed LFRU policy with respected to the existing ones of LRU/LFU [2] and LR+5LF [7]. Let, X, Y, Z, A, Z, Y, X, Y, B, D, Z, D, A, Z, X, B, A be the sequenced data blocks being referred. Fig. 8 and fig. 9 gives us an idea of how access and replacements take place when policy [2] and [7] are in control of the cache. Whereas, Fig. 10 pictorially represents the frequency and recency values against each cache lines for each of the access and replacements as per the newly proposed cache replacement policy.

2016 International Conference on VLSI Systems, Architectures, Technology and Applications (VLSI-SATA)

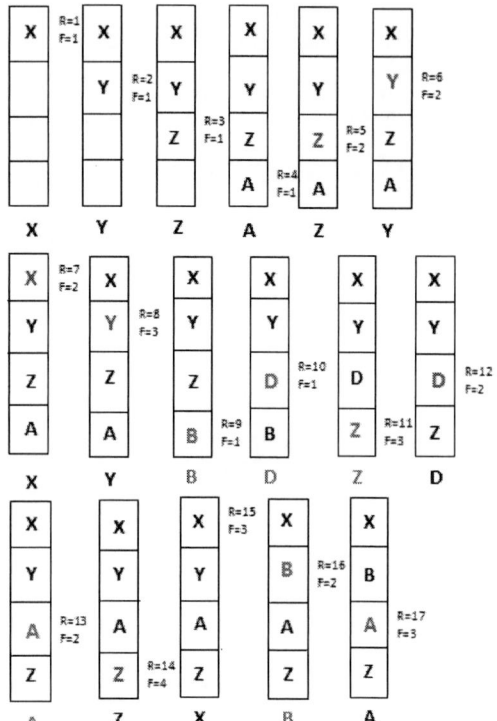

Fig. 8   Access and Replacements as per LRU/LFU [2] policy

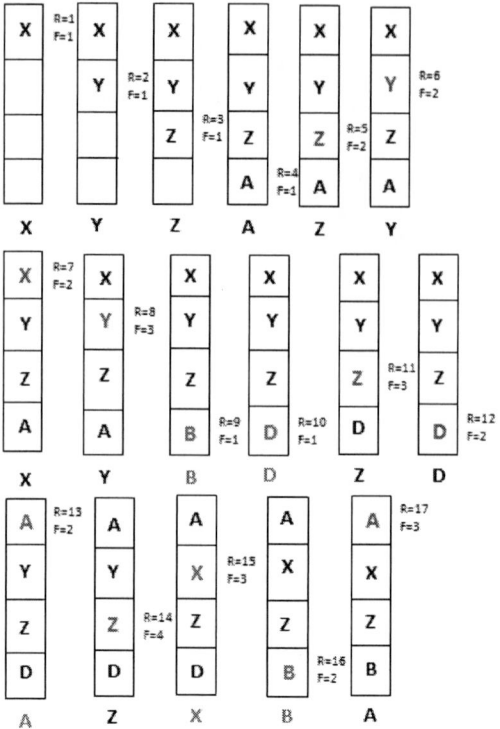

Fig. 10   Access and Replacements as per the newly proposed LFRU policy

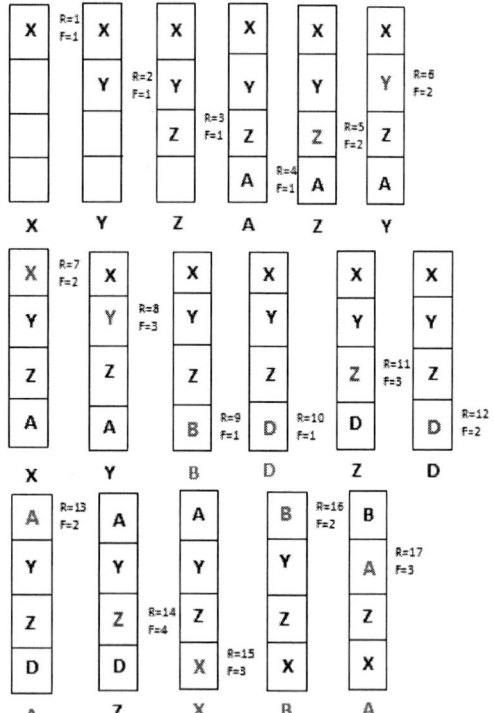

Fig. 9   Access and Replacements as per LR+5LF [2] policy

For the LRU/LFU policy [2] a CRF (Combined Recency and Frequency) value is calculated for each cache line of a set and as and when a replacement is inevitable, the cache line with minimum CRF value is selected for replacement. Fig. 8 shows the referred data blocks that are accessed but not replaced, in blue and those data blocks which are loaded in a cache line due to replacement of the old ones in red color.

Similarly, for LR+5LF policy [7] a WLRFU (Weighted Least Recently Least Frequently Used) value is calculated for each cache line of a set and as and when a replacement is inevitable, the cache line with minimum WLRFU value is selected for replacement. Fig. 9 shows the referred data blocks that are accessed but not replaced, in blue and those data blocks which are loaded in a cache line due to replacement of the old ones in red color.

As per the newly proposed LFRU (Least Frequently and Recently Used) cache replacement policy is concerned, it mainly deals in computations of F-R (frequency-recency) product values for each cache lines during loading, accessing and replacing of data blocks into it. For any particular instance when a replacement of cache line is required the line with minimum F-R product value bears the brunt of getting replaced. Fig. 10 shows the referred data blocks that are accessed but not replaced, in blue and those data blocks which are loaded in a cache line due to replacement of the old ones in red color.

978-1-5090-0037-1/16 $31.00 © 2016 IEEE

VI. RESULT ANALYSIS

Replacement parameters of already existing replacement policies [2] [7] have been calculated for example data block set of section V and compared along with the values of F-R products as proposed in the new LFRU policy (shown in Table I). Further analyses of parametric results are shown in fig. 11 and fig. 12.

TABLE I.   COMPARISON OF REPLACEMENT POLICIES

Replacement Policies	Sequenced Referral of Blocks of Data								
	B	D	Z	D	A	Z	X	B	A
LRU/LFU[2]	16	20	81	100	48	198	84	144	234
LR+5LF[7]	5	10	14	11	18	42	28	30	33
Newly Proposed LFRU	4	9	10	10	14	33	24	24	26

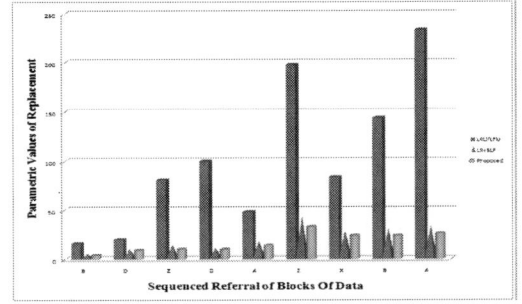

Fig. 11 Comparative Chart for Sequence of Blocks of Data being Referred

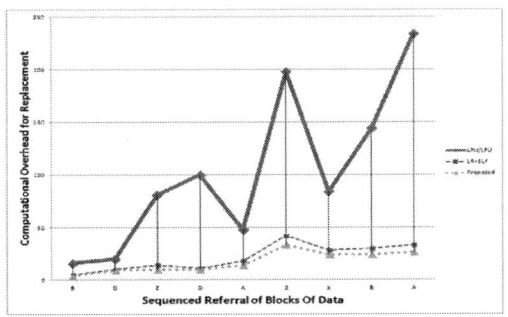

Fig. 12 Graphical View of Relative Computational Overhead

VII. CONCLUSION

Comparing the newly proposed LFRU cache replacement policy with the already existing methods [2] [7] suggests that the proposed policy is computationally less intensive. Replacements at cache lines with minimum frequency-recency product values also ensure better results in hit and miss rates. The proposed policy has been successfully implemented for set-associative cache memories. Implementation of the proposed policy on multi-level cache hierarchical structures remains as future scope of research.

REFERENCES

[1]   A. J. Smith, "Design of CPU Cache Memories," Proceedings of the IEEE TENCON, Seoul, Korea, August, 1987.

[2]   D. Lee, J. Choi, J. Kim, S. Noh, S. Min, Y. Cho and C. Kim, "LRFU: A Spectrum of Policies that Subsumes the Least Recently Used and Least Frequently Used Policies," *IEEE Transaction on Computers*, vol. 50, no. 12, pp. 1352-1361, 2001.

[3]   K. Qureshi, D. Thompson and N. Patt, "The V-Way Cache: Demand Based Associativity via Global Replacement," Proceedings of the 32th International Symposium on Computer Architecture, USA, pp. 544-555, 2005.

[4]   D. Swain, B. Nidhi Dash, D. O. Shamkuwar and D. Swain, " Analysis and Predictability of Page Replacement Techniques towards Optimized Performance," International Journal of Computer Applications, pp. 12-16, 2011.

[5]   A. Wong and L. Baer, "Modified LRU Policies for Improving Second-Level Cache Behavior," Proceedings of 6th International Symposium on High-Performance Computer Architecture, France, pp. 49-60, 2000.

[6]   J. Alghazo, A. Akaaboune and N. Botros, "SF-LRU Cache Replacement Algorithm," Proceedings of International Workshop on Memory Technology Design and Testing, USA, pp. 19-24, 2004.

[7]   A. Abdel Fattah, A. Abu Samra, "Least Recently Plus Five Least Frequently Replacement Policy (LR+5LF)," *The International Arab Journal of Information Technology*, Vol. 9, No. 1, January 2012.

[8]   N. Megiddo and D. Modha, "ARC: A Self-Tuning, Low Overhead Replacement Cache," Proceedings of the 2nd USENIX Symposium on File and Storage Technologies, USA, pp. 115-130, 2003.

[9]   L. Zhansheng, L. Dawei, and B. Huijuan, "CRFP: A Novel Adaptive Replacement Policy Combined the LRU and LFU Policies," Proceedings of IEEE 8th International Conference on Computer and Information Technology Workshops, Sydney, pp. 72-79, 2008.

Clock skew optimized VLSI Architecture for Zero Frequency Filter

K.R.Radhakrishnan*, S.Subha Rani,Faculty,Department of Electronics and Communication Engineering ,
PSG College of Technology Coimbatore,Tamilnadu,India
radhabe@gmail.com, krr@ece.psgtech.ac.in

Abstract—In today's world, mobile phones play an important role in communication. People using mobile phones often face the problem of noise signals affecting the quality of speech. Passive noise control suppresses the higher frequency acoustic noise. It is a technique that provides sound reduction by noise-isolating materials such as insulation, sound- absorbing tiles, or a muffler rather than a power source. In lower frequencies, passive techniques require material that is too bulky and heavy. So an alternative method called active noise cancellation that separates noise signal and speech signal is chosen. Zero frequency filter is a technique used for active noise cancellation of noisy speech signals. This filter is used for the characterization of glottal activity from speech signals thereby cancelling the noise. The main advantage of this method when compared to other noise cancellation methods is that noise need not be modeled separately. In this paper is to present the VLSI implementation architecture of zero frequency filter with useful clock skew optimization. This architecture can be used as voice processor in mobile applications. As clock frequency increases and the technology move towards sub-nanometer process, handling timing violations in the design becomes an increasingly complex and challenging task. Traditional approaches target for global zero skew in the process of timing closure costs in area and power and also limits the maximum achievable operating frequency. Useful clock skew optimization is an emerging technique that helps achieve timing closure. The work presented in this paper achieves timing closure with an area overhead of about 15.87% through useful clock skew optimization.

Keywords—Clock skew, useful skew, active Noise cancellation, zero frequency filtering.

I. INTRODUCTION

The most frequent problem in speech processing is the consequence of interference noise in speech signals. Interference noise in a certain way modulates the speech signal and reduces its clarity. Interference noise is produced from acoustical sources such as ventilation equipment, echoes, crowds, and in general with any type of signal that interferes with the speech signals. Noise Cancellation is a much sought after process nowadays with the increase in environmental noise due to various reasons. Initially Passive Noise Cancellation was followed everywhere because of its effectiveness. But with the increase in environmental noise, the methods of passive noise cancellation just began to become a burden to the user. This led to invent of the concept of Active Noise Cancellation which uses phase cancellation techniques to cancel out the noise.

Clock skew is the difference in clock arrival times at different sequential elements in the clock-distribution network. A lot of work has been done in the past to minimize clock skew [1][2]. Targeting global zero skew not only costs in area and power, but also limits the achievable operating frequency to the maximum data path delay in the circuit. This has led to a paradigm shift from skew minimization to useful skew optimization as the latter has the potential to significantly improve design performance [3][4][5][6].

As technology move towards sub-nanometer process, interconnects play an important role in timing [7]. Several works have focused on timing optimization during placement and routing as well [8][9][10]. But in spite of all these efforts, timing violations still exist after detail routing in MCMM designs. So the designers have to intervene manually to analyze and fix the timing violations considering every mode and process variation altogether in an iterative and non- convergent way, whereas the verification engineers need to run timing analysis for each scenario.

Engineering change order (ECO) is always used after detail routing in order to fix existing timing violations by incremental adjustment of pertaining cells and nets [11]. These ECO adjustments, focused mainly on data path optimization, are not sufficient to handle all timing violations. So data path aware clock scheduling becomes an important step for timing closure, as it allows modifications in the clock tree which is towards timing closure. Several works study the clock scheduling problem [12][13][14]. [13] formulates an LP problem to optimize clock period in the post-CTS stage by bounded delay buffering at the leaves of the clock tree. But since this work only considers inserting delay but not speeding up clock arrival at the leaves, the scope of the optimization is limited and buffering at the leaf level introduces a high area overhead in clock tree. A recent work [14] focuses on the realization of the useful skew on industrial-scale designs at post-routing stage. It also performs local transformations at the leaf-level by inserting/removing buffers to minimize negative D-slack/Q-slack violations.

To tackle these issues, a novel clock tree re-synthesis methodology is presented in [15]. Instead of estimating clock schedule at the leaf level registers, the author consider offsets in clock arrival at the clock tree driver pins of any placed design with already synthesized and routed clock tree.

The organization of this paper is as follows: Section II presents the zero frequency filtering of speech signal. Section III presents the Useful Clock Optimization technique. Section IV provides a implementation of zero frequency filtering & Useful Clock Optimization technique.

II. ZERO FREQUENCY FILTER

The great majority of current voice technology applications rely on acoustic features, such as the widely used MFCC or LP parameters. Adaptive filters are used with various algorithms such as LMS, NLMS for active noise cancellation [16][17]. The major source of excitation, namely the glottal flow, is expected to convey useful complementary information. The glottal flow is the airflow passing through the vocal folds at the glottis. Various methods such as group delay method, Hilbert envelope method are used for extracting glottal activity from speech signals [18][19][20][21].

Epoch is the instant of significant excitation of the vocal-tract system during production of speech. For most voiced speech, the most significant excitation takes place around the instant of glottal closure. Extraction of epochs from speech is a challenging task due to time-varying characteristics of the source and the system. Zero frequency filter is used in [22] for Epoch extraction from speech signals. The block diagram of IIR implementation is shown in Fig. 1.

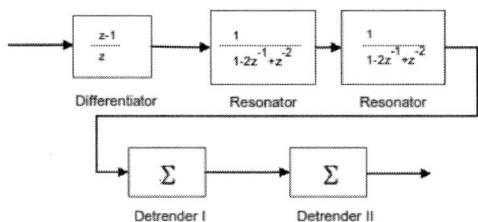

Fig.1.Zero frequency filter block diagram.

Zero Frequency Filtering technique is used to filter out the necessary instants of significant excitation which are not affected by time varying resonances from vocal tract. Voiced/Non-voiced (V/NV) detection involves identifying the regions of speech when there is significant glottal activity (i.e., the vibration of vocal folds) [23]. Such regions of speech are generally referred to as voiced speech. The non-voiced regions of speech include both silence (and background noise) as well as unvoiced speech.

III. USEFUL CLOCK SKEW OPTIMIZATION

Targeting zero skew limits the achievable operating frequency to the maximum data path delay in the circuit. If clock is skewed intentionally to resolve violations, it is called useful skew. The timing slacks on critical paths can be increased while still satisfying the timing constraints on non- critical paths. Timing can be met on net list with "impossible" timing constraints, such as an input delay longer than the clock cycle time. This can be done by skewing the receiving flip flop a little as long as it does not create a hold time problem. This can be explained using Fig.2.

Consider the setup requirement of flipflop2 is 2 ps. So the required time is 10 ps – setup which equals 8 ps. But the actual arrival time is 12 ps. This results in slack of -4 ps. This timing violation can be handled in two ways. (i) To make the

data arrive early either by removing buffers in data path or by upsizing the gates. Upsizing the gates result

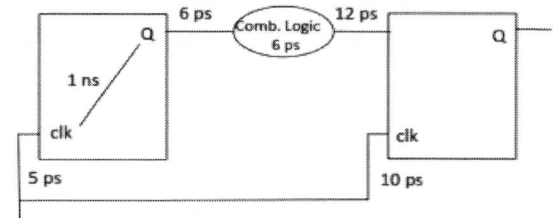

Fig.2.Example for timing violation

is increase in area and power. (ii) To increase the clock period so that the capture clock arrives at 14 ps thereby meeting setup requirement. This reduces the operating frequency.

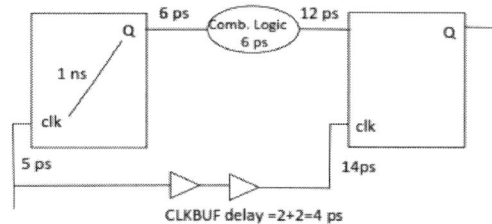

Fig.3. Timing violation handled useful clock skew optimization method

By using the useful clock skew optimization concept Skew can be introduced by inserting clock buffers in the clock path to handle the timing violation thereby overcoming

The drawbacks of the previous methods. This is shown in Fig.3.

IV. IMPLEMENTATION

Cadence RTL Compiler and Encounter RTL-to-GDSII system is used for implementing zero frequency filter with useful clock skew optimization. The conventional zero-skew implementation flow and the useful clock skew flow are shown in Fig. 4

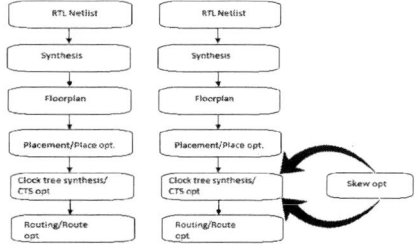

Fig.4. Zero skew vs useful clock skew implementation flow.

The difference as seen in the figure is performing clock skew optimization after clock tree synthesis optimization. This is done in cadence encounter RTL-to-GDSII system. After optimization static power analysis is done and the values are presented in the next section.

V. RESULTS AND DISCUSSION

RTL functionality of zero frequency filter design is synthesized using Cadence RTL Compiler by using 180nm standard cell libraries with a target frequency of 600 MHz. The synthesized structure is shown in Fig. 5.

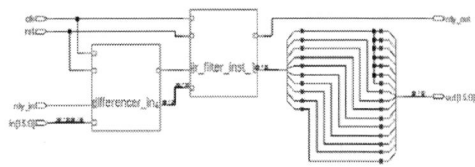

Fig. 5. Schematic of zero frequency filter

The worst path timing after synthesis is shown in Fig. 6. It can be seen that the slack is a positive value. The schematic of the worst path is also shown in the figure

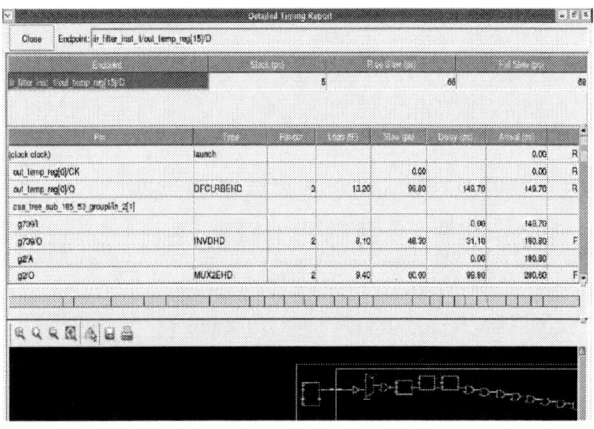

Fig. 6. Timing worst path after synthesis

Physical design from Floorplan to Routing is done in Cadence Encounter RTL-to- GDSII system with a starting utilization of 0.6. Floorplanning, powerplanning and clock tree synthesis is performed. The design after clock tree synthesis stage is shown in Fig. 7. The clock tree is highlighted in the figure.

The die co-ordinates of the design is [118.4 115] in

µm. The total area of core is 12206.272 µm^2

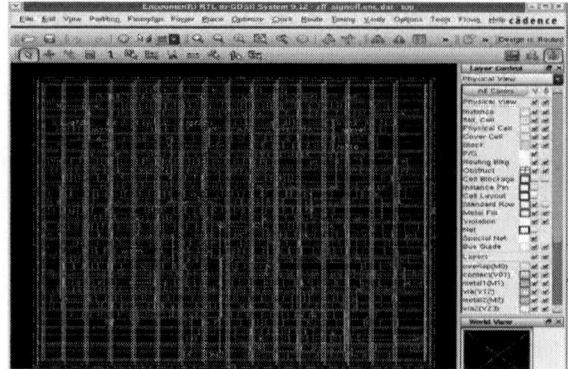

Fig.7.Design after clock tree synthesis

The timing debug results after clock tree synthesis optimization is shown in Fig. 8. It can be seen that the number of violating paths are 134 out of 236. The Worst Negative Slack (WNS) and Total Negative Slack (TNS) values are also shown in the figure. The values are negative indicating that there are timing violations.

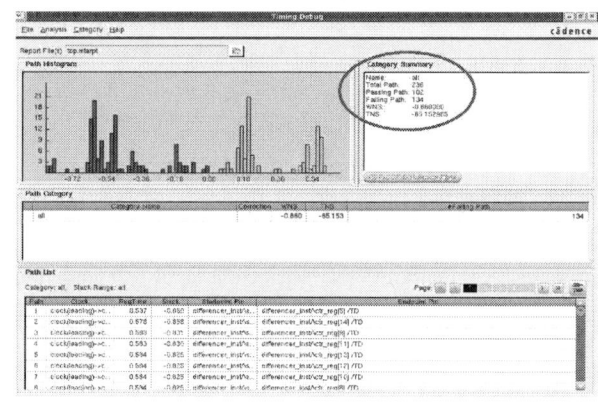

Fig.8. Timing debug window after CTS opt.

Fig. 9 shows the timing analyzer window after CTS optimization. It shows the skew is initially -0.001 ns. The worst path has a negative slack of about -0.86 ns. These violations could not be handled using datapath optimization. So useful clock skew optimization in driver level is performed.

To reduce the timing violations after CTS optimization, useful skew optimization is initially done in driver level by inserting clock buffers. The timing debug window after one iteration is shown in Fig. 10. It can be seen that the number of violating paths have reduced to 109.

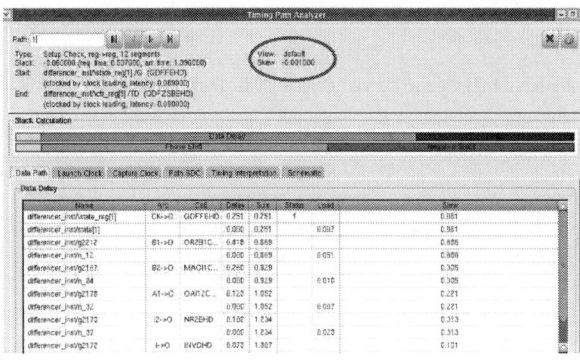

Fig .9. Timing analyzer after cts opt

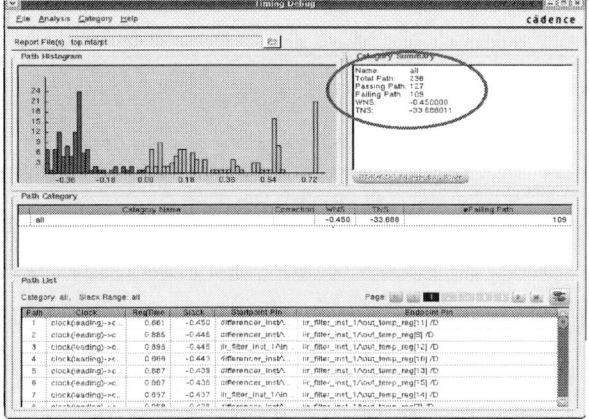

Fig.10.Timing debug window after one iteration of skew opt.

The timing analyzer after one iteration of skew optimization is shown in Fig. 11. It can be seen that the skew has increased to 0.0012 ns. WNS value is also reduced to -0.45 ns from the previous value of -0.86 ns.

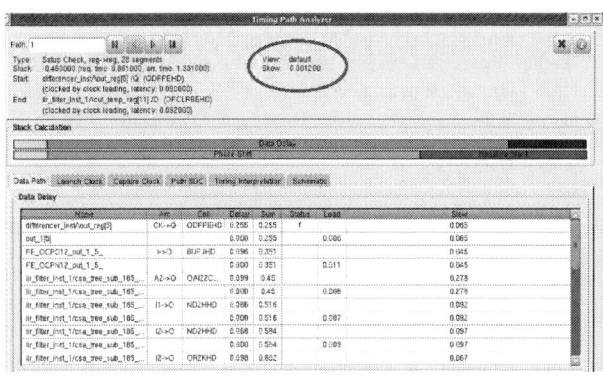

Fig. 11. Timing analyzer after one iteration of skew opt

Since driver level optimization yields lesser area overhead, more iterations are performed in driver level. It reduces the number of violating paths to 7. It can be shown in Fig. 12.

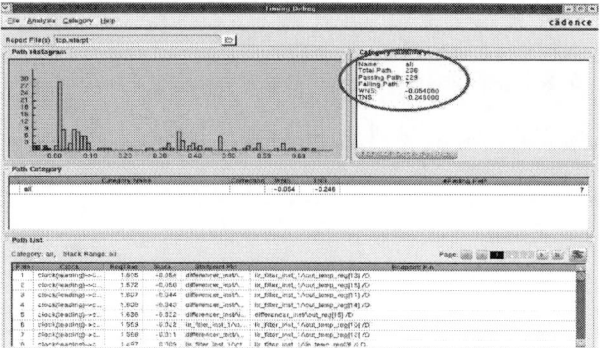

Fig.12.Timing debug window after driver level skew opt

In Fig. 13 the timing analyzer window shows the worst path timing after driver level skew optimization. The skew has increased to 0.026 ns and WNS is -0.054 ns.

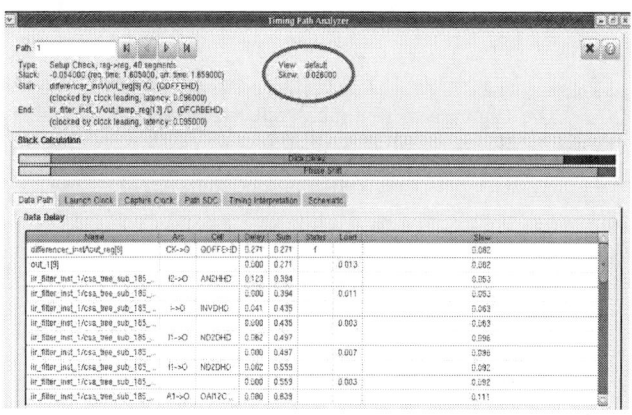

Fig. 13. Timing analyzer window after driver level skew opt

After this stage, performing driver level optimization is not recommended because it introduced more timing violations in the design. So leaf level optimization is performed as mentioned in Timing debug window after leaf level optimizations is shown in Fig. 14. It can be seen that there are zero violating paths. TNS and WNS values are zero. It means that the slack of the worst timing path is positive implying that timing has been met for the design.

The timing analyzer window is shown in Fig. 15 It shows that the skew is 0.0288 ns. So by introducing a small skew, timing violations are handled.

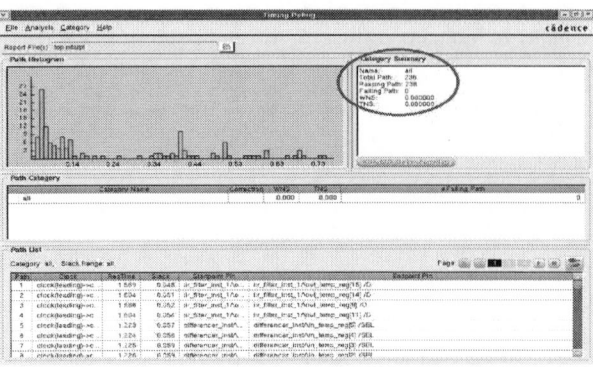

Fig.14. Timing debug window after leaf level skew opt.

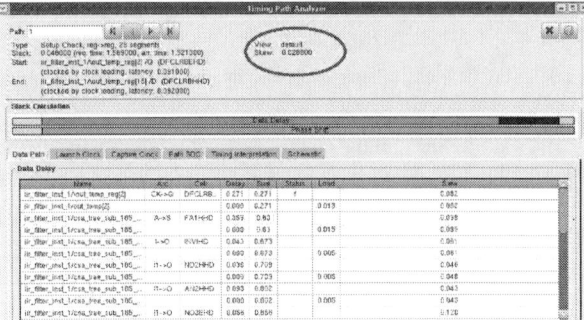

Fig.15.Timing analyzer window after leaf level skew opt.

The comparison of timing values before and after skew optimization is shown in TABLE I.

TABLE I. Timing values before and after skew opt.

Stage	Failing paths	WNS (ns)	TNS (ns)	Skew (ns)
Pre skew opt	134	-0.86	-65.153	-0.001
Post skew opt	0	0	0	0.0288

The increase in skew is 0.0288 ns. The design after Routing is shown in Fig. 16.

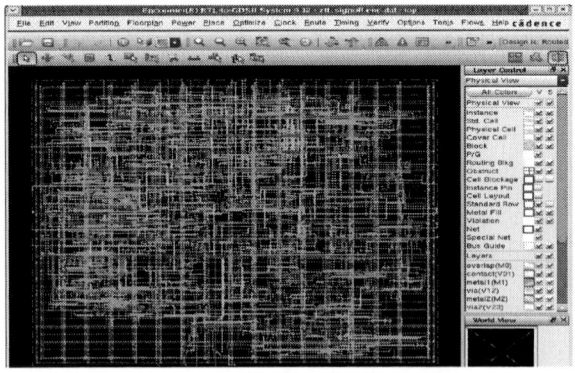

Fig.16. Design after routing

The number of standard cells and the power values at each stage are listed in TABLE II

TABLE II Area and power values before and after skew opt

Stage	No. of Std. cells	Std. cell area (μm2)	Leakage Power (mW)	Internal Power (mW)	Switching Power (mW)	Total Power (mW)
Pre-CTS	420	6103	0.005115	0.9599	0.1447	1.11
Post-CTS (Before Skew Opt)	711	8407	0.009027	1.137	0.48	1.626
After Skew Opt	806	9741	0.008869	1.189	0.5482	1.746

From TABLE II., it can be noted that the increase in standard cells is 13.36% and the increase in standard cell area is 15.87%. Distribution of power is shown in Fig. 17.

Fig. 17. Power distribution in the design

Switching activity is generated in the form of VCD (Value Change Dump) file using Synopsys VCS. Dynamic and Peak power analysis is done using Synopsys PrimeTime-PX tool. The results are shown inFig.18.

The switching activity is taken from the input signal that contains male voice and vehicle noise. The figure also shows the power distribution among the various power components.

978-1-5090-0037-1/16 $31.00 © 2016 IEEE

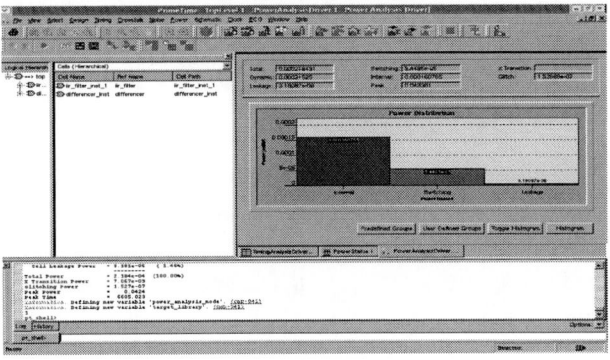

Fig.18. Dynamic and peak power analysis results

V.CONCLUSION

The most frequent problem in speech processing is the consequence of interference noise in speech signals. Noise cancellation is a much sought after process nowadays with the increase in environmental noise due to various reasons. People using mobile phones often face the problem of noise signals affecting the quality of speech. To tackle this issue, an active noise cancellation architecture based on zero frequency filter is realized and voiced/nonvoiced classification is done. The functionality is initially verified in Matlab and the VLSI implementation of the same has been carried out in 180 nm technology using Cadence RTLCompiler and Encounter RTL-to-GDSII system with an operating frequency of 600 MHz. In the physical design process, timing closure is achieved through useful skew optimization concept in a hybrid method of combining driver level and leaf level optimization. The skew optimization resulted in an area overhead of about 15.87%.

REFERENCES

[1] R. Tsay, "Exact zero skew clock routing algorithm," Computer Aided Design of Integrated Circuits and Systems, pp. 242–249, 1993.

[2] J. L. Tsai, T. H. Chen, and C. C. Chen, "Zero skew clock-tree optimization with buffer insertion/sizing and wire sizing,"Computer Aided Design of Integrated Circuits and Systems, pp. 565–572, 2004.

[3] J. P. Fishburn, "Clock skew optimization," IEEE Trans. On Computers, pp. 945–51, 1990.

[4] R. Deokar and S. Sapatnekar, "A graph-theoretic approach to clock skew optimization," ISCAS, pp. 407–10, 1994.

[5] I. S. Kourtev and E. G. Friedman, "Clock skew scheduling for improved reliability via quadratic programming," ICCAD, pp. 239–43, 1999.

[6] V. Nawale and T. W. Chen, "Optimal useful clock skew scheduling in the presence of variations using robust ILP formulations," ICCAD, pp. 27–32, 2006.

[7] V. Mehrotra and D. Boning, "Technology scaling impact ofvariation on clock skew and interconnecet delay," Interconnect Tech. Conference, pp. 4–6, 2001.

[8] K. Rajagopal, T. Shaked, Y. Parasuram, T. Cao, A. Chowdhury, and B.Halpin, "Timing driven force directed placement with physical net constraints," ISPD, pp. 60–66, 2003.

[9] Y. Liu, R. S. Shelar, and J. Hu, "Delay-optimal simultaneous technology mapping and placement with applications to timing optimization,"ICCAD, pp. 101–106, 2008.

[10] S. W. Hur, A. Jagannathan, and J. Lillis, "Timing driven maze routing,"TCAD, pp. 234–241, 2000.

[11] K. Sato, H. E. M. Kawarabayashi, and N. Maeda, "Post-layout optimization for deep sub-micron design," DAC, pp. 740–745, 1996.

[12] M. Ni and S. O. Memik, "A revisit to the primal-dual based clock skew scheduling algorithm," ISQED, pp. 755–764, 2010.

[13] J. Lu and B. Taskin, "Post-CTS clock skew scheduling with limited delay buffering," International Midwest Symposium on Circuits and Systems, pp. 224–227, 2009.

[14] W. Shen, Y. Cai, W. Chen, Y. Lu, Q. Zhou, and J. Hu, "Useful clock skew optimization under a multi-corner multi-mode design framework," ISQED, pp. 62–68, 2010.

[15] Subhendu Roy, Pavlos M. Mattheakis, Laurent Masse-Navetteand David Z. Pan, "Clock Tree Resynthesis for Multi-corner Multi-mode Timing Closure," IEEE Transactions on Computer-Aided Design of Integrated Circuits and Systems, April 2014.

[16] J. M. Gorriz, Javier Ramirez, S. Cruces-Alvarez, Carlos G. Puntonet, Elmar W. Lang and DenizErdogmus, "A Novel LMS Algorithm Applied to Adaptive Noise Cancellation ," IEEE Signal Processing Letters., vol.16, no. 1,pp. 34 – 37, January 2009

[17] M.M. Dewasthale. andR.D.Kharadkar, "Acoustic Noise Cancellation Using Adaptive Filters: A Survey ," International Conference on Electronic Systems, Signal Processing and Computing Technologies (ICESC),2014.

[18] Kruthiventi S. S. Srinivas and Kishore Prahallad, "An FIR Implementation of Zero Frequency Filtering of Speech Signals," IEEE Trans. Audio, Speech, Lang. Process., vol. 20, no.9, pp. 2613-2617, Nov. 2012.

[19] R. Smits and B. Yegnanarayana, "Determination of instants of significant excitation in speech using group delay function," IEEETrans. Speech Audio Process., vol. 3, no. 5, pp. 325–333, Sep. 1995.

[20] K. SreenivasaRao, S. R. MahadevaPrasanna and B. Yegnanarayana, "Determination of Instants of Significant Excitation using Hilbert Envelope and Group Delay Function," IEEE Signal Processing Letters.,vol. 14, no. 10,pp. 762 – 765, October 2007.

[21] K. S. R. Murty, B. Yegnanarayana, and M. A. Joseph, "Characterization of glottal activity from speech signals," IEEE Signal Process. Lett.,vol.16, no. 6, pp. 469 – 472, June 2009.

[22] B. Yegnanarayana, R. K. Swamy, and K. S. R. Murty, "Determining mixing parameters from multispeaker data using speech-specific information,"IEEE Trans. Audio, Speech, Lang. Process., vol. 17, no.6,pp. 1196–1207, Aug. 2009.

[23] K. Sri Rama Murty and B. Yegnanarayana, "Epoch extraction from speech signals," IEEE Trans. Audio, Speech, Lang. Process., vol. 16, no.8, pp. 1602–1613, Nov. 2008.

[24] N. Dhananjaya and B. Yegnanarayana, "Voiced/Nonvoiced detection based on robustness of voiced epochs," IEEESignal Processing Letters,vol. 17, no. 3, March 2010.

Effect of Split Manufacturing on Power Supply Requirements

Sharath K Rangan
Student
Dept. of Electronics
and Communication Engg.
M.S. Ramaiah Inst. of Tech.
Bangalore-560054, INDIA
sharathrangan@gmail.com

Shazia Afreen
Student
Dept. of Electronics
and Communication Engg.
M.S. Ramaiah Inst. of Tech.
Bangalore-560054, INDIA
shaziaafreen26@gmail.com

Raghuram Srinivasan*
Associate Professor
Dept. of Electronics
and Communication Engg.
M.S. Ramaiah Inst. of Tech.
Bangalore-560054, INDIA
raghuram@msrit.edu

* Corresponding Author

Abstract—Split manufacturing has become one of the most important methods to effectively control malware insertion and IP piracy. 3D IC technologies are still in the nascent stage, but 2.5D IC technology is currently adopted by most fabrication facilities. While recent studies on split manufacturing have focused on graph theoretic algorithms to effect a minimum cut split, none of them have discussed the effect of circuit performance due to the splitting. In this paper we show that split manufacturing has an adverse effect on power supply requirement, if circuit performance metrics have to stay the same before and after splitting. In particular, we have shown that multiple power supplies are necessary if circuit performance metrics have to be satisfied before and after the splitting procedure.

Keywords—Security, Split manufacturing, 3D IC, Supply Voltage

I. INTRODUCTION

As Moore's law continues its march deep into the submicron region, the Integrated Circuit design and manufacturing details continue to increase in complexity. With these complexities are also associated costs related to maintaining expensive design and fabrication teams. One of the most important fallouts of this cost is the new trend of design piracy [1]. Thus, it has become harder for companies that create Intellectual Property to ensure that their designs are not revealed to a competitor. Unfortunately, most IP manufacturers are fabless semiconductor companies, and are forced to depend on third parties for manufacturing capabilities. This is due to the large costs involved in fabrication - for instance the cost of owning a foundry is currently around $5 billion [2]. So, a design becomes susceptible to many types of misuse when an untrusted third party is given the complete design of an IP block, e.g., piracy, malware and trojans insertion etc [3]. Trojan insertion can be, in certain cases, detected by comparison with a golden chip [4]. However, piracy cannot be directly dealt with. An indirect method of dealing with piracy is to distribute the design among multiple untrusted foundries in such a way that a certain amount of obfuscation disallows the overall

functionality to be understood by any single foundry [5]. This type of manufacturing is called as split manufacturing and is discussed briefly in the next section.

II. BACKGROUND AND RELATED WORK

A. Split Manufacturing

Split manufacturing was initially proposed by research agencies like International Advanced Research Projects Agency (IARPA) and prime fabless companies like AMD [6], [7]. The principle of split manufacturing is to split the layout of the design into Front End of Line (FEOL) and Back End of Line (BEOL) layers which are fabricated in separate foundries. This method revolves around the idea that the attacker never has complete access to the full design. Hence, the probability of an engineer at the foundry inserting a fatal trojan, or re-engineering the design to get the full netlist is significantly reduced. FEOL is defined as the layer where there are transistors and lower level metal layers (M1 to M4) and BEOL is defined as all the metal layers about M4 [8]. The idea behind split manufacturing is to fabricate the FEOL and BEOL separately and then align them together as a single IC chip. If the FEOL and BEOL are fabricated separately, then no single fabrication facility will know the full details of the design. Fig. 1 shows possible stages in split manufacturing. [8]

B. Front End Of Line (FEOL) and Back End Of Line (BEOL)

The FEOL involves the fabrication of all the active components like transistors, passive components like resistors and capacitors and hence is an expensive process. It involves processes like Local Oxidation of Silicon (LOCOS), demarcation of source and drain [9]. This is the layer that is sent to another (possibly untrusted) foundry.

In the BEOL, all the interconnections are made between the various components that are obtained after the FEOL process. Thus, this stage involves processes like creation of metal layers, passivation layers and making interconnections between transistors, and other components. This is not as expensive as fabricating the active components. Thus, this part is given to small and (possibly trusted) foundries.

978-1-5090-0037-1/16 $31.00 © 2016 IEEE

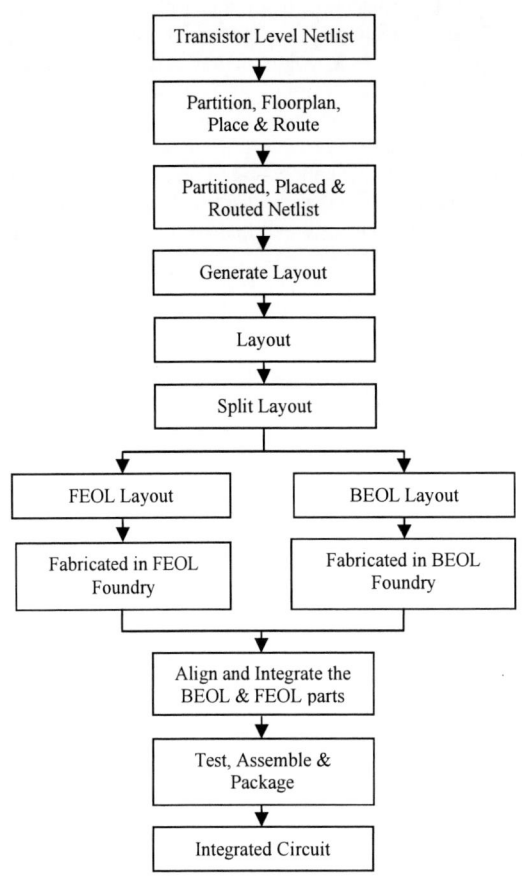

Fig. 1: Stages in Split Manufacturing

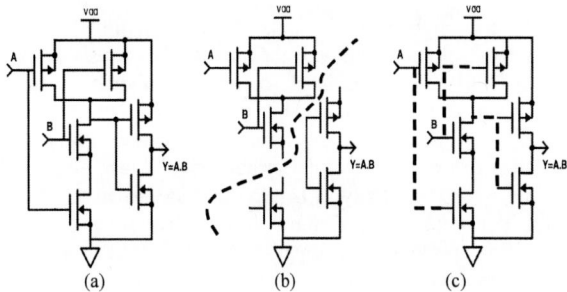

Fig. 2: (a) Transistor Level AND gate. (b) 3D Split. (c) 2.5D split.

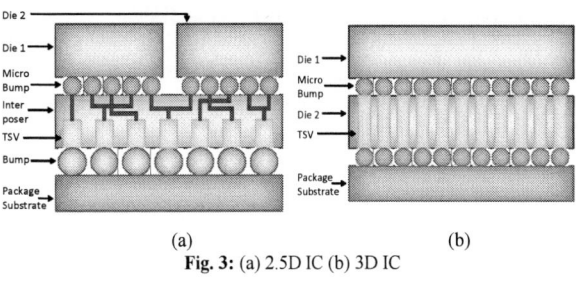

Fig. 3: (a) 2.5D IC (b) 3D IC

C. 3D IC and 2.5D IC

In order to reduce the form factor and delays due to length of interconnects, two IC architectures have emerged. The first approach which is currently in production is the 2.5D IC. In this paradigm, the individual dice are placed beside each other. There is a silicon interposer present whose purpose is to route the signals between the two dice. Through Silicon Vias (TSV) are used to help with the connections in the overall package [10].

Hence, there is a possibility to realize interconnections in the vertical dimension through multiple types of these TSVs. The objective of this approach is to allow for more FEOL areas to be occupied by logic. For example, we could have a memory and logic chip placed beside each other and the interposer doing the job of routing the signals between the two blocks and the TSVs could be used to connect these blocks to the power and ground rails. The wires in an individual die can also be connected using TSVs in the interposer layer.

Another approach is to stack these two blocks on top of each other and making use of the third dimension to conserve on area and reduce delays. Owing to the use of the third dimension for logic circuit, it is called 3D IC.

To illustrate splitting of a circuit with respect to 2.5D and 3D architecture we have taken the example of AND gate as shown in Fig. 2. The layouts of 2.5D IC and 3D IC are shown in Fig. 3. It is seen from the figures that the 3D implementation shares transistors between the two layers and how only the wires are lifted in the 2.5D implementation.

Connections between the various stacked on layers are made through TSVs. One of the crucial requirements for the formation of 3D layers is a single defect-free crystalline semiconductor layer. 3D IC manufacturing is challenging especially because it is very hard to deposit a single crystalline layer on top of amorphous layers. 3D ICs can be used to provide better obfuscation as we can now place the transistors in many layers.

As per our knowledge, there is a dearth of research on the power requirements due to split manufacturing. Security can be provided to a circuit containing logic gates through a wire lifting algorithm as described in Ref. [11]. When it comes to IP in the field of analog circuits, Ref. [12] proposes techniques to improve security like adding spare transistors, evening up the various form factors so that different transistors almost lean in on each other to provide better obscurity to the attacker.

We chose an analog circuit for two reasons. Firstly because, power requirements could play a much more crucial role in analog circuits as it could drive a transistor into saturation, cutoff or triode region. Secondly, any fluctuations in the power rails could add unnecessary noise and cause distortions. Thus, in this paper, we try to analyze the effect on power supply by taking the example of an analog circuit. We split it by mimicking its 2.5D and 3D IC architecture.

978-1-5090-0037-1/16 $31.00 © 2016 IEEE

III. MODELING SPLIT MANUFACTURING

A few parameters that must be kept in mind when we intend to split a circuit are the number of transistors to be present on the various layers and the number of vias. It is important to ensure the number of transistors present on the various layers (in our case, only 2 layers) are evenly shared because this could affect performance and cause unnecessary timing issues. Secondly, it is important to implement the split with the minimum of transmission lines to avoid signal distortion and loss of information. Since this paper deals with split manufacturing for analog circuits, it is important to take care of the signal strength.

One possible approach to split the circuit is to use minimum cut algorithm [13]. Given an undirected graph with vertices and edges G=(V, E), the algorithm comes up with the smallest set of edges that need to be removed so that we get two disjoint graphs. In our circuit, we implemented algorithm but the resulting split was very poorly balanced. Hence, we chose to go with another approach. As defined earlier, in a 3D IC the transistors are spread over various layers and suitable interconnects are made. We implemented a novel way of splitting all the NMOS and PMOS in two different layers. This improves security because the attacker can see only two different types of transistors spread across two layers and the interconnections between these two layers would be best obfuscated this way. Any other split would have both PMOS and NMOS transistors and would represent some logical portion of the circuit. Also, fabrication becomes a less expensive process as only one type of transistors is present in a single layer at a given time.

The two separate layers are then connected using vias. The vias were modeled using RC coupled lossy transmission lines as shown in Fig. 4. R represents the attenuation in the line and C represents the time delay. We chose to go with RC model for a transmission line because the TSV has an inherent resistance which is represented by R and there is no need for inductance as we are not operating in the high frequency region.

In 2.5D IC obfuscation of connections is done by lifting wires. Some connections are removed so that an attacker will not be able to insert a trojan in the chip. An attacker can exploit the typicalities present in floor planning, placement and routing to figure out connections between transistors. In fact, previous researchers have developed attacks capable of predicting 96% of the connections correctly. Thus, to avoid this we removed wires that were not so obvious to the attacker's eye. We chose wires that were long and connected many transistors.

IV. EXPERIMENTAL RESULTS

We conduct our experimental study on how split manufacturing affects power supplies using a MOSFET cascode current source [14] as shown in Fig. 6(a). The output of this circuit is logarithmic in nature and attains a value of 10µA when a voltage of 0.2V is given between QP1 and QN1 as shown in Fig. 6(b). We split this circuit as described in the previous section for 2.5D and 3D as shown in Fig. 7(a) and 8(a) respectively. All experimental results are obtained using LTSpice IV.

Looking at results of 2.5D simulation in Fig. 7(b), we observe that the current at even 1V is only in the order of a few hundred nano amperes. To achieve the desired current of 10µA, the supply voltage had to be increased to 2.67V as shown in Fig. 7(c).

The results of the 3D split as seen in the Fig. 8(b) and 8(c) show that the current is zero even until a voltage of 1.6V and then begins to increase exponentially. The current then reaches 10µA at 1.91V

In order to try and decrease the supply voltage to get the desired current, we removed a few transmission lines. But the results turned out to suggest that removing transmission lines need not necessarily correspond to a decrease in output voltage. For example, removing 3 transmission lines showed the desired current of 10uA at 2.75V, leading to an increase of eighty millivolts as shown in Fig. 9.

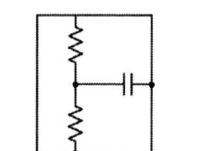

Fig. 4: Transmission Line Model.

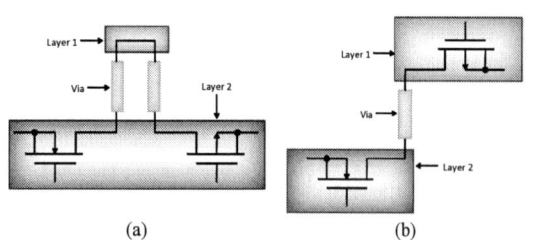

(a)   (b)

Fig.5: (a) Via connected in a 2.5D architecture (b) Via connected in a 3D architecture

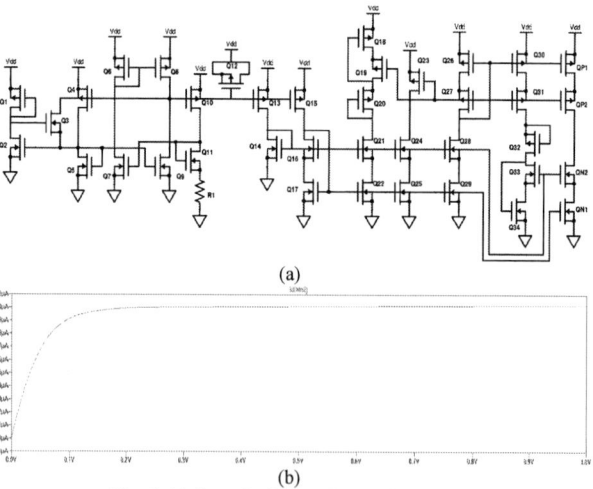

(a)

(b)

Fig. 6: (a) Cascode Current Source (b) Output.

978-1-5090-0037-1/16 $31.00 © 2016 IEEE   48

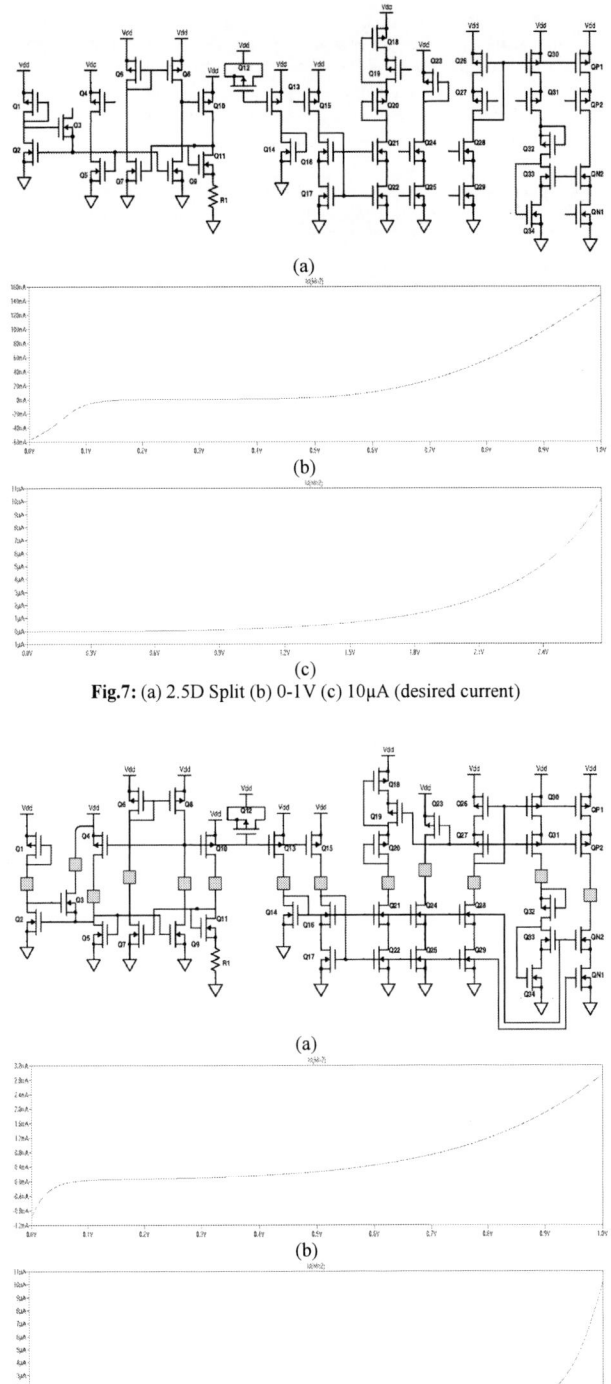

(a)

(b)

(c)

Fig.7: (a) 2.5D Split (b) 0-1V (c) 10µA (desired current)

(a)

(b)

(c)

Fig. 8: (a) 3D Split (b) 0-1V (c) 10µA (desired current)

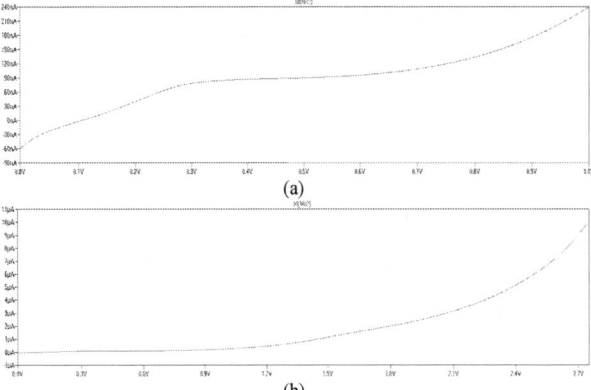

(a)

(b)

Fig. 9: Output for modified 2.5D split (a) 0-1V (b) 10µA

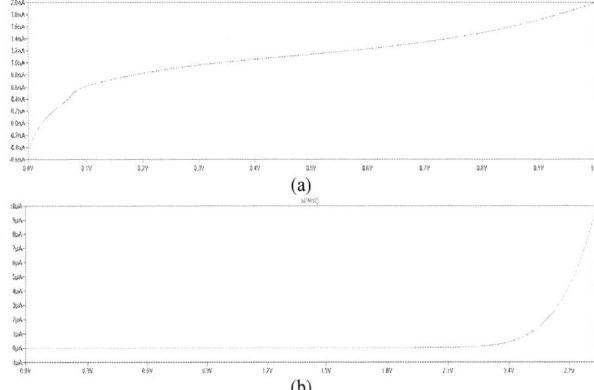

(a)

(b)

Fig. 10: Output for modified 3D split (a) 0-1V (b) 10µA

TABLE I COMPARISION OF VOLATGE VALUES OF DIFFERENT SPLITTING METHODS

Split Implemented	Voltage Required	Current Desired
NA	0.20 V	10µA
2.5D	2.67 V	10µA
3D	1.91 V	10µA
2.5D Modified	2.75 V	10µA
3D Modified	2.83 V	10µA

Continuing from the previous 3D architecture of separating all the NMOSs and PMOSs, we went one step further by pushing an arbitrary NMOS or/and PMOS up or down respectively by removing appropriate transmission lines. This produced a current of 10uA only at 2.83V increasing the supply voltage even further as shown in Fig. 10.

V. CONCLUSION AND FUTURE WORK

Techniques like wire lifting and splitting of transistors into NMOSs and PMOSs are simulated to evaluate the effect of split manufacturing on circuit performance. We have reasoned that a lossy RC line can reliably model a TSV. With the unmodified analog circuit as a reference, we have analyzed various ways of splitting the transistors, both in 2.5D and in 3D. From the results, it is shown that logic encryption comes at the cost of increased circuit complexity, in particular the requirement for multiple power supplies and the associate overhead of routing power and ground signals. Also, it can be observed that the best split is obtained from partitioning the two types of MOSFETs in different layers. Another important conclusion of this work is that a min-cut type graph theoretic placement methods might not always lead to the best splitting possible, since circuit performance has to also be considered.

Scope for future investigations in this area would be to analyze transient responses of the system and measuring parameters like jitter and response time. Also, research can be carried out on how assembly process variations are affected due to split manufacturing. New ways of splitting circuits that don't just focus on the transistor/via count, but also other performance parameters changed due to the new topology can be studied as an optimal means to perform splitting.

ACKNOWLEDGEMENT

The authors would like to thank the referees for their valuable feedback.

REFERENCES

[1] Yousra Alkabani, Farinaz Koushanfar, and Miodrag Potkonjak, "Remote activation of ICs for piracy prevention and digital right management,"*Proceedings of the 2007 IEEE/ACM International Conference on Computer-aided design*, pp. 674-677, 2007.

[2] DIGITIMES Research. "Trends in the global IC design service market," http://www.digitimes.com/Reports/Report.asp?datepublish=2012/3/13&pages=RS&seq=400&read=toc.

[3] Jeyavijayan Rajendran, Ozgur Sinanoglu, and Ramesh Karri, "Regaining Trust in VLSI Design: Design-for-Trust Techniques," *Proceedings of IEEE*, Vol. 102, No. 8, 2014.

[4] L. Xiu, "VLSI circuit design methodology demystified: a conceptual taxonomy," *Wiley-IEEE Press*, 2008. [cited at pp. 164]

[5] M. Young, The Technical Writer's Handbook. Mill Valley, CA: University Science, 1989.

[6] Intelligence Advanced Research Projects Activity, "Trusted Integrated Circuits Program," https://www.fbo.gov/utils/view?id=b8be3d2c5d5babbdffc6975c370247a6.

[7] R.W Jarvis and M. G. McIntyre, " Split manufacturing method for advanced semiconductor circuits," *US Patent no. 7195931*, 2004.

[8] Jeyavijayan Rajendran, Ozgur Sinanoglu, and Ramesh Karri, "Is Split Manufacturing Secure," *Design, Automation & Test in Europe Conference & Exhibition*, pp. 1259-1264, 2013.

[9] Karen A. Reinhardt and Werner Kern, " Handbook of Silicon Cleaning Technology," William Andrew, 2008.

[10] Meng-Jen Wang, Chang-Ying Hung, Chin-Li Kao, Pao-Nan Lee, Chi-Han Chen, Chih-Pin Hung, and Ho-Ming Tong, " TSV Technology for 2.5D IC Solution," *Electronic Components and Technology Conference (ECTC),IEEE*, pp. 284-288, 2012.

[11] Frank Imeson, Ariq Emtenan, Siddharth Garg, and Mahesh V. Tripunitara, "Securing Conputer Hardware Usng 3D Integrated Circuit Technology and Split Manufacturing for Obfuscation," *Proceeding of the USENIX Security Symposium*, 2013.

[12] Kaushik Vaidyanathan, Renzhi Liu, Ekin Sumbul, Qiuling Zhu, Franz Franchetti, and Larry Pileggi, "Efficient and Secure Intellectual Property(IP) Design with split Fabrication," *IEEE Internation Symposium on Hardware-Oriented Security and Trust (HOST)*, 2014.

[13] "An Implementation of Min-Cut Algorithm," http://in.mathworks.com/matlabcentral/fileexchange/37726-an-implementation-of-min-cut-algorithm/content/mincut.m.

[14] R. Jacob Baker, "CMOS Circuit, Design, Layout and Simulation," *IEEE Press Series on Microelectronic Systems, 2011.* [cited at pp. 650-651]

2016 International Conference on VLSI Systems, Architectures, Technology and Applications (VLSI-SATA)

Design of CMOS Programmable Output Binary and Fibonacci Switched Capacitor Step-down DC-DC Converter

Mahesh Zanwar[*], Subhajit Sen[†]

International Institute of Information Technology,

Bangalore, India

Email: [*]mahesh.zanwar@iiitb.org, [†]subhajit.sen@iiitb.ac.in,

Abstract—**This paper describes the CMOS implementation of an open-loop variable output voltage switched capacitor step-down DC-DC converter with a large number of target voltages. The number of target voltages generated using n-flying capacitors are of the order of 2^n. A switch selection scheme is presented that optimizes silicon area. Expressions for equivalent series resistance R_{eq}, conduction, switching power loss and efficiency are obtained and compared with the spice simulation results. The Digital Switch Controller is designed to switch between various target voltages and simulated in Cadence Analog-Mixed Signal flow. The 3/4 step-down converter circuit is described and analysed by varying switching frequency and load for different values of bottom plate parasitic capacitance. The optimum value of switching frequency and switch sizes is obtained for a switched capacitor converter. An efficiency of about 78.4% is achieved with 5% bottom plate parasitic capacitance for a load current of 1.35 mA and input voltage of 1.8 V at 20 MHz of switching frequency.**

Index Terms—**Binary, DC-DC converter, Fibonacci, Flying Capacitor, Power Efficiency, Switched Capacitor Converter (SCC), Digital Switch Controller, Target Voltages**

I. INTRODUCTION

Energy harvesting networks are being used to provide power supply to the modern VLSI applications like Internet of Things (IoT), wireless sensor networks (WSN), and other low power circuits. The input of these networks varies largely from few milliVolt to a few Volt and we desire a constant voltage at the output. This can be achieved by switching regulators that can generate a large number of target voltages from the same circuit and also ensures better efficiency.

Switching regulators can be divided into two groups: Inductive and Capacitive. Inductor-based converter has been the design for most of the switching voltage regulators for moderate to high power applications. Generally, inductors in such converters are of relatively large values & are not realisable on-chip. The SCC's, on the other hand, are widely used in applications requiring low power as these converters exhibit low noise, minimal radiated EMI and allow integration of the entire converter circuit including capacitors. In the literature, two types of SCC are discussed: (a) with limited number of conversion ratios [1], [2] with a feedback loop to regulate output voltage. However, [1] uses a complex switch network for conversion & [2] uses different switch network

and flying capacitor values for each conversion ratio. (b) with large number of conversion ratios, which can be generated using Binary and Fibonacci sequences [3]. A SCC circuit with n-flying capacitors can generate target voltages of the order of 2^n. For example, with n=2 (two-flying capacitor) we can generate 5 target voltages (1/4, 1/3, 1/2, 2/3, 3/4) as shown in the Figure 1.

It is proved that the SCC exhibits maximum efficiency when its output voltage is close to the target voltage,

$$V_{TRG} = MV_{in} \tag{1}$$

where M is the no-load conversion ratio. The efficiency of a SCC is given by,

$$\eta = \frac{V_o}{V_{TRG}} \tag{2}$$

which shows that efficiency has dependency on the value of M.

The work presented in [3], [4] uses discrete complementary CMOS switches for the implementation of the SCC circuit. In the analysis of a circuit ideal switches with constant on-resistance are assumed. This is not the case of MOS switches integrated in CMOS VLSI process because of the dependency on PVT (process, voltage and temperature) parameters. This work also neglects losses due to bottom plate (stray) capacitance of a flying capacitor and switching power losses due to parasitic capacitance of switches. These losses play a very important role in the degradation of efficiency at high frequencies.

This paper describes the CMOS implementation of a SCC

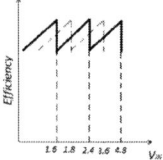

Fig. 1.   Expected total efficiency for multiple target voltages

with generalized switch network of the type proposed by [3],

978-1-5090-0037-1/16 $31.00 © 2016 IEEE   51

[4] that uses Binary & Fibonacci sequences. A methodology for appropriate switch selection is discussed. In addition, the expression for conduction, switching power loss and efficiency are obtained for a converter. The Digital Switch Controller is designed to switch between various target voltages for 2-flying capacitor case. The optimum value of switching frequency and switch size is obtained using MATLAB tool to achieve maximum efficiency for given load current.

The advantage of having large number of target voltages is that any desired output voltage can be obtained with very small ripple by interpolating between two target voltages and with a very simple control loop or even in open loop. The maximum efficiency is calculated and plotted for capacitor with different values of bottom plate capacitances by varying load.

In the next section, switched capacitor circuit configuration with switch selection scheme. In section III operation of a SCC with Digital Switch Controller and comparison of efficiency for different integrated capacitor is presented. In section IV efficiency and loss analysis is explained for the case of 3/4 step-down conversion. Section V describes the optimization of switching frequency and switch size. Section VI presents simulated results of efficiency by varying switching frequency & load for a 3/4 converter.

II. SWITCHED-CAPACITOR CIRCUIT CONFIGURATION

The general switch network shown in Figure 2(a) uses n-flying capacitors and generate target voltages of the order of 2^n. For example, with n=3 (3-flying capacitors) we can generate 13 target voltages (1/8, 2/8, 3/8, 4/8, 5/8, 6/8, 7/8, 1/5, 2/5, 3/5, 2/3, 4/5). The equivalent circuit for a generalized converter is shown in Figure 2(b) where, R_{eq} is the equivalent series resistance due to on-resistance of switches and charging/discharging of flying capacitors.

A. Switch Selection Criterion

The switches are realised using PMOS or NMOS transistor or a combination of them depending on the location of switch in the generalized switch network. This switch selection plays very important role in the reduction of gate driving loss at high switching frequencies of operation.

III. RECONFIGURABLE STEP-DOWN SCC

The methodology to determine multiple target voltages from different topologies in the design of a SCC is discussed in [3]. In this section we explain the 2-flying capacitor SCC. The main advantage of having switched capacitor converter is that we can integrate entire circuitry on-chip including flying capacitors. The integrated capacitor selection issue is addressed and Digital Switch Controller is explained.

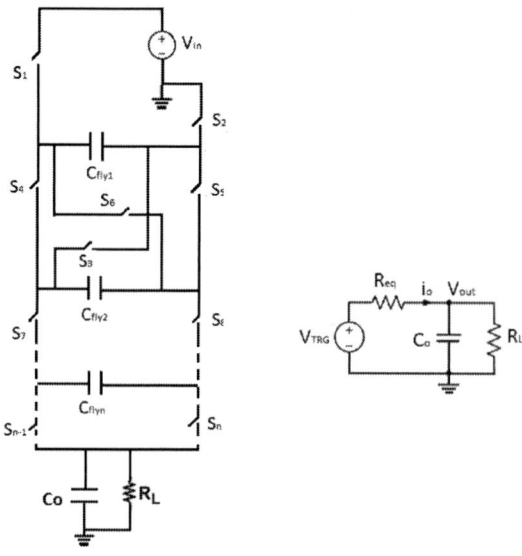

(a) General Switch Network   (b) Equivalent Circuit

Fig. 2.   Equivalent circuit representation of a SCC

A. Operation of SCC

The operation of 3/4 step-down converter as shown in Figure 3. This topology switches between three phases using different sets of switches as shown in Figure 4. In phase 1 switches S_1,S_4,S_8 will be ON, in phase 2 switches S_1,S_5,S_7 will be ON and in phase 3 switches S_2,S_6,S_7 will be ON.

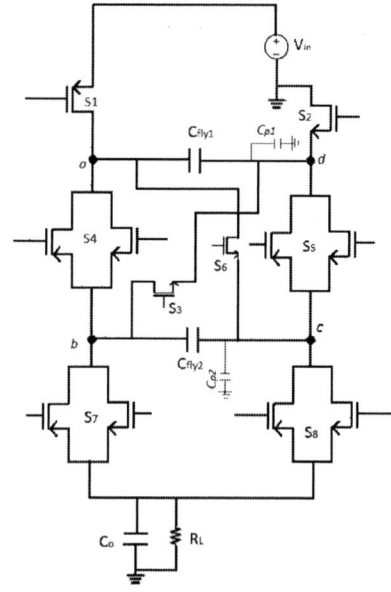

Fig. 3.   3/4 Step-Down SCC using switch selection criteria

978-1-5090-0037-1/16 $31.00 © 2016 IEEE   52

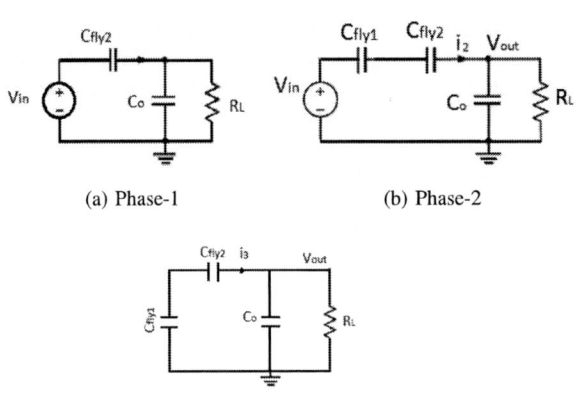

(a) Phase-1　　　　　(b) Phase-2

(c) Phase-3

Fig. 4. Topologies of 3/4 down converter

TABLE I
SWITCH ON AND OFF FOR VARIOUS TARGET VOLTAGES

Gain	Phase 1	Phase 2	Phase 3
1/4	S_1,S_3,S_8	S_2,S_4,S_8	S_2,S_5,S_7
1/3	S_1,S_5,S_8	S_2,S_4,S_8	S_2,S_5,S_7
2/4	S_1,S_5,S_8	S_2,S_4,S_7	-
2/3	S_1,S_4,S_8	S_1,S_5,S_7	S_2,S_4,S_7
3/4	S_1,S_4,S_8	S_1,S_5,S_7	S_2,S_6,S_7

As the switching cycle is typically much smaller than the charge/discharge time constant, the ramp rate of voltages across the capacitor is relatively constant and therefore the load can be approximated as a constant current source. The ripple voltage has a direct implication on the power losses and hence achievable efficiency of the converter so it is desirable to keep the output capacitor value to be large [1]. The Table I shows the set of ON switches of a converter using 2-flying capacitors which switch between multiple phases to obtain desired target voltage.

B. Integrated Capacitor

The most important component of the SCC from efficiency point of view, are the capacitors. If they are integrated, they always have stray capacitance to the ground. Due to limitation of capacitance per unit area, capacitors always occupy a considerable chip area in the whole circuit layout. Therefore, saving the chip area is important consideration in capacitor selection of CMOS ICs. Nowadays, three kinds of capacitors are commonly used in IC applications, which are MOS capacitor, metal-insulator-metal (MIM) capacitor and metal-oxide-metal (MOM) capacitor [6]. MOS capacitor has the highest capacitance density per unit area, but it shows non-linear behaviour. The capacitance density per unit area of MOS capacitor is 8 times greater than MIM capacitor and 2 times greater than that of MOM capacitor.

Table II shows simulated efficiency of converter with two

TABLE II
EFFICIENCY OF SCC FOR VARIOUS INTEGRATED CAPACITOR

Gain	Efficiency (%)			
	Ideal cap	PMOS cap	NMOS cap	MIM cap
1/4	80.73	79.29	78.78	80.54
3/4	75.79	73.35	73.20	75.81
2/3	80.51	80.11	80.21	80.52
1/3	82.96	82.67	82.75	82.88
1/2	81.95	81.76	81.84	81.92

TABLE III
CODE FOR DIFFERENT GAINS CONFIGURATIONS FOR 2-CAP CONVERTER

Gain	A	B	C
1/4	0	0	1
3/4	0	1	0
2/3	0	1	1
1/3	1	0	0
1/2	1	0	1

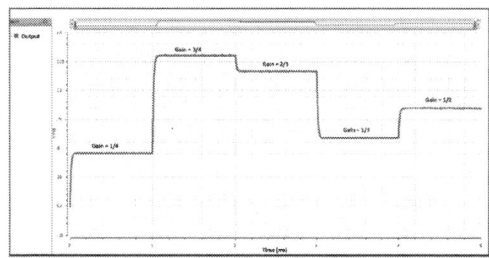

Fig. 5. Output of switch controller for various gains

flying capacitor for various integrated capacitors in standard 180 nm CMOS technology. The results are obtained for maximum load current of 1.35 mA, Vin = 1.8 V, switching frequency of 20 MHz, and $C_{fly1} = C_{fly2} = 100$ pF. As seen from the table above, the efficiency of a SCC is unaffected by the non-linearity of MOS capacitor.

C. Switch Controller

To switch between different target voltages for two flying capacitor, Digital Switch Controller is developed. Particular topology can be selected among 5 possible target voltages using A,B,C inputs as shown in Table III. The output of a controller are clocks which controls ON and OFF states of switches. The waveform in Figure 5 shows simulated result of the converter with Digital Switch Controller for various target voltages in Cadence Analog-Mixed flow.

IV. SCC EFFICIENCY AND LOSS ANALYSIS

Efficiency of a SCC is a key metric for battery operated electronics and energy starved systems. The principle contributors to efficiency loss in SCC are switching power losses and conduction power losses. The loss analysis is explained in this section is for 3/4 conversion and it can be extended with little effort to other topologies of 2-flying

978-1-5090-0037-1/16 $31.00 © 2016 IEEE

capacitors as shown in Table IV.

The efficiency of the converter is affected by two types of power losses: conduction losses and switching losses. Conduction power loss is generated from the charging /discharging of the flying capacitor along with the on-resistance of the MOS switches (R_{on}). These losses can be modelled by the equivalent series resistance (R_{eq}) which is derived for 3/4 step-down converter shown in Figure 3,

$$R_{eq_{3/4}} = \frac{1}{8 F_{sw} C_{fly}} \left\{ coth(\frac{\beta_1}{2}) + \frac{1}{2} coth(\frac{\beta_2}{2}) + \frac{1}{2} coth(\frac{\beta_3}{2}) \right\} \tag{3}$$

$$\beta_1 = \frac{t_1}{(R_{s1} + R_{s4} + R_{s8}) C_{fly}} \tag{4}$$

$$\beta_2 = \frac{t_2}{(R_{s1} + R_{s5} + R_{s7}) \frac{C_{fly}}{2}} \tag{5}$$

$$\beta_3 = \frac{t_3}{(R_{s2} + R_{s6} + R_{s7}) \frac{C_{fly}}{2}} \tag{6}$$

Where, F_{sw} is the switching frequency and R_s is switch on-resistance. For the given load current of I_{out}, the conduction power loss in the SCC is given by

$$P_c = I_{out}^2 R_{eq_{3/4}} \tag{7}$$

Switching power loss originates from the bottom plate capacitance of flying capacitor and gate driving loss of the switches. The bottom plate capacitance is assumed to be a fixed value of the flying capacitor,

$$C_p = \alpha C_{fly} \tag{8}$$

Thus, switching power loss can be estimated by the total parasitic capacitance at a node and the voltage swing (ΔV) across that node.

$$P_{sw} = F_{sw} C (\Delta V^2) \tag{9}$$

Total switching loss is the summation of the switching losses over all the nodes of the converter and it is given as,

$$P_{sw} = F_{sw} V_{in}^2 \{ C_G + C_{p1}/16 + C_{p2}/4 \} \tag{10}$$

where, C_G, C_{p1} and C_{p2} are gate oxide capacitances of ON switches, bottom plate parasitic capacitance of C_{fly1} and C_{fly2} flying capacitors respectively as shown in Figure 3. The total power loss is the sum of conduction and switching power losses. From Figure 2, we can write an expression of V_{out} in terms of target voltage and equivalent series resistance (R_{eq}). Hence the resultant efficiency can be expressed as,

$$\eta = \frac{R_L}{R_L + R_{eq_{3/4}}} \left\{ \frac{1}{1 + \frac{16}{9} F_{sw} C_{eq_{3/4}} (R_L + R_{eq_{3/4}})} \right\} \tag{11}$$

where, $C_{eq_{3/4}}$ is

$$C_{eq_{3/4}} = C_G + C_{p1}/16 + C_{p2}/4 \tag{12}$$

At low frequencies R_{eq} becomes purely a function of switching frequency and flying capacitor value therefore conduction power losses dominate switching power losses. However at higher frequencies R_{eq} becomes constant and a function of the on-resistance (R_{on}) of MOS, as shown in Figure 6. Here the switching power losses dominate conduction power losses.

V. OPTIMIZATION OF SWITCHING FREQUENCY AND SWITCH SIZE

There is a well known trade off between efficiency of a SCC and switching frequency & switch size. For a given switching frequency, if we increase size of switch then conduction loss will reduce but switching loss will increase as gate oxide capacitance will increase and vice versa. For a given switch size, if we reduce switching frequency then conduction losses will increase but switching losses will decrease and vice versa. Hence, it is desirable to obtain optimum value of switch size and switching frequency.

We calculated optimum values of switching frequency and switch size using a MATLAB code by dividing switch sizes into multiple regions. For every region of switch size, we calculated optimum value of switching frequency where we get maximum efficiency. The region of switch size in which we get maximum efficiency is selected.

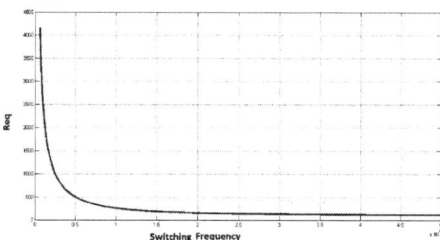

Fig. 6. Plot of Equivalent resistance vs. Switching Frequency

TABLE IV
EFFICIENCY FOR VARIOUS GAIN

Gain	Efficiency
1/4	$\eta = \frac{R_L}{R_L + R_{eq_{1/4}}} \left\{ \frac{1}{1 + F_{sw} C_{eq_{1/4}} (R_L + R_{eq_{1/4}})} \right\}$
2/4	$\eta = \frac{R_L}{R_L + R_{eq_{2/4}}} \left\{ \frac{1}{1 + F_{sw} C_{eq_{2/4}} (R_L + R_{eq_{2/4}})} \right\}$
3/4	$\eta = \frac{R_L}{R_L + R_{eq_{3/4}}} \left\{ \frac{1}{1 + \frac{F_{sw}}{9} C_{eq_{3/4}} (R_L + R_{eq_{3/4}})} \right\}$
1/3	$\eta = \frac{R_L}{R_L + R_{eq_{1/3}}} \left\{ \frac{1}{1 + F_{sw} C_{eq_{1/3}} (R_L + R_{eq_{1/3}})} \right\}$
2/3	$\eta = \frac{R_L}{R_L + R_{eq_{2/3}}} \left\{ \frac{1}{1 + \frac{F_{sw}}{4} C_{eq_{2/3}} (R_L + R_{eq_{2/3}})} \right\}$

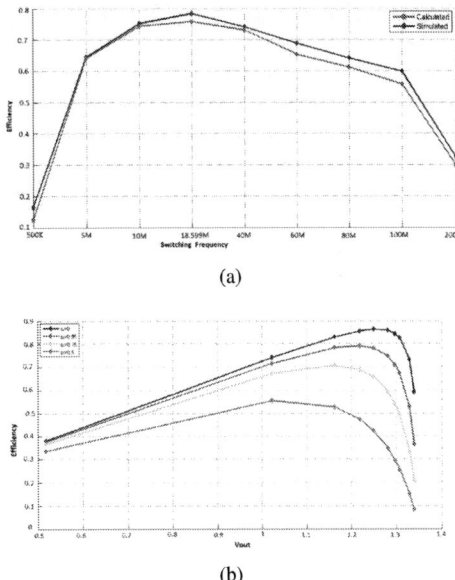

(a)

(b)

Fig. 7. (a) Plot of Efficiency vs. Switching Frequency ; (b) Plot of Efficiency vs. Output Voltage for 3/4 step down converter

VI. SIMULATED RESULTS AND DISCUSSION

To illustrate the loss analysis mechanism presented in the paper, we compare the simulation results with calculated results for the case of 3/4 step-down converter using two flying capacitors. We have used C_{fly1} = C_{fly2} = 100 pF and V_{in}=1.8 V. The switch selection is done based on the scheme presented. The efficiency was measured and plotted for various switching frequencies (F_{sw}) ranging from 500 KHz to 200 MHz as shown in Figure 7(a). The plot shows a good agreement (approximately 5% error in the calculated and simulated values) with the calculated efficiency values. The maximum efficiency of 78.5% can be achieved with 5% bottom plate capacitance at a optimum switching frequency of 20 MHz.

The efficiency is calculated and simulated for different bottom plate capacitance values for $\alpha = 0$, $\alpha = 0.05$, $\alpha = 0.15$ and $\alpha = 0.5$. We plotted these simulated values of efficiency vs. output voltage by varying load as is shown in Figure 7(b). As seen from the plot, efficiency reaches the maximum value of 86.30% with $\alpha = 0$ and decreases for values of $\alpha = 0.05$, $\alpha = 0.15$ and $\alpha = 0.5$. This explains the effect of the bottom plate capacitance of the flying capacitor on efficiency.

A load step from 1.35 mA to 2.7 mA is performed at an input voltage of 1.8 V to check load transient behaviour of the converter, 7% drop in the efficiency is observed as shown in Figure 8. The input voltage is changed from 1.8 V to 1 V to check line transient behaviour of the converter, there is 7.8% drop in the efficiency.

TABLE V
PERFORMANCE SUMMARY AND COMPARISON

	Ref:[1]	Ref:[5]	Ref:[2]	This work
Technology	32nm	0.7μm	0.18μm	0.18μm
Architecture	Step Down	Step up	Step Down	Step Down
Conversion Ratios	1/2,2/3 1/3	2/1	1,1/2,1/3 2/3,3/4	1/4,1/2,1/3 2/3,3/4 for n=2
Input Voltage	2V	1.1V	1.2V	1.8V
Flying Cap	-	100pF(2)	200pF(12)	100pF(2)
Frequency	10MHz-800MHz	100KHz-10MHz	15MHz	500KHz-200MHz
Peak Efficiency	79.76 %	75%	74%	78.4%
Load Current	3.4mA-6.8mA with η>70%	20μA-2mA with η>70%	100μA-0.25mA with η>70%	0.45mA-1.35mA with η>74%

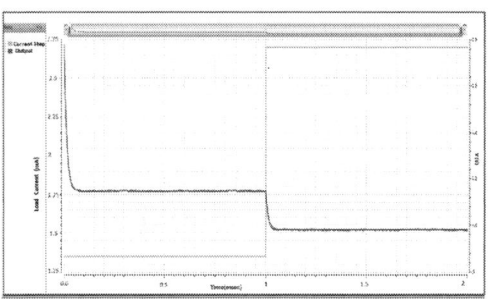

Fig. 8. Load Transient Behaviour

The performance comparison of different SCC is shown in Table V.

VII. CONCLUSION

CMOS implementation of SCC with generalized switch network is presented. The optimization of switching frequency and switch size is discussed. The Digital Switch Controller is developed to switch between all possible target voltages for two-flying capacitor converter. Future work includes the implementation of simple digital feedback control mechanism to obtain desired output voltage by interpolating between two target voltages.

REFERENCES

[1] Le, Hanh-Phuc, Seth R. Sanders, and Elad Alon. "Design techniques for fully integrated switched-capacitor DC-DC converters." Solid-State Circuits, IEEE Journal of 46.9 (2011): 2120-2131.

[2] Ramadass, Yogesh K., and Anantha P. Chandrakasan. "Voltage scalable switched capacitor DC-DC converter for ultra-low-power on-chip applications." Power Electronics Specialists Conference, 2007. PESC 2007. IEEE. IEEE, 2007.

2016 International Conference on VLSI Systems, Architectures, Technology and Applications (VLSI-SATA)

[3] Kushnerov, Alexander. "High-efficiency self-adjusting switched capacitor DC-DC converter with binary resolution." arXiv preprint arXiv:1003.4301 (2010).

[4] Kushnerov, Alexander, and Sam Ben-Yaakov. "Algebraic synthesis of Fibonacci switched capacitor converters." Microwaves, Communications, Antennas and Electronics Systems (COMCAS), 2011 IEEE International Conference on. IEEE, 2011.

[5] Favrat, Pierre, Philippe Deval, and Michel J. Declercq. "A high-efficiency CMOS voltage doubler." Solid-State Circuits, IEEE Journal of 33.3 (1998): 410-416.

[6] Chiu, Po-Yen, and Ming-Dou Ker. "Metal-layer capacitors in the 65nm CMOS process and the application for low-leakage power-rail ESD clamp circuit." Microelectronics Reliability 54.1 (2014): 64-70.

Design and Analysis of Novel Fuzzifer Circuits in CMOS Current Mode Approach

Abdullah Gubbi
Dept of Electronics and Communication,
P A College of Engineering,
Mangalore, Karnataka, India
abdullahgubbi@yahoo.com

Deeksha.M
Dept of Electronics and Communication,
P A College of Engineering,
Mangalore, Karnataka, India
deekshamohan9@gmail.com

Abstract—In this paper, the hardware realization of the basic blocks of Fuzzy Inference System (FIS) using simplified inference mechanism circuits are designed and tested in Complementary Metal Oxide Semiconductor (CMOS) Current Mode (CM). These circuits are useful in fuzzy and neuro-fuzzy systems. FIS consists of three main functional blocks. The fuzzification block using Membership Function Generator Circuit (MFGC), rule evaluation and defuzzification. The circuits are designed using the Cadence Virtuoso Design environment in 180nm technology and tested using the Spectre tool. The responses of the circuits, for variations in different signal values are represented using characteristics obtained from spectre tool. The circuit delays and average power are calculated from transient responses with simulation matching the mathematical calculation.

Keywords—*Fuzzy Logic; Membership Functions; Current Mirror; S-shaped MFGC; Gaussian MFGC; Min-Max; Schematic; Aspect Ratio; Delay.*

I. INTRODUCTION

Fuzzy Logic (FL) theory is the extension of conventional (crisp) set theory. In broad sense, fuzzy logic refers to fuzzy sets, which are sets with blurred boundaries. It handles the concept of partial truth (truth values between 1 (completely true) and 0 (completely false)). It was introduced by Prof. Lotfi A. Zadeh of UC Berkely in 1965 as a mean to model the vagueness and ambiguity in complex systems [1]. For the complicated and vague processes the FL is effectively used to get the solutions for which there is no mathematical model or the mathematical model is highly nonlinear. Since its inception, FL has got numerous revenues in fields of commercial and non commercial applications. The concept has touched almost all fields. To name a few, Computer Vision, Control Engineering, Medicine, Signal processing, Forecasting, Robotics and so on. As FL is gaining applications in all fields, hence researchers are motivated to find novel methods to implement the same.

In the literature, FIS are implemented using reconfigurable Field Programmable Gate Array (FPGA) [2], Microcontroller and Computer Aided-Design (CAD) tools of Very Large Scale Integration (VLSI). Yamagawa and Mikki [3] have implemented analog CMOS CM circuits for FIS. J.L. Huertas et.al has implemented fuzzy microcontroller using analog current CMOS circuit and also a voltage interface is provided with the external world, combining analog and digital technique [4]. Adapting FIS concepts to fit microelectronics constraints, both in terms of architecture and circuitry, has been the major issues in this field. In the literature it is reported that FIS are implemented by massively parallel structure for both evaluating inference rules and handling the fuzzy consequents.

Analog VLSI ((CM) and voltage mode) circuits are suited for FIS, as they are easily implementable for the arithmetic operations. The analog circuits design methodologies; especially in CM techniques are best suited for direct implementation of the arithmetic operations needed by the fuzzy operators. The advantage of CM method is that the inputs and output values are defined as currents only. One of the features of CM is an unwanted current to voltage and/or voltage to current intermediate converters can be avoided. By doing so, the delay through cascaded operator may even be shortened to gain higher speed. Using the CM, the dynamic range problem can be handled very flexibly in a manner as if it does not depend upon bias voltage values. The important feature is that existing FIS system can be trained to different environment with the expert knowledge. Parameter tuning capability of FIS exists in such way that it can be adapted to solve complex problems without demanding complex adjusting procedures. In order to work at a reasonable speed with lesser number of transistors, analog CM FIS elements are built which can be used for fuzzification and defuzzification. In general the sensor data is available in the form of voltage with noise. Hence the primary task is to remove the noise and use voltage to current converter for further processing.

The objective of this paper is to describe the hardware realization of basic blocks of FIS using CM. These elements can be used for applications such as process control, home appliances, automotive and robotics etc. Analog circuits occupy less area in contrast to digital circuits and also allow parallelisms to be employed. They are susceptible to process variations and temperature. They also suffer from noise and interferences. There is a need for robust analog design with low cost, low power and high operational speed. The rest of the paper is presented as follows.

Section II, III, IV, V presents the fuzzification using S-shaped Membership Function Generator (MFG), Z-shaped MFG and Gaussian MFG, rule evaluation circuits like minimum and maximum selector circuits, simulations and results of the respective circuits, comparison of MFG's and conclusion.

II. PROPOSED CIRCUITS FOR FUZZIFICATION

In any FIS, the first task is to convert the crisp input to the fuzzy membership values (crisp to fuzzy set). This operation is known as Fuzzification. These MFGC are implemented in CM and voltage mode using CMOS design. In general this block makes use of S-shaped or Z-shaped or Triangular or Gaussian MFGs or combinations of any of the above. These MFGCs are selected based on expert's knowledge.

A. S-shaped Membership function generator

One of the commonly used membership function is the S-shaped membership function. The S-shaped membership function $\mu_S(i)$ is given by,

$$\mu_S(i) = \begin{cases} 0 & -\infty < i \leq a \\ \dfrac{i-b}{b-a} + 1 & a \leq i \leq b \\ 1 & b \leq i \leq +\infty \end{cases} \qquad (1)$$

Where i represents the input current, the parameters a and b locate the extremes of the sloped portion of the curve.

The Fig 1 shows schematic of the S-shaped MFGC. It has one crisp input I_{IN} and two constant inputs I_{REF} and I_X. The current I_X is used to control the location of the membership function in horizontal axis and I_{REF} defines the upper limit of the output current. This effect is pictorially shown in the simulation results.

TABLE I   Transistor sizing for S-shaped MFGC.

Transistors	Aspect ratio(μm/μm)
N1,N5,N6,N7,N8,P3	0.4/0.18
N2	2.3/0.36
N3	0.56/0.36
N4,P1,P2	2/0.18
P4	1/0.36

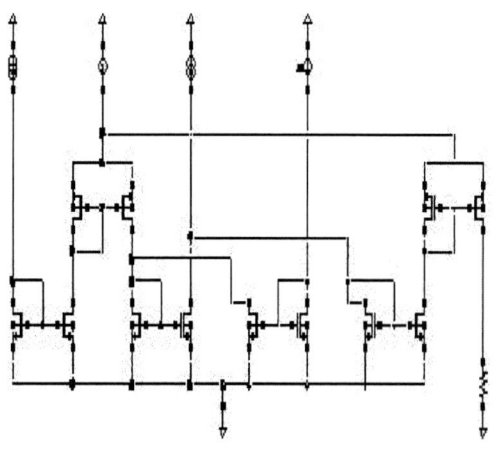

Fig 1   S-shaped MFGC.

The S shaped MFGC is built on the concept of current mirror. There are six current mirrors using only 4 Positive channel Metal-Oxide Semiconductor (PMOS) and 8 Negative channel Metal-Oxide Semiconductor (NMOS) transistors. The working of S-shaped MFGC is as follows. As long as transistor N4 conducts, the transistor N7 remains in the cut-off state, as the path exist from gate to ground via N4. Gradually the current in N3 $(I_x - I_{IN})$ reduces to a certain point and N4 goes to cut-off state. This gives room for transistor N7 to conduct and the current in N7 gradually reaches to I_{REF}. The individual branch currents are depicted in the Fig 2.

Since the current mirrors are used, the negative current cannot flow through the braches. Hence the currents evaluated to negative values are Zero in the Circuit. The current I_{N7} is the output which is stabilized through mirrors N7-N8 and P3-P4.

$$I_{OUT} = I_{N7} = I_{REF} - (I_X - I_{IN}) \qquad (2)$$

1) Results and discussions.

The currents used for the simulation are given in TABLE II.

The V_{DD} is set to 1.8V. The transfer characteristic of the circuit is shown in Fig 3. The DC analysis done from 0-10μA indicates that the unity gain point is at 5μA, the lower cut-off for the S-shaped curve is approximately at $I_{IN} = 2.5$μA and output attains its peak value at 7.5μA.

The transient response is the important parameter, to characterize the circuit. The transient response is computed to calculate the speed of the circuit. The average delay obtained is 361.6 Pico-second. The power dissipation of the circuit is 40.23μW. It is also proved that the DC response and Transient response are same if ramp current is given as an input. The comparison is done in the next section. The proposed method uses less number of transistors and provides more speed.

TABLE II   Current values for Fig.1

Current	Value
I_{REF}	10μA
I_X	5μA
I_{IN}	(0μA-10μA)

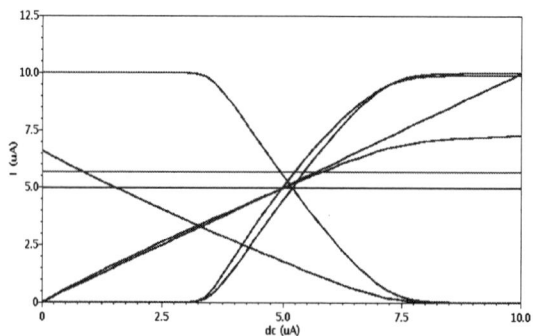

Fig 2   Individual Branch Currents of S-shaped MFGC.

2016 International Conference on VLSI Systems, Architectures, Technology and Applications (VLSI-SATA)

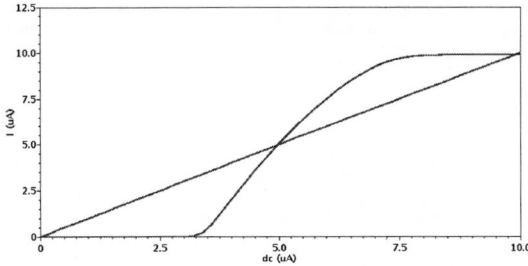

Fig 3   Transfer curve of S-shaped MFGC.

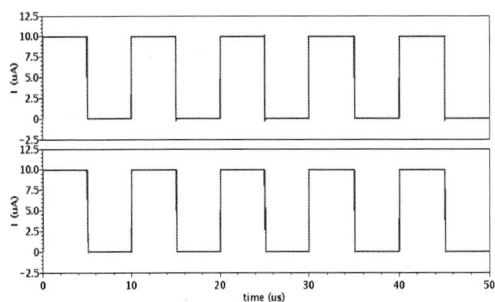

Fig 4   Transient response of S-shaped MFGC.

(a)Output current waveform (b) Input current waveform.

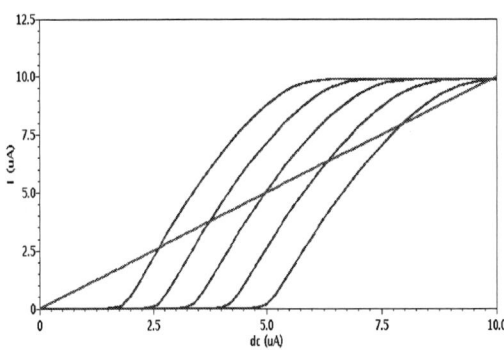

Fig 5   Response of S-shaped MFGC for variation in I_X.

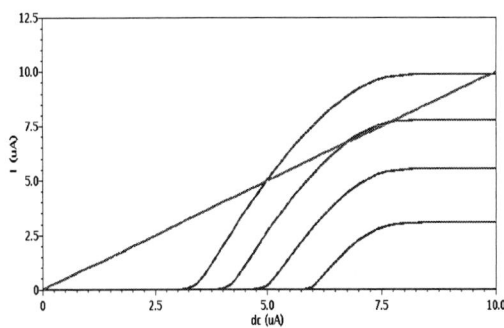

Fig 6   Response of S-shaped MFGC for variation in I_{REF}.

B. Z-shaped membership function generator

When the inputs I_X and I_{IN} are interchanged, the shape of the transfer curve resembles the Z-shape. Hence, it can be used as Z-shaped MFGC. The waveform in Fig 7 shows the transfer curve of Z-shaped MFGC.

C. Gaussian membership function generator

Gaussian MFGC is also built on current mirror concept. We use 6 PMOS and 12 NMOS transistors as current mirrors. Gaussian membership functions are the most adequate option for representing the uncertainty of the situations. The Gaussian membership function $\mu_G(i)$ is given by,

$$\mu_G(i) = \begin{cases} 0 & -\infty < i \le a - \delta \\ \exp\left(-(i-1)^2/\delta^2\right) & a - \delta \le i \le a + \delta \\ 0 & a + \delta \le i \le +\infty \end{cases} \quad (3)$$

Where the value of a represents the position of the peak point of the Gaussian curve and δ represents the width of the curve.

In the schematic of the Gaussian MFGC we use one crisp input I_{IN} and two constant inputs I_{REF} and I_X. The current I_X is used to control the location of the membership function in horizontal axis and I_{REF} defines the upper limit of the output current. The effect of I_{ref} and I_x is shown in the simulation results.

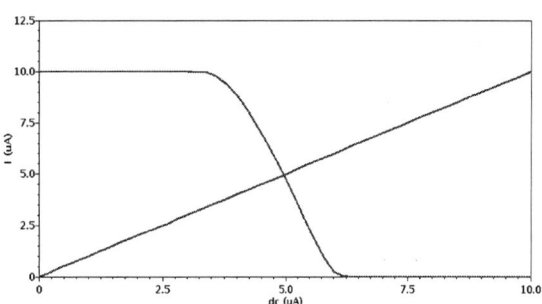

Fig 7   Transfer curve of Z-shaped MFGC.

TABLE III   Transistor sizing for Gaussian MFGC

Transistors	Aspect ratio(μm/μm)
N1,N10,N11,N12,P5	0.4/0.18
N2	1.8/0.36
N3	3/0.18
N4,N7	0.56/0.36
N8	2.2/0.18
N9	2.3/0.18
N5,N6,P1,P2,P3,P4	2/0.18
P6	1/0.36

978-1-5090-0037-1/16 $31.00 © 2016 IEEE   59

Gaussian MFGC circuit is a combination of S and Z-shaped MFGCs. The transistors P1, P2, N1, N2, N4, N5, N8 and N10 form the S-shaped MFGC giving output at drain of N5. Similarly transistors P3, P4, N1, N3, N6, N7, N9 and N10 form the Z-shaped MFGC giving output at drain of N6. These two outputs are then added and then subtracted by the current I_{REF} to get the Gaussian output. The outputs at different nodes of the circuit are shown in the Fig 9.

2) Results and discussions

The V_{DD} is set to 1.8V. The transfer characteristic of the circuit is shown in Fig 10. The DC analysis done from 0-10μA indicates that maximum current is reached at 5μA, the lower cut-off and upper cut-off for the Gaussian curve are approximately at $I_{IN} = 2.5\mu A$ and $I_{IN} = 7.5\mu A$ respectively.

The transient response is performed to calculate the speed of the circuit. The average delay obtained is 994.4psec. The power dissipation of the circuit is 53.35 μW.

The effect of I_X on the position of the Gaussian curve on the input axis is shown in Fig 11. It indicates that the slope remains same for a given design irrespective of the I_X value.

In the similar way, the effect of I_{REF} is shown in Fig 12. From the figure it is clear that, as the I_{REF} decreases the maximum current attained by the circuit decreases.

TABLE IV   Current values for simulation of Gaussian MFGC

Current	Value
I_{REF}	10μA
I_X	5μA
I_{IN}	(0μA-10μA)

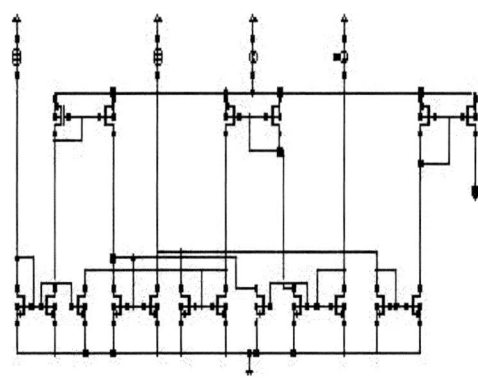

Fig 8   Gaussian MFGC.

Fig 9   Individual Branch Currents of Gaussian MFGC

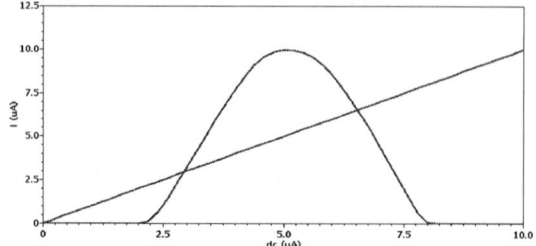

Fig 10   Transfer characteristics of Gaussian MFGC

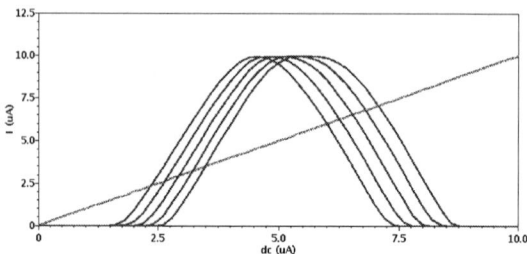

Fig 11   Response of Gaussian MFGC for variation in I_X

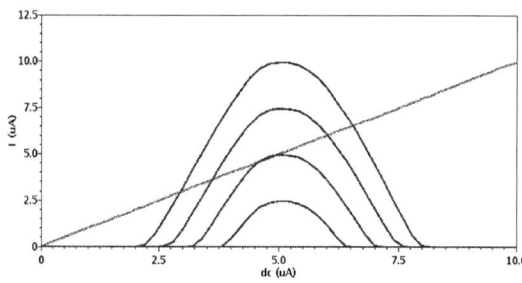

Fig 12   Response of Gaussian MFGC for variation in I_{REF}

III.   FUZZY RULE EVALUATION

The two main circuits used for the manipulation of fuzzy rules are minimum and maximum selector circuits. These perform fuzzy operations of conjunction and disjunction respectively.

A. Max current selector circuit

Fig 13 shows the schematic of the maximum current selector circuit. The principal function of this circuit is to detect the maximum current among all the inputs given to it and generate this maximum current at its output terminal. The following expression describes the input and output relationship of this circuit.

$$Imax = \max(I1, I2) \qquad (4)$$

Where I_1 and I_2 are the two input current signals and $Imax$ is the output, which is equal to the maximum of I_1 and I_2.

The transient response for the maximum current selector circuit is as shown in Fig 15 given below.

B. Min current selector circuit

Fig 14 shows the schematic of the minimum current selector circuit. The main function of this circuit is to detect the minimum current among all the input currents given to it and generate the minimum current at its output terminal. The following expression describes the input and output relationship of this current.

$$Imin = \min(I1, I2) \qquad (5)$$

Where I_1 and I_2 are the two input current signals and $Imin$ is the output, which is equal to the minimum of I_1 and I_2.

The V_{DD} is set to 1.8V. The transfer characteristics of the circuit are shown in Fig 16. Fig 15 and Fig 16 shows the simulation results for max current selector circuit and min current selector circuits. In this case, two types of signals are used, a square wave and a sinusoidal waveform.

TABLE V   Current values for simulation of max selector.

Current	Value
I1 (I_{sin})	20µA
I2 (I_{pulse})	10µA

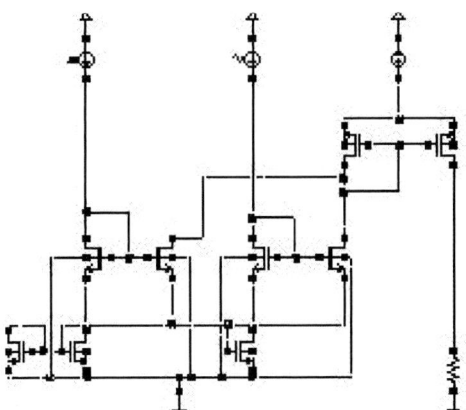

Fig 13   Max current selector circuit

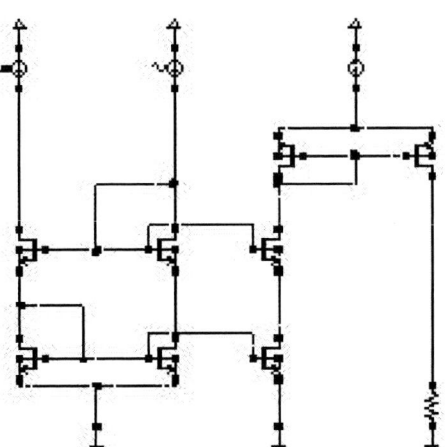

Fig 14   Min current selector circuit.

TABLE VI   Current values for simulation of min selector.

Current	Value
I1 (I_{sin})	20µA
I2 (I_{pulse})	10µA

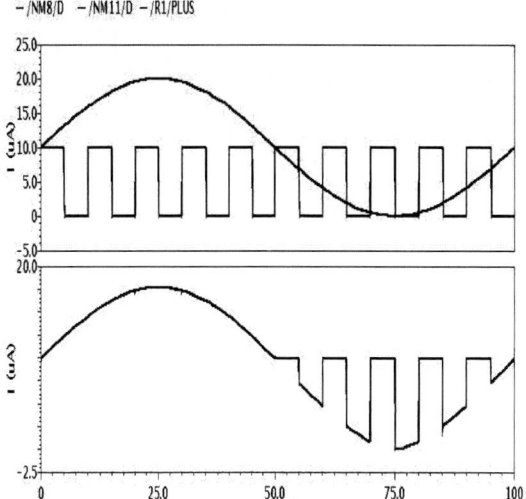

Fig 15   Transient characteristics of max current selector circuit

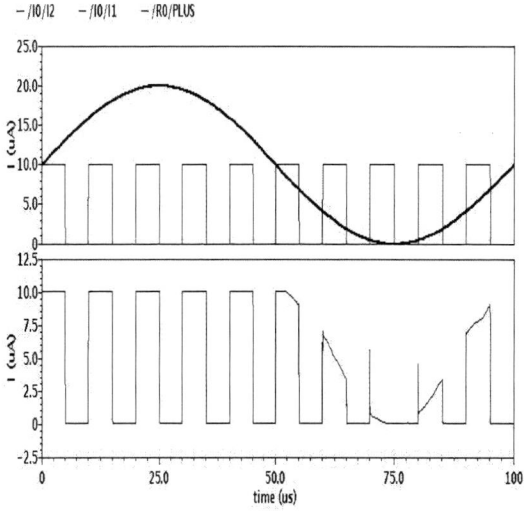

Fig 16   Transient characteristics of max current selector circuit

IV.   COMPARISON OF REPORTED MEMBERSHIP FUNCTION GENERATOR

This section shows the comparison between the proposed MFGC's and the existing MFGC's. The proposed method is found to be effective for low power and high speed design applications. TABLE VII shows the comparisons of existing and proposed MFGC's.

TABLE VII   Comparison of reported MFGC

Ref	Tech	Supply	Transistors count or area.	Parameters controlled	Controlled by	Power or *Ibias*	Control	In	Out
[4]	1.5μm	-	variable	3	Variable trans-conductance	-	Non-independent	V	I
[6]	-	10V	38+4 *Ibias*	6	2 voltage source 2 current sources	25 μA	Slope: non-linear	V	I
[7]	0.6μm	5V	21+2 *Ibias* 0:03 mm2	4	2 voltage source 2 current sources	20 μA	Only position is independent. Slope: non-linear	V	I
[4]	1.5μm		variable	4	D/A converters	-	Digital linear control	V	I
[8]	0.35 μm	1.5V	42+13 *Ibias*	5	2 voltage source 3 current sources	-	Slope: non-linear and non-independent	V	I
[9]	0.8 μm	-	84 0:07 mm2	5	2 voltage source 3 current sources	80 μA	Non-linear slope control	V	I
[10]	0.5 μm	3V	28	5	Switched branches 3 current sources	90 μA	All linear and non-independent	V	I
[12]	0.7 μm	5V	>50	3		1.7mW	Non-independent	V	I
[11]	0.35 μm	3.3V	>132	3	2 voltage source 1 current source	100 μA	Non linear slope control	I	I
[13]	0.5 μm	1.5V	61 (+)	4	4 voltage source	200 μW,5μA	All linear including slope	I	I
This work	0.18 μm		variable	4	1 voltage source 3 current source	10 μA	Non independent	I	I

V.   CONCLUSION

A detailed design and analysis of the two mostly used MFGC's, S-shaped and Gaussian shaped are carried out. The circuits are designed with minimum number of transistors. The design parameters provided. The detailed analysis is provided with graphical representation. Changing the transistor size results change in the slope of curves. The functionality is verified and good result is achieved.

These circuits are suited for high speed, high accuracy fuzzy and neuro-fuzzy applications.

REFERENCES

[1] L. A. Zadeh, "Fuzzy logic," Institute of Electrical and Electronics Engineers (IEEE) Computer, vol. 21, no. 4, pp. 83–93, 1988.

[2] An Implementation of Fuzzy Logic Controller on the Reconfigurable FPGA System, Daijin Kim, Member, IEEE

[3] T. Yamakawa and I. Miki, "The current mode fuzzy logic integrated circuits fabricated by the standard CMOS process," *IEEE Trans. Compute.*, vol. 35, pp. 161–167, Feb. 1986.

[4] J.L. Huertas "Building blocks for current mode implementation on VLSI fuzzy microcontroller", 5th international fuzzy system association, world congress Vol. ii, pp, 929-932, Seoul- Korea July 4-9-1993

[5] Chuen-Yau Chen, Yuan-Ta Hsieh and Bin-Da Liu, "Circuit implementation of linguistic-hedge fuzzy logic controller in current mode approach", IEEE Transactions on Fuzzy Systems, Vol. 11, No. 5, pp.624-645, October 2003.

[6] Tokmakci, M., Alkci, M., Kilic R. Simple CMOS-based membership function circuit. Analog Integrated Circuits and Signal Processing, 2002, vol. 32, p. 83 - 88.

[7] Shan, W., LU, Y., Sun, H., Liu J. A novel analog circuit design and test of a triangular membership function. Lecture Notes in Electrical Engineering, 2011, vol. 97, p. 727 - 733.

[8] Ghanavati, B. A 1.5 V CMOS fuzzifier. International Journal of Multidisciplinary Sciences and Engineering, 2012, vol. 3.6, p. 52-55 .

[9] Carvajal, R. G., et al. Mixed-signal CMOS fuzzifier with emphasis in power consumption. IEEE 42nd Midwest Symposium on Circuits and Systems. Las Cruces (USA), 2000, p. 929-933.

[10]   Kachare, M., Ram´Irez-Angulo, J., Carvajal, R. G Lo'pez-Marti'n, C. New low voltage fully programmable CMOS triangular /trapezoidal function generator circuit. IEEE transactions on circuits and systems- I, 2005, vol. 52, no. 10, p. 2033-2042.

[11] Khalilzadegan, A., Khoei, A., Hadidi, K. A fully programmable MFG with a new analog programmable current mirror for slop tuning of membership functions. In 19th Iranian Conference on Electrical Engineering (ICEE). Tehran (Iran), 2011, p. 1 - 6.

[12] Bouras, S., Kotronakis, M., Suyama, K., Tsividis, Y. Mixed analog - digital fuzzy logic controller with continuous – amplitude fuzzy inferences and defuzzification. IEEE Transactions on Fuzzy Systems, 1998, vol. 6, no. 2, p. 205 - 215.

[13] Carlos Mun˜ IZ-Montero1, Luis A. SA´Nchez-Gaspariano1, Jose´ M. Rocha-PE´REZ2, Jesu´s E. Molinar-Soli´S3 and Carlos SA´Nchez-LO´Pez4 1Universidad Polit´ecnica de Puebla, Tercer Carril del Ejido "Serrano" s/n, Juan C. Bonilla, Puebla 72649, Mexico.

2016 International Conference on VLSI Systems, Architectures, Technology and Applications (VLSI-SATA)

Performance of Asymmetric Gate Oxide on Gate-Drain Overlap in Si and $Si_{1-x}Ge_x$ Double Gate Tunnel FETs

S.Poorvasha, *Student Member, IEEE* & B.Lakshmi, *Member, IEEE*
School of Electronics Engineering
VIT University, Chennai, India
Email: poorvasha.s2014@vit.ac.in, lakshmi.b@vit.ac.in

Abstract—This paper studies the performance of asymmetric gate oxide on gate-drain overlap for Si and $Si_{1-x}Ge_x$ based double gate (DG) Tunnel FETs (TFETs). For the first time, asymmetric gate oxide is introduced in the gate-drain overlap and compared with that of DG TFETs. For the different values of the mole fraction (x), $Si_{1-x}Ge_x$ is optimized to get ON current (I_{ON}) enhancement. $Si_{1-x}Ge_x$ based DG TFETs with gate-drain overlap offers a very good I_{ON} of 232 μA with the subthreshold swing (SS) of 26 mV/dec. This is achieved because of the high tunneling rate of electrons occurring at the source side of $Si_{1-x}Ge_x$.

Keywords—*DG TFETs; asymmetric gate oxide; gate drain overlap; I_{ON}; SS; TCAD.*

I. INTRODUCTION

CMOS technology faces major challenges, mainly due to high leakage current, high power consumption and non-scalability of sub-threshold swing (SS). In order to overcome these challenges many novel devices have been reported with different principles. Tunnel field effect transistors (TFETs) being one of the promising devices of MOSFETs are widely preferred due to its SS limit of 60 mV/dec, less leakage current (I_{OFF}) and low threshold voltage [1-3]. TFET works on the principle of band-to-band tunneling (BTBT) mechanism.

Comparing to single gate TFETs (SG TFETs), Double Gate TFETs (DG TFETs) shows improved on-current and less threshold voltage [4-6]. It has been reported that TFETs with the lower bandgap material, Silicon-Germanium (SiGe) for the source [7, 8], in the channel region [9], in the entire region [10], gives better ON-current (I_{ON}) comparing to Silicon (Si) based TFETs. Ambipolar conduction also plays a vital role in affecting the performance of TFET [11]. This kind of conduction can be suppressed by overlapping the gate on drain in DG TFET structure [12].

In this paper, asymmetric gate oxide structure is introduced in gate-drain overlap region and compared with the DG TFET. This study is done for both Si and $Si_{1-x}Ge_x$ source channel TFETs. Section II describes the device structures of Si and $Si_{1-x}Ge_x$ based DG TFET along with the simulation methodology. The results are discussed in Section III. Finally Section IV provides the conclusion.

II. DEVICE STRUCTURE AND SIMULATION METHODOLOGY

Technology CAD (TCAD) simulator from Synopsys [13] is used to perform all the simulations. The following modules are used in this study.

- Structure Editor (SDE): The 2D device structure is created using this module.

- Sentaurus Device (SDEVICE): The DC characteristics of the device are analyzed using this module.

- Inspect and Svisual: These modules are used to view the results and device respectively.

A. Si based TFETs

Fig. 1(a) and 1(b) shows the 2D structure of Si based DG TFET and DG TFET with gate drain overlap respectively. Fig. 2 shows the schematic of the device with all the parameters. The device simulator includes appropriate models for doping dependence mobility, effects of high and normal electric fields on mobility and velocity saturation. A non-local Hurkx band-to-band tunneling model is used along with Fermi-dirac statistics and Shockley-Read-Hall recombination model. Supply voltage used in this study is 1 V with the gate voltage of 1.8 V. The dimensions of the device are given in Table I. DG TFETs with gate drain overlap is optimized to give I_{ON} with reduced I_{OFF} by tuning the work function at the gate and gate-drain overlap.

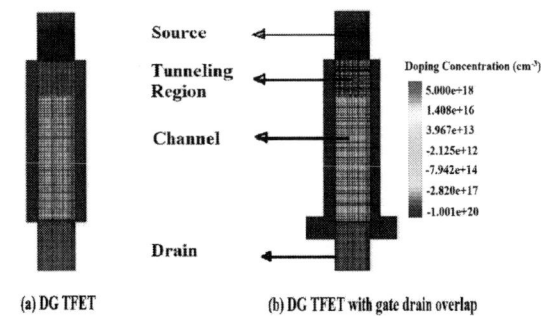

Fig. 1. Simulated structure of Si based DG TFETs and DG TFETs with gate drain overlap.

978-1-5090-0037-1/16 $31.00 © 2016 IEEE

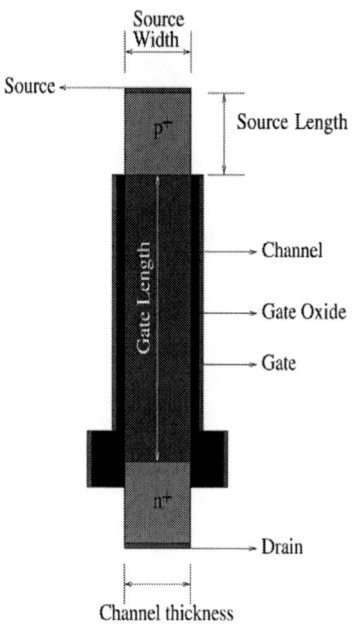

Fig. 2. Schematic structure of DG TFET with gate drain overlap.

B. $Si_{1-x}Ge_x$ based TFETs

The structural and doping dimensions of $Si_{1-x}Ge_x$ DG TFETs is same as in Table I. Fig. 3(a) and 3(b) shows the simulated structure of $Si_{1-x}Ge_x$ DG TFET and DG TFET with gate drain overlap. $Si_{1-x}Ge_x$ TFETs is compared with Si based TFETs for their performance enhancement. Compared to Si, the smaller bandgap in $Si_{1-x}Ge_x$ yields higher I_{ON} in TFETs [10].

TABLE I.   PARAMETER SPACE

Parameters	Nominal Value
Gate length (L_g)	50 nm
Gate oxide thickness (t_{ox})	3 nm
Channel thickness (t_{ch})	10 nm
Channel doping concentration (N_a)	1e17 cm^{-3}
Drain doping concentration (N_d)	5e18 cm^{-3}
Source doping concentration (N_s)	1e20 cm^{-3}
Gate electrode work function	4.5 eV

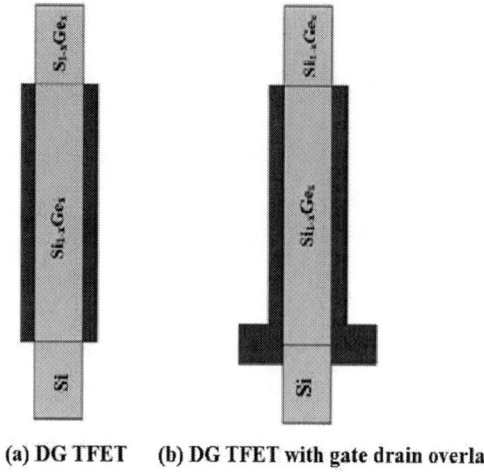

(a) DG TFET   (b) DG TFET with gate drain overlap

Fig. 3. Simulated structure of $Si_{1-x}Ge_x$ based DG TFET and DG TFET with gate drain overlap.

III. RESULTS AND DISCUSSION

A. I_d - V_g Characteristics of DG TFET for different mole fractions

Fig. 4 depicts the I_d–V_g curve for different mole fraction of $Si_{1-x}Ge_x$ based DG TFET. For x=0, I_{ON} is found to be 31.5 μA and for x=0.4, it increases to 58.5 μA. The composition of Germanium (Ge) content in $Si_{1-x}Ge_x$ improves I_{ON} for DG TFETs. This is because of the reduction in the tunneling width at the source side [14]. In this study, x=0.4 is used for the $Si_{1-x}Ge_x$ based DG TFETs.

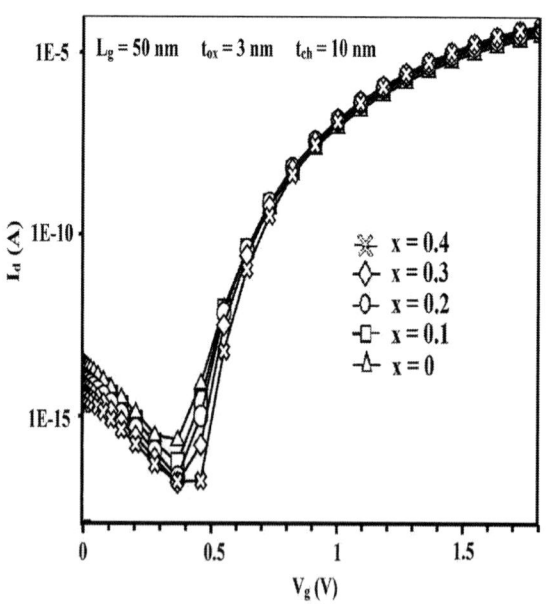

Fig. 4. I_d-V_g characteristics for different mole fraction of $Si_{1-x}Ge_x$ based DG TFET.

B. I_d - V_g Characteristics of Si and $Si_{0.6}Ge_{0.4}$ DG TFETs

I_d-V_g characteristics of Si and $Si_{0.6}Ge_{0.4}$ based DG TFETs and DG TFETs with gate drain overlap are plotted in Fig. 5. I_{OFF} has been matched to 29 fA for all the devices which is done by tuning the work function at the gate and gate-drain overlap. It can be inferred from Fig. 5 that $Si_{0.6}Ge_{0.4}$ based DG TFETs gives more I_{ON} compared to Si based DG TFETs. The SS is extracted and found to be less for $Si_{0.6}Ge_{0.4}$ based DG TFETs with gate drain overlap.

The increased I_{ON} for $Si_{0.6}Ge_{0.4}$ based TFETs can be reasoned out with Fig. 6. Fig. 6(a) and 6(b) depicts the electron barrier tunneling along the channel for DG TFETs and DG TFETs with gate drain overlap respectively. This plot gives the rate at which electrons are generated because of the tunneling at the source side. It can be observed that higher tunneling rate of electrons are achieved for $Si_{0.6}Ge_{0.4}$ based TFETs compared to that of Si based TFETs. Similarly DG TFETs with gate drain overlap offers more electron tunneling over DG TFETs.

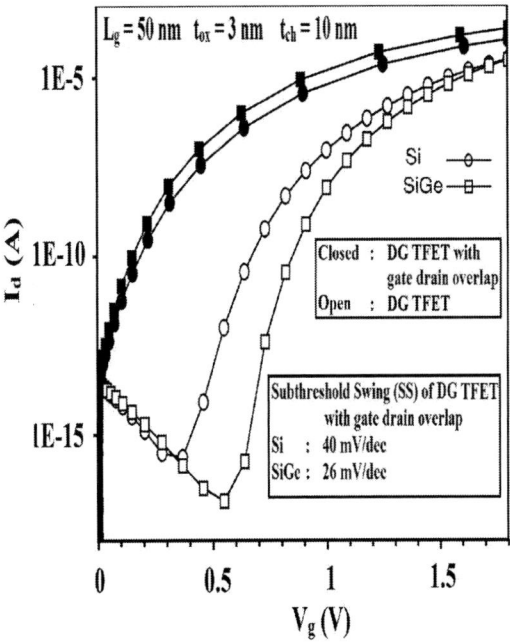

Fig. 5. I_d-V_g characteristics of Si and $Si_{0.6}Ge_{0.4}$ based DG TFETs and DG TFETs with gate drain overlap with I_{OFF} matching =29 fA.

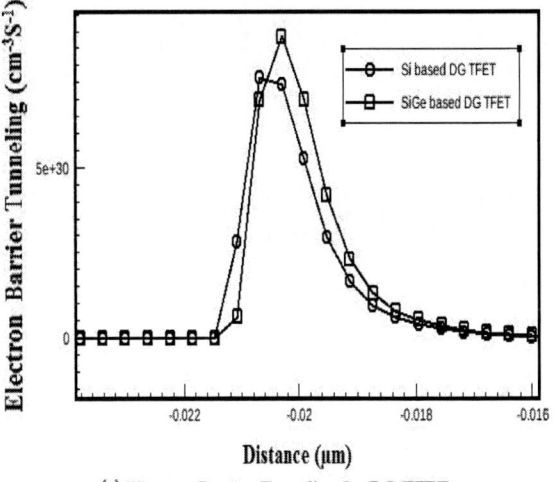

(a) Electron Barrier Tunneling for DG TFETs

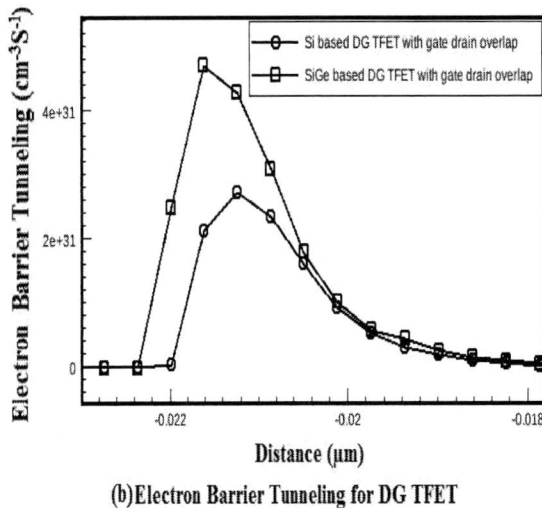

(b) Electron Barrier Tunneling for DG TFET with gate drain overlap

Fig. 6. Electron barrier tunneling for Si and $Si_{0.6}Ge_{0.4}$ based DG TFETs and DG TFETs with gate drain overlap.

IV. CONCLUSION

In this study asymmetric gate oxide is introduced on gate-drain overlap of DG TFETs and its performance is compared with that of DG TFETs. This comparison is done for both Si and $Si_{1-x}Ge_x$ based devices. Higher I_{ON} is achieved for $Si_{1-x}Ge_x$ based DG TFETs for a mole fraction of x=0.4. It is also found that $Si_{1-x}Ge_x$ DG TFETs with gate-drain overlap offers an excellent I_{ON} enhancement with the SS reduction of 26mV/dec. Hence $Si_{1-x}Ge_x$ DG TFETs with gate drain overlap is considered to be superior to that of DG TFETs and seems to be a promising candidate for future RF/ analog or mixed signal circuit applications.

ACKNOWLEDGMENT

This work is supported by Department of Science and Technology, Government of India under SERB scheme (Grant No: SERB/F/2660).

REFERENCES

[1] Yasin Khatami, and Kaustav Banerjee, "Steep subthreshold slope n- and p- type tunnel FET devices for low power and energy efficient digital circuits," IEEE Trans. Electron Devices, vol. 56, no. 11, pp. 2752-2760, Nov. 2009.

[2] Partha Sarathi Gupta, Sayan Kanungo, Hafizur Rahaman, Kunal Sinha, and Partha Sarathi Dasgupta, "An extremely low sub-threshold swing UTB SOI tunnel-FET structure suitable for low-power applications," Int. J.Applied Physics and Mathematics, vol. 2, pp. 240-243, July 2012.

[3] Alan C. Seabaugh, and Qin Zhang, "Low-voltage tunnel transistors for beyond CMOS logic," Proc. IEEE, vol. 98, no. 12, pp. 2095–2110, Dec 2010.

[4] Kathy Boucart, and Adrian Mihai Ionescu, "Double-gate tunnel FET with high-k gate dielectric," IEEE Trans. Electron Devices, vol. 54, no. 7, pp. 1725–1733, July 2007.

[5] Kuo-Fu Lee, et al., "Characteristic optimization of single and double gate tunneling field effect transistors," NSTI-Nanotech, vol. 2, 2010.

[6] Lining Zhang, Mansun Chan, and Frank He, "The impact of device parameter variation on double gate tunneling FET and double gate MOSFET," International Conference on Electron Devices and Solid-State Circuits (EDSSC), pp. 1-4, IEEE, Dec. 2010.

[7] Eng-Huat Toh, Grace Huiqi Wang, Lap Chan, Dennis Sylvester, Chun-Huat Heng, Ganesh S. Samudra, and Yee-Chia Leo, "Device design and scalability of a double-gate tunneling field-effect transistor with Silicon-Germanium source," *Jpn. J. Appl. Phys.*, vol. 47, no. 4, pp. 2593–2597, April 2008.

[8] Q. T. Zhao, J. M. Hartmann, and S. Mantl, "An improved Si tunnel field effect transistor with a buried strained $Si_{1-x}Ge_x$ source," *IEEE Electron Device Lett.*, vol. 32, no. 11, pp. 1480–1482, Nov. 2011.

[9] Hyun Woo Kim, Jang Hyun Kim, Sang Wan Kim, Min-Chul Sun, Euyhwan Park, and Byung-Gook Park, "Tunneling field-effect transistor with Si/SiGe material for high current drivability," Jpn. J. Appl. Phys., vol. 53, Article ID 06JE12-1 -06JE12-4, May 2014.

[10] S. Richter, et al., "Silicon-germanium nanowire tunnel-FETs with homo and heterostructure tunnel junctions," Solid-State Electronics, vol. 98, pp. 75-80, April 2014.

[11] A. Hraziia, C. Andrei, A. Vladimirescu, A. Amara, and C. Anghel, "An analysis on the ambipolar current in Si double-gate tunnel FETs," Solid-State Electron, vol. 70, pp. 67–72, April 2012.

[12] Dawit B. Abdi and M. Jagadesh Kumar, "Controlling ambipolar current in tunneling FETs using overlapping gate-on-drain," Journal of Electron Device Society, vol. 2, no. 6, pp. 187-190, Nov. 2014.

[13] Synopsys Sentaurus Device User Guide, version J-2014.09.

[14] S. Saurabh and M. J. Kumar, "Impact of strain on drain current and threshold voltage of nanoscale double gate tunnel field effect transistor (TFET): Theoritical investigation and analysis," Jpn. J. Appl. Phys., vol. 48, Article ID 064503-1 – 064503-7, June 2009.

2016 International Conference on VLSI Systems, Architectures, Technology and Applications (VLSI-SATA)

CORDIC-based VLSI Architecture for Implementing CI-OFDM and Its FPGA Prototype

Vikas Kumar, Kailash Chandra Ray and Preetam Kumar
Electrical Engineering Department
Indian Institute of Technology Patna,
India 800013
Email: {vikaskumar,kcr,pkumar}@iitp.ac.in

Abstract—Since decades, orthogonal frequency division multiplexing (OFDM) has been drawing its attention in the area of wireless and satellite communication systems such as IEEE 802.11 a/g/n, ADSL, WiMAX and DVB-T/SH. In these applications, OFDM has drawbacks of high Peak-to-Average Power Ratio (PAPR) and Inter carrier symbol interference (ISI). Recently, Carrier Interferometry OFDM (CI-OFDM) as an alternative to OFDM is being studied to minimize the aforesaid drawbacks. To the knowledge of authors, there is no Field Programmable Gate Array (FPGA) prototyping of CI-OFDM is addressed for real time wireless and satellite communication system. In this context, authors in this paper have proposed a new hardware efficient and flexible CO-ordinate Rotational Digital Computer (CORDIC) based CI-OFDM architecture. The novelty of this proposed architecture is its capability to change the number of orthogonal subcarriers and data symbols upto the maximum 32K, that shall be suitable for most of the standards of wireless and satellite communication system and prototyped using commercially available FPGA device XC3S500E-5FG320. The implementation and experimental results of this proposed scheme are highlighted and validated with the result from MATLAB simulation.

Keyword: *CI-OFDM; CORDIC; VLSI Architecture; FPGA Prototyping.*

I. INTRODUCTION

Orthogonal frequency division multiplexing/ multiple access (OFDM/OFDMA) scheme has been adopted in IEEE 802.11 a/g/n WLAN, HIPERLAN/2, DVB [1], WiMAX [2], ADSL, VDSL [3]. In these applications, OFDM converts a frequency-selective channel into a parallel collection of frequency-flat subchannels. In present trend, a novel orthogonal complex spreading code referred as carrier interferometry (CI) code is being used in OFDM [4], where each parallel data stream is simultaneously modulated onto all carriers via unique orthogonal complex phase spreading sequences. CI codes present unique features which allow multicarrier systems to have orthogonal spreading sequences of any length N and support N additional data symbol using pseudo-orthogonal CI (PO-CI) codes ($\pi/2$ shifted) with no expansion in bandwidth [5], [6]. Recent literature [5]–[9] states that CI based OFDM (CI-OFDM) scheme provides improved bit-error-rate (BER) performance and lower peak-to-average-power-ratio (PAPR) compared to conventional OFDM and multiple access schemes. By spreading all information symbols over all N subcarriers with CI spreading codes, CI-OFDM exploits frequency diversity and suppresses narrowband interference [10].

Real time computation of CI-OFDM demands the efficient complex arithmetic units where co-ordinate rotation digital computer (CORDIC) is an obvious choice because of its less hardware complexity and is used for real-time computation of linear, trigonometrical and transcendental functions [11]–[13].

To the knowledge of the authors, no hardware efficient CORDIC based architecture of CI-OFDM has been proposed in the literature and prototyped on Field Programmable Gate Array (FPGA). In this paper, a new CORDIC based VLSI architecture for real time implementation of CI-OFDM is proposed that will be suitable for wireless baseband modulator and further this proposed architecture is designed using verilog HDL and prototyped on commercially available FPGA device XC3S500E-5FG320, whose implementation and experimental results are presented and compared with MATLAB simulation.

The rest of the paper is organized as follows,section II highlights the CI-OFDM system model and CORDIC used for proposed architecture. Section III presents the proposed VLSI architecture of CORDIC based CI-OFDM. FPGA prototyping and experimental results of proposed architecture are reported in Section IV. Finally, Section V concludes the paper.

II. CI-OFDM AND CORDIC

This section briefs about CI-OFDM including CORDIC as arithmetic unit used for proposed architecture.

A. CI-OFDM System Model

Carrier Interferometry (CI) spreaded Orthogonal Frequency Division Multiplexing (OFDM) is being used to reduce intercarrier interference (ICI) and to reduce Peak-to-Average Power Ratio (PAPR) in multicarrier system. It also utilizes the spectrum more efficiently by enabling closer spacing of the channels [14], [15]. In CI-OFDM, each parallel information symbols are simultaneously modulated on all N subcarriers via unique orthogonal complex phase spreading sequences using processing element (PE) as shown in Fig. 1. CI code for the kth information symbol corresponds to $\{\beta_k^i\}_{i=0}^{N-1} = \{0, e^{\frac{j2\pi k}{N}}, e^{\frac{j2\pi 2k}{N}}, e^{\frac{j2\pi 3k}{N}}, \ldots e^{\frac{j2\pi(N-1)k}{N}}\}$, which ensures orthogonality among all N transmitted symbol over same subcarriers at same time [5], [6], [16], [17]. The data vector is given by $\mathbf{a} = [a(0), a(1), a(2), \ldots a(N-1)]^T \in \mathbb{C}^{N \times 1}$ where $a(k)$ denotes kth data symbol and $\mathbb{C}$ represents the set

978-1-5090-0037-1/16 $31.00 © 2016 IEEE

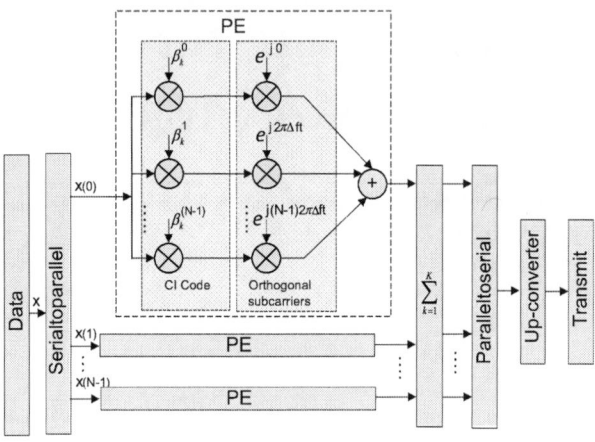

Fig. 1. CI-OFDM system model [14].

of complex numbers. In CI-OFDM, the CI signal for kth data is expressed as

$$c_k(t) = \sum_{i=0}^{N-1} \beta_k^i e^{j2\pi i \Delta f t} \qquad (1)$$

where the frequency of ith subcarrier is $\{f_i = f_c + i\Delta f\}$ with center frequency f_c, where Δf is selected such that the orthogonality between carrier frequencies can be maintained, typically $\Delta f = \frac{1}{T}$, where T is time duration of Nyquist pulse shape $p(t)$. The CI-OFDM transmitted signal is expressed as

$$s(t) = \sum_{k=0}^{N-1} s_k(t) e^{j2\pi f_c t} \qquad (2)$$

where $s_k(t)$ is the CI-OFDM baseband signal and f_c is the frequency of carrier signal. Equation (2) can be further rewritten in terms of equation (1) as below

$$s(t) = \frac{1}{\sqrt{N}} \sum_{k=0}^{N-1} \left[a(k) c_k(t) p(t-kT) \right] e^{j2\pi f_c t} \qquad (3)$$

$$= \frac{1}{\sqrt{N}} \sum_{k=0}^{N-1} \left[a(k) \sum_{i=0}^{N-1} \beta_k^i e^{j2\pi f_i t} p(t-kT) \right]$$

where $a(k)$ is the input data vector and $c_k(t)$ is CI signal for kth data. CI-OFDM transmitted signal, $s(t)$, in terms of equation (3) can be expanded as below.

$$s(t) = \frac{1}{\sqrt{N}} \sum_{k=0}^{N-1} a(k) \sum_{i=0}^{N-1} e^{\frac{j2\pi ik}{N}} e^{j2\pi i \Delta f t} p(t-kT) e^{j2\pi f_c t} \qquad (4)$$

Based upon equation (4) system model for CI-OFDM transmitter is presented in Fig. 1 for N orthogonal subcarriers, where serial input data, **a**, is converted into parallel stream using serial to parallel converter then these parallel bits are spreaded over each orthogonal subcarriers by CI code block shown as Processing Element (PE) in Fig. 1. Transmitted signal, $s(t)$ is obtained by summing up all the modulated subcarriers, converting them from parallel to serial stream and finally up converting this base band signal.

B. CORDIC (Co-ordinate Rotation Digital Computer)

CORDIC algorithm [18], [19] is an efficient, reliable method used in linear, trigonometrical and hyperbolic functions implementation. The CORDIC algorithm [18] is a vector rotation algorithm carried out by micro-rotation stages in terms of a set of prefixed angles, α_i and evaluated by add and shift operation which is presented as follows

$$\begin{aligned} x_{i+1} &= x_i - d_i y_i 2^{-i} \\ y_{i+1} &= d_i x_i 2^{-i} + y_i \\ z_{i+1} &= z_i - d_i \alpha_i \end{aligned} \qquad (5)$$

where (x_{i+1}, y_{i+1}) is the resulting vector when a vector (x_i, y_i) is rotated through an angle $\alpha_i = \tan^{-1}(2^{-i})$ and i varies from 0 to $(m-1)$, where m is an integer equals to the bit precision or number of micro-rotations [18], [19]. z_i is the residual angle and $d_i \in (+1, -1)$ is the sign bit of the residual angle z_i. Any vector (x_0, y_0) rotated through an angle $z_0 = \theta$ results in a vector (x_m, y_m) after m number of iteration stages. θ is expressed in terms of prefixed micro-rotation angles α_i as $\theta = \sum_{i=0}^{m-1} d_i \alpha_i$. A hardware efficient iterative CORDIC architecture designed using based on equation (5) is presented in Fig. 2 which is one of the building block in our proposed architecture of CI-OFDM.

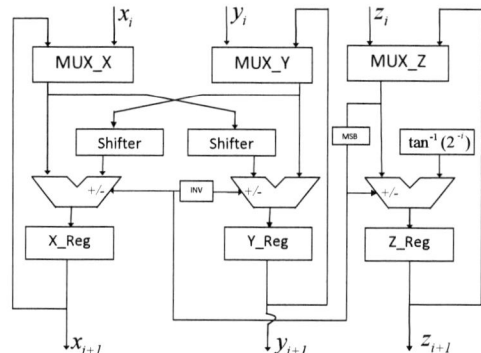

Fig. 2. Iterative CORDIC architecture

III. PROPOSED CORDIC BASED CI-OFDM ARCHITECTURE

In this section, the proposed architecture of CORDIC based CI-OFDM along with constituent blocks shown in Fig. 3 are discussed. In this proposed hardware efficient CI-OFDM architecture, one circular CORDIC block along with one angle generator unit are being used with reference to equation (4). Angle generator unit as shown in Fig. 3(c) is used to generate angle argument of baseband signal shown in equation (4) i.e z_{in} to the circular CODIC block of CORDIC based CI-OFDM symbol generator. The two blocks i.e CORDIC based CI-OFDM symbol generator and an angle accumulator of the proposed CI-OFDM transmitter architecture shown in Fig. 3(a) consists the input and output signals as *master clock*, z_{in}, *data_in*, *EOC*, *sel1*, *sel2*, *clk1* and *clk2* which are discussed in next subsections. These signals help in synchronisation and

2016 International Conference on VLSI Systems, Architectures, Technology and Applications (VLSI-SATA)

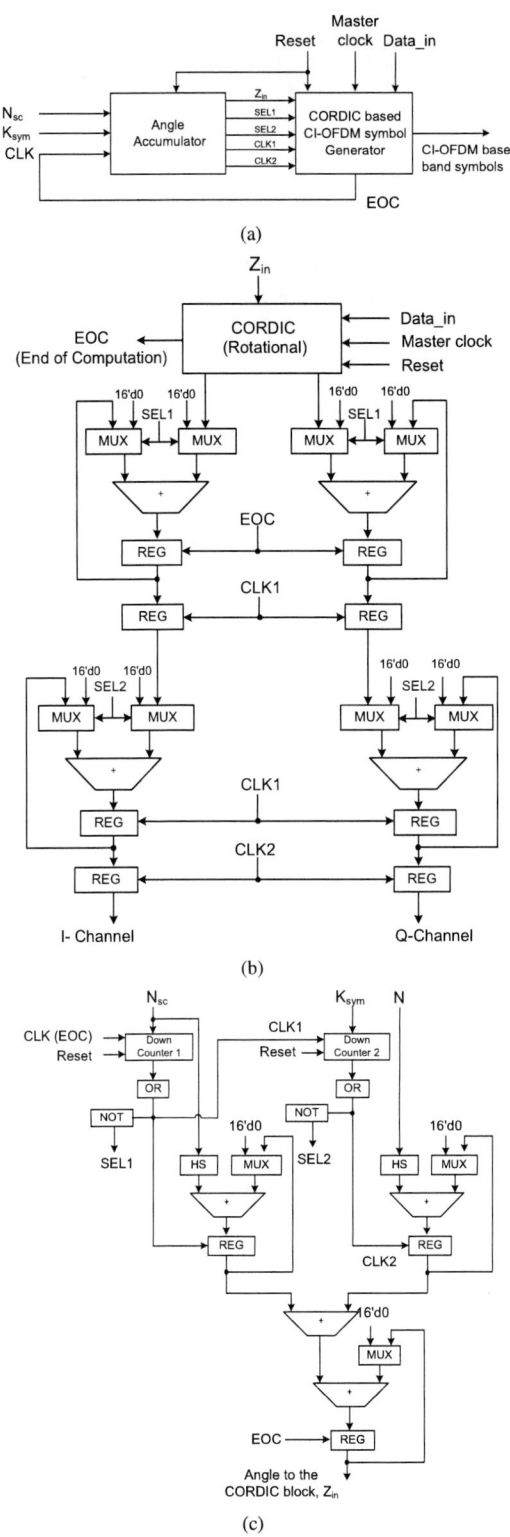

Fig. 3. Proposed (a) CI-OFDM transmitter architecture, (b) CORDIC based CI-OFDM symbol generator and (c) Angle generator architecture.

control of desired outputs i.e I-channel and Q-channel outputs of CI spreaded baseband signal.

A. CORDIC based CI-OFDM symbol generator

CORDIC based CI-OFDM symbol generator generates the CI spread signal whose inputs are *master clock*, z_{in} and *data_in* and outputs are *CI-OFDM baseband signal* and *EOC* as shown in Fig. 3(b). *master clock* is the primary clock which drives the proposed architecture, z_{in} is the input angle to the CORDIC block inside the CORDIC based CI-OFDM symbol generator and *data_in* is the information bit which modulates the N orthogonal subcarriers of the CI spread baseband signal. *EOC* is the output signal which represents the completion of CORDIC iteration stages for each z_{in} and also used as clock signal for the *counter1* shown in Fig. 3(c) taking care of number of subcarriers being used in CI spread signal. Other inputs *sel1*, *sel2*, *clk1* and *clk2* are discussed in next subsection.

B. Angle Generator unit

Angle generator unit generates all the necessary arguments required in equation (4). It has two down counters i.e *counter1* and *counter2* having their inputs as N_{sc}, number of subcarriers, and K_{sym}, number of orthogonal symbols, as required by CI spreads codes referring equation (4). Angle generator unit has a new hardwired shifter (HS) whose input is number of sample points N used in CI spread codes and produces output equal to $\frac{2\pi}{N}$ as shown in Fig. 4. Output of HS is weighted as $-\pi, \frac{\pi}{2}, \frac{\pi}{2^2} \ldots \frac{\pi}{2^{w-1}}$ and the same weighting has been used in the input angle to the CORDIC block i.e z_{in}. Here w is the word size used in the proposed architecture. Different argument required for the CI spread codes is generated by using output of HS i.e $\frac{2\pi}{N}$ as shown in Fig. 3(c). The signals *sel1*, *sel2*, *clk1* and *clk2* are generated by using two down counters *counter1* and *counter2*. *EOC* is the driving clock for *counter1*. *counter1* controls the number of subcarriers being used within CI spread OFDM symbol. *clk1* is generated by using output of *counter1*, which shows the completion of computation of sample point of individual CI spread code. *clk1* is the input clock signal for *counter2* whose other input is K_{sym} decides the number of CI spread symbol used in CI baseband signal. *clk2* is generated with the help of *counter2* which shows the completion of computation of individual sample point of CI baseband signal i.e completion of sample point of combined CI spread code. The signals *sel1* and *sel2* generated by using two counters is the selection line for the MUX as shown in Fig. 3(b) which helps in clearing the register value to zero after arithmetic computation of sample point of individual CI symbol and sample point of CI baseband signal respectively. CORDIC block generates the cosine (*in phase component*) and sine (*quadrature component*) of the argument z_{in}. These output are summed up for the generation of I-channel and Q-channel outputs of baseband CI spread codes. This proposed architecture has been prototyped on FPGA and experimental results are discussed in next section.

978-1-5090-0037-1/16 $31.00 © 2016 IEEE

2016 International Conference on VLSI Systems, Architectures, Technology and Applications (VLSI-SATA)

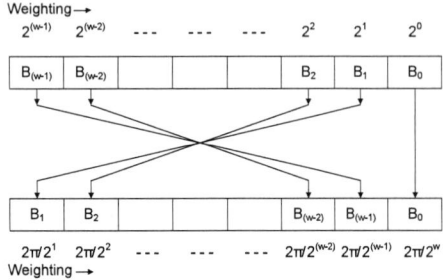

Fig. 4. Block diagram of hard wired shifter.

IV. FPGA Prototyping experiments and Results

The proposed VLSI architecture of CI-OFDM transmitter has been designed using Verilog hardware description language (Verilog HDL) for 16-bit data path size and simulated using XILINX ISE11.1 tool. Further, the design is prototyped on a FPGA chip XC3S500E-5FG320. The output of proposed architecture has been captured through mixed signal oscilloscope Tektronix MSO2024 with the help of an external digital-to-analog-converter (DAC) Digilent Pmod-DA1 AD7303. The experiment has been performed with master clock frequency of 120 MHz with $N = 256$, $N_{sc} = 16$, $K_{sym} = 16$.

The experiment for proposed design and implementation is carried out for input vector $\mathbf{a} = [a(0), a(1), a(2), \ldots, a(15)]$ as in equation (4) by taking a BPSK data pattern as follows.

$$
\begin{aligned}
\mathbf{a} = \ & [a(0), a(1), a(2), \ldots, a(15)] \\
= \ & [(1111100110000010), (1111001100000101), \\
& (1110011000001010)(1100110000010100), \\
& (1001101000101000), (0011011001010000), \\
& (0110111010100001), (1101110101000011), \\
& (1011100010000110), (0111001100001100), \\
& (1110011000011000), (1100111000110001), \\
& (1001111001100010), (0011110011000101), \\
& (0111100110001011), (1111001100010111)]
\end{aligned}
\tag{6}
$$

For the validation and comparison of the real-time output with MATLAB simulation result, FPGA physical output for individual CORDIC based CI spread code has been generated and shown in Fig. 5(a) and (b). The experimental set up is shown in Fig. 6. BPSK modulated CI-OFDM baseband signal has been generated and found comparable with MATLAB simulation result as shown in Fig. 7(a) and (b). I-channel and Q-channel outputs of baseband CI spreaded signals in time domain using MATLAB simulation is shown in Fig. 7(a) which is obtained by adding up all the 16 CI spreading codes for the given bit pattern, i.e input data vector, shown in equation (6). Further, for each bit of the kth input vector, $a(k)$, individual CI spreading code for 16 subcarriers is generated and summed up to get the CI-OFDM signal of the kth input vector which is the output of the proposed architecture shown in Fig. 7(b). From Fig. 7(a) and (b), it is clear that the envelope of CI-OFDM FPGA output closely matches with the MATLAB simulation. Spectrum of CI-OFDM baseband signal

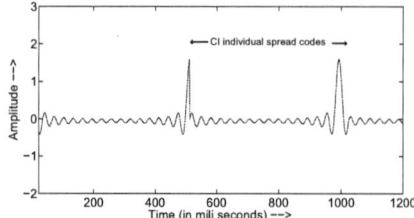

(a) Matlab simulation output of CI spread codes.

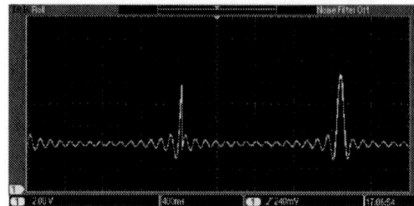

(b) Experimental output of CI spread codes captured through Tektronix MSO2024.

Fig. 5. Matlab simulation and FPGA physical output of Indivisual CI codes.

obtained from FPGA output is shown in Fig. 7(c). Peaks of $N_{sc} = 16$ subcarriers used in CI-OFDM modulator are clearly observed in spectrum analyzer Tektronix RSA3303B.

Hardware utilization report is found based on the physical synthesis report using XILINX XST of the proposed architecture targeting the FPGA device XC3S500E-5FG320. In summary, the proposed architecture uses total 620 slices, 232 Flip Flops, 1186 number of 4 inputs look-up-tables (LUTs), 22 number of input-output-blocks (IOBs) and 4 global clocks. The experimented 16-bit data path size of the proposed architecture has the flexibility to generate maximum up to $32K$ number of subcarriers (N_{sc}) and information symbols (K_{sym}). The CORDIC based proposed design provides the flexibility in terms of number of orthogonal subcarriers and orthogonal information symbol upto $32K$. The critical path of the proposed architecture is limited to adder and shifter delay. The work is concluded in the next section.

V. Conclusion

In this paper, a new hardware efficient and flexible CORDIC based architecture for implementing CI-OFDM is proposed with the novelty of having the flexibility of changing the

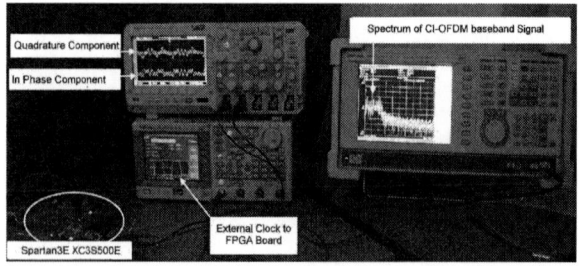

Fig. 6. Experimental set up and FPGA prototyping of proposed CI-OFDM architecture.

978-1-5090-0037-1/16 $31.00 © 2016 IEEE

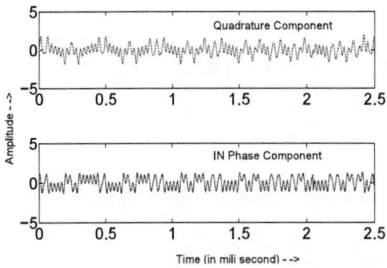

(a) Matlab simulation output of CI-OFDM Symbol for In phase and Quadrature component.

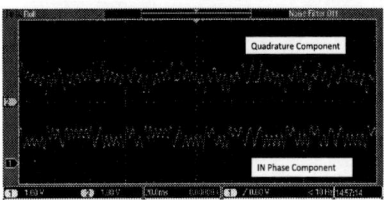

(b) Experimental output of CI-OFDM Symbol captured through Tektronix MSO2024 for In phase and Quadrature component.

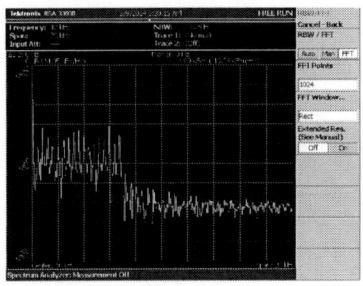

(c) Spectrum of CI-OFDM baseband signal captured through Tektronix RSA3303B (zoomed view as in Fig. 6).

Fig. 7. (a) and (b) CI-OFDM base-band transmitted signal (I-channel and Q-channel) corresponding to equation (4) with $N_{sc} = 16$ subcarriers, $K_{sym} = 16$ information symbol, data vector $\mathbf{a} = [a(0),a(1),a(2),\ldots] = [(1111001100000010),(1111001100000101),(1110011000001010),\ldots]$ and (c) Base-band signal spectrum.

number of orthogonal subcarriers and data symbols depending upon the applications. This proposed design is prototyped on commercially available FPGA device and further the implementation and experimental results of this baseband modulation scheme are highlighted and validated with the simulation results obtained using MATLAB. The work in this paper can be extended to design a wireless MODEM using CI-OFDM suitable for satellite communication systems. The authors are working towards this end in one of their sponsored research project.

ACKNOWLEDGMENT

The authors would like to thank DEITY, Govt. of India for funding of this research under the project entitled "FPGA

Prototyping of Multicarrier Multiple Access Schemes for Variable Rate Multimedia Satellite Communication", Project Reference No. R&D/SP/EE/DEIT/DFP/20 13-14/61.

REFERENCES

[1] K. Fazel and S. Kaiser, *Multicarrier spread spectrum: for future generation wireless systems*, 4th international workshop, Germany, September 17-19, 2003.

[2] H. Arslan, *Cognitive Radio, Software Defined Radio, and Adaptive Wireless Systems*, Springer, 2007.

[3] A. Jamalipour, T. Wada, and T. Yamazato, "A tutorial on multiple access technologies for beyond 3G mobile networks," *IEEE Commun. Mag.*, vol. 43, no. 2, pp. 110–117, Feb. 2005.

[4] P. N. Whatmough, M. R. Perrett, S. Isam, and I. Darwazeh, "VLSI architecture for a reconfigurable spectrally efficient FDM baseband transmitter," *IEEE Trans. Circuits Syst. I, Reg. Papers*, vol. 59, no. 5, pp. 1107–1118, May 2012.

[5] B. Natarajan, C.R. Nassar, S. Shattil, M. Michelini, and Z. Wu, "High-performance MC-CDMA via carrier interferometry codes," *IEEE Trans. Veh. Technol.*, vol. 50, no. 6, pp. 1344–1353, Nov. 2001.

[6] D. A. Wiegandt, Z. Wu, and C. R. Nassar, "High-throughput, high-performance OFDM via pseudo-orthogonal carrier interferometry spreading codes," *IEEE Trans. Commun.*, vol. 51, no. 7, pp. 1123–1134, Jul. 2003.

[7] D. A. Wiegandt and C. R. Nassar, "High-throughput, high-performance OFDM via pseudo-orthogonal carrier interferometry coding," in *Proc. IEEE PIMRC*, Sept./Oct. 2001, vol. 2, pp. G–98 –G–102 vol.2.

[8] D. A. Wiegandt and C. R. Nassar, "High-performance 802.11a wireless LAN via carrier-interferometry orthogonal frequency division multiplexing at 5 GHz," in *Proc. IEEE GLOBECOM*, 2001, vol. 6, pp. 3579–3582.

[9] F. Bortoletto, M. D'Alessandro, E. Giro, D. Magrin, L. Corcione, D. Pelusi, C. Giuliani, and A. di Cianno, "Control system architecture for amica: the antarctic nir/mir camera for irait," 2006.

[10] Z. Wu and C. R. Nassar, "Narrowband interference rejection in OFDM via carrier interferometry spreading codes," *IEEE Trans. Wireless Commun.*, vol. 4, no. 4, pp. 1491–1505, Jul. 2005.

[11] Ray Andraka, "A survey of cordic algorithms for fpga based computers," in *Proceedings of the 1998 ACM/SIGDA Sixth International Symposium on Field Programmable Gate Arrays*, New York, NY, USA, 1998, FPGA '98, pp. 191–200, ACM.

[12] Min-Woo Lee, Ji-Hwan Yoon, and Jongsun Park, "Reconfigurable cordic-based low-power dct architecture based on data priority," *IEEE Trans. Very Large Scale Integr. (VLSI) Syst*, vol. 22, no. 5, pp. 1060–1068, May 2014.

[13] V. Kumar, K. C. Ray, and P. Kumar, "Cordic-based VLSI architecture for real time implementation of flat top window," *Microprocessors and Microsystems*, vol. 38, no. 8, Part B, pp. 1063 – 1071, 2014.

[14] M. Mukherjee and P. Kumar, "Design and performance of WH-spread CI/MC-CDMA with iterative interference cancellation receiver," *Phys. Commun.*, vol. 5, no. 3, pp. 217 – 229, Sept. 2012.

[15] P. Kumar and P. Kumar, "A comparative study of spread OFDM with transmit diversity for underwater acoustic communications," *Wireless Personal Communications*, pp. 1–18, 2015.

[16] X. Li, V. D. Chakravarthy, B. Wang, and Z. Wu, "Spreading code design of adaptive non-contiguous SOFDM for dynamic spectrum access," *IEEE J. Sel. Topics Signal Process.*, vol. 5, no. 1, pp. 190–196, Feb. 2011.

[17] M. Mukherjee and P. Kumar, "Variable rate transmission schemes for CI/MC-CDMA system," *IEEE Commun. Lett.*, vol. 16, no. 7, pp. 1137–1139, July 2012.

[18] J. S. Walther, "A unified algorithm for elementary functions," in *Proceedings of the May 18-20, 1971, Spring Joint Computer Conference*, New York, NY, USA, 1971, pp. 379–385, ACM.

[19] J. E. Volder, "The CORDIC trigonometric computing technique," *IRE Trans. Electron. Comput.*, vol. EC-8, no. 3, pp. 330 –334, Sept. 1959.

2016 International Conference on VLSI Systems, Architectures, Technology and Applications (VLSI-SATA)

A Hardware optimized Low power RNM Compensated three stage Operational amplifier with Embedded Capacitance Multiplier Compensation

Karan Raj Singh
Department of Electrical and Electronics Engineering
Birla Institute of Technology and Science, Pilani
Pilani, India
singhkaranraj049@gmail.com

Anu Gupta
Department of Electrical and Electronics Engineering
Birla Institute of Technology and Science, Pilani
Pilani, India
anug@pilani.bits-pilani.ac.in

Abstract—This paper proposes a hardware optimized low power three stage compensated operational amplifier with a capability of driving a wide range of capacitive loads ranging from 200pF to 5nF. The amplifier is compensated by implementing Embedded Capacitance Multiplier (CM) Compensation on the outer Miller capacitor of traditional Reverse Nested Miller Compensation (RNMC) with a feed forward stage. This provides a unity gain bandwidth (UGB) greater than 1MHz and phase margin greater than 60° for the range of loads mentioned above. The circuit has a 100uW of DC power dissipation for a 2V supply. The proposed technique uses two compensation capacitances of 1pf and 500fF only. The design achieves a unity gain bandwidth of 9.227MHz at 500pF capacitive load. The simulation is carried for 180nm CMOS technology in Cadence Virtuoso environment.

Keywords— Reverse Nested Miller Compensation; Embedded Capacitance Multiplier Compensation; Feedforward Stage; Phase Margin; Unity Gain Bandwidth

I. INTRODUCTION

Operational Amplifiers (Op-Amp) are the basic building blocks in most of the analog and mixed signal electronic applications. Modern day operational amplifiers are required to operate at low power and simultaneously provide a high gain and good bandwidth along with improved stability. Improved stability is achieved by employing efficient frequency compensation techniques which typically use Miller capacitors. Since, these compensation capacitances take up most of the chip area, it is preferred to use lower values of these capacitances.

Embedded CM (capacitance multiplier) technique shown in Fig.1, is used to increase the effective value of Miller capacitance C_b by enhancing it to $(g_{mb}R_b-1)$ C_b, as seen at the output node V_2 [1], [2].

Typically traditional Reverse Nested Miller Compensation (RNMC) with feed forward stage (-g_{mf}) shown in Fig. 2, has been used to compensate the three stage amplifier. Feedback capacitors C_{C1} and C_{C2} are used compensate the op-amp.

RNMC proves efficient in driving heavy output loads with good bandwidth for three stage amplifiers as only C_{C1} loads the output due to its Miller effect on it [3]. One drawback of RNMC is that an RHP zero is introduced due to forward signal paths in the circuit through the Miller capacitances which adversely affect phase margin and stability of the amplifier. Numerous techniques have been developed to either completely eliminate the RHP zero by use of nulling resistor [3] or convert RHP zero to an LHP zero [4], [5]. Feed forward stage at the output acts as a pseudo class AB output stage that is capable of driving C_L with a current much higher than output branch quiescent current [5]. This enhances slew rate of the op-amp.

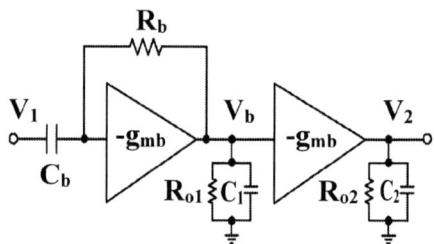

Fig. 1. Embedded Capacitor Multiplier Compensation

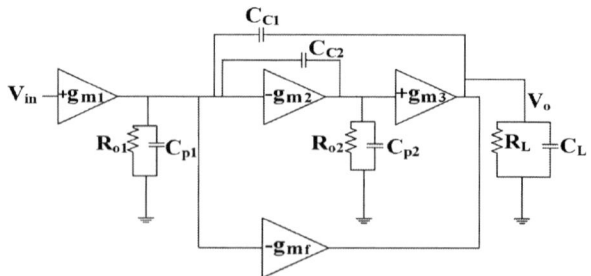

Fig. 2. Traditional RNMC with feedforward stage

978-1-5090-0037-1/16 $31.00 © 2016 IEEE

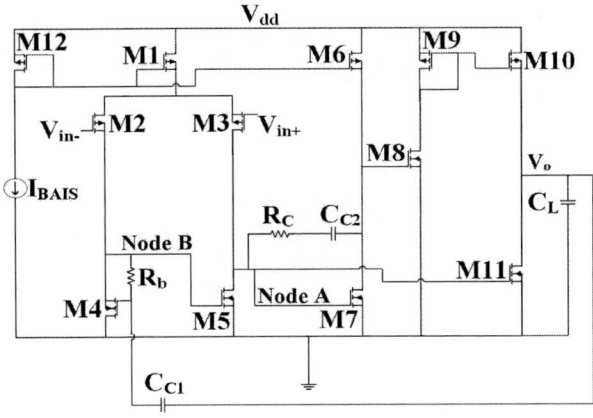

Fig. 3. Schematic of three stage op amp with RNMC and embedded CM compensation

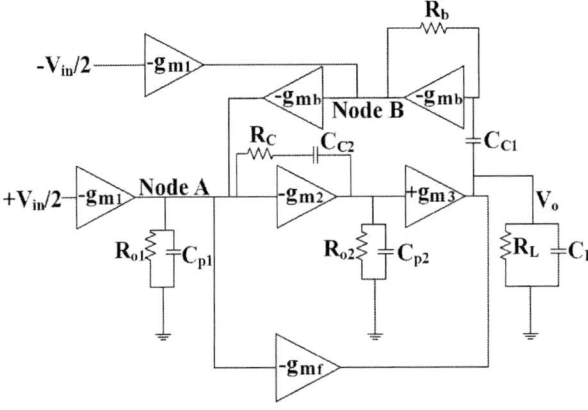

Fig. 4. Small signal model of op amp with proposed compensation

The technique proposed in this paper implements embedded CM technique over C_{C1} and enhances its effective value for performing compensation. The novelty of the proposed technique is this unique combination of traditional RNMC with embedded CM. The technique has been analyzed and explained by simulation graphs whose results have been discussed later. Performance parameters such as phase margin, unity gain bandwidth (UGB) and power dissipation have been improved compared to previous variants of RNMC techniques. Unlike previous variants of RNMC, the proposed technique can drive a wide range of capacitive loads ranging from 200pF to 5nF. A figure of merits (FOMs) table has been presented in a later section to demonstrate the advantages of the proposed technique over previous variants of RNMC.

II. DESIGN AND ANALYSIS OF PROPOSED TECHNIQUE

A. Description of the technique

Fig.3 shows the schematic of the proposed technique that has been used for performing simulation. Transistors M1 to

M5 comprise the first stage, M6 and M7 comprise the second stage and M8 to M10 comprise the third stage. M11 acts as the feed forward stage and M12 acts as a current mirror used for biasing M1 and M6. I_{BAIS} is the constant current flowing through M12 which is mirrored in the first and second stages. C_L is the load capacitance. Resistor R_C and capacitors C_{C1} and C_{C2} are used for performing RNMC on the amplifier.

The circuit for capacitance enhancement of C_{C1} is implemented by reusing current mirror M4 and M5. Since, the current mirror part of differential input stage is reused for implementing embedded CM on C_{C1}, no additional biasing circuitry or power dissipation is incurred [2]. R_b is used for adjusting the multiplication factor of ($g_{mb}R_b - 1$) of capacitor C_{C1}. Fig. 4 shows the small signal model of the proposed design of Fig. 3. g_{m1}, g_{m2} and g_{m3}, here are transconductances of the first, second and third stages, respectively. g_{mf} is the transconductance of the feed forward stage. g_{mb} is the transconductance of the transistors M4 and M5. R_{oi} and C_{oi} are the resistance and parasitic capacitance for corresponding i-th stage for $i \leq 2$. R_L is output resistance of the op-amp.

B. Analysis of nature of zeroes

In Fig.4, a Miller capacitance of value ($g_{mb}R_b$-1) C_{C1} is effectively realized at node A. The NMOS transistor M5 blocks the forward signal path from node A through C_{C1} to V_o, while maintaining the feedback path for compensation intact. This removes the contribution of C_{C1} towards formation of the RHP zero [2]. The presence of R_C-C_{C2} across the input and output of the second stage introduces only an LHP zero.

In Fig.4, one half of input signal at node B gets divided into two forward signals. One part of the half input signal flows through R_b and C_{C1} to V_o. The other part of this signal flows through g_{mb} and gets divided into two signals at node A. One part of the signal at node A goes through second and third stage of the op amp to V_o, and the other part goes through feed forward stage (g_{mf}) to V_o. Since, the three signals at V_o are in phase, hence a resultant second LHP zero is formed in the system [2]. Hence, the RHP zero is completely eliminated in the system.

C. AC analysis of the proposed technique

Simplified transfer function of the small signal model of the proposed op amp is given in (3) with DC gain and expression for dominant pole given in (1) and (2), respectively. The transfer function is calculated by taking the approximations stated as under:

$$g_{m1}R_{o1}, g_{m2}R_{o2}, g_{m3}R_L \gg 1$$

$$C_L, C_{C2}, C_{C1} \gg C_{p1}, C_{p2}$$

$$A_{DC} = g_{m1}r_{o1}\left(g_{m2}r_{o2}g_{m3}R_L + g_{mf}R_L\right) \quad (1)$$

$$\omega_{p1} \approx \frac{1}{\left(g_{mb}R_b - 1\right)C_{C1}R_{o1}g_{m2}R_{o2}g_{m3}R_L} \quad (2)$$

$$A(s) = \cfrac{A_{DC}\left(1 + \dfrac{g_{mb}}{g_{m1}} + \left(C_{C1}\left(\dfrac{g_{mb}R_b}{g_{m1}} - \dfrac{1}{g_{mb}}\right) + C_{C2}R_C\left(1 + \dfrac{g_{mb}}{g_{m1}}\right)\right)s + \left(C_{C1}C_{C2}R_C\left(1 + \dfrac{g_{mb}}{g_{m1}}\right)\left(\dfrac{g_{mb}R_b}{g_{m1}} + \dfrac{1}{g_{mb}}\right)\right)s^2\right)}{\left(\begin{array}{l}1 + (g_{mb}R_b - 1)C_{C1}R_{o1}\left(g_{m2}R_{o2}g_{m3}R_L + g_{mf}R_L\right)s \\[2mm] + \left((g_{mb}R_b - 1)C_{C1}R_{o1}C_{C2}R_Cg_{m2}R_{o2}g_{m3}R_L + C_LR_L\left(\dfrac{C_{C1}}{g_{mb}} + g_{m2}C_{C2}R_{o2}R_{o1}\right)\right)s^2 + \left(\dfrac{g_{m2}C_{C1}R_{o1}C_{C2}R_{o2}C_LR_L}{g_{mb}}\right)s^3\end{array}\right)} \quad (3)$$

Compared to the expression of dominant pole for traditional RNMC [6], the pole is divided by a factor of $(g_{mb}R_b - 1)$. This indicates the effect of capacitance multiplication on C_{C1}.

The op amp has three LHP poles and two LHP zeroes in the system. The values of poles and zeroes can be adjusted by using R_C and C_{C2} such that the dominant pole (ω_{p1}) is followed by the first non-dominant pole (ω_{p2}), followed by the first zero (ω_{z1}), followed by the second zero (ω_{z2}) and finally followed by the second non-dominant pole (ω_{p3}). This allows the two consecutive zeroes to increase the phase until ω_{p3} is encountered. The same is displayed by the projected Bode phase plot given in Fig. 5. A good phase margin is achieved for a unity gain bandwidth (UGB) that lies between ω_{z1} and ω_{p3}.

III. SIMULATION AND RESULTS

Simulation of the op-amp is carried out at a power supply of 2V for 180nm CMOS technology. The values of R_b, R_C, C_{C1} and C_{C2} used for simulation are 27KΩ, 500KΩ, 1pF and 0.5pF, respectively, for g_{m1}, g_{m2}, g_{m3}, g_{mb} and g_{mf} of 176μA/V, 115μA/V, 176μA/V, 115μA/V and 341μA/V, respectively, and I_{BIAS} of 5μA.

Fig. 6 and Fig. 7 display variations in AC Gain with varying values of C_{C1} and R_b, respectively. Fig. 8 and Fig. 9 display AC gain and bode phase plots for varying values of C_{C2} and R_C, respectively. Fig. 10 shows the AC gain and bode phase plot for different C_L loads for the op-amp. AC gain and phase plots are displayed for C_L values of 200pF, 500pF, 1nF and 5nF stating the UGB and phase margin (PM) for each load. Fig. 11 shows the transient response of the amplifier in unity gain feedback configuration corresponding to an input pulse of 2V for different C_L values. All the graphs shown here are plotted at TT process corner at a temperature of 27°C. Legends are indicated in all the figures from Fig. 6 to Fig.11, which correspond to value of the parameter being varied. Simulations for varying parameter are carried out by keeping the rest of the design parameters as constant.

The values of performance parameters of op amp at 'TT', 'SS' and 'FF' process corners are presented in Table I. The results indicate that the amplifier gives good performance at the three process corners.

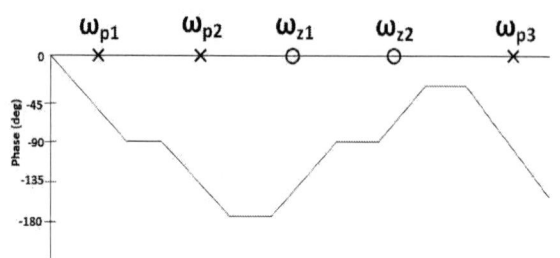

Fig. 5. Projected Bode phase plot

A. Graphs and Tables

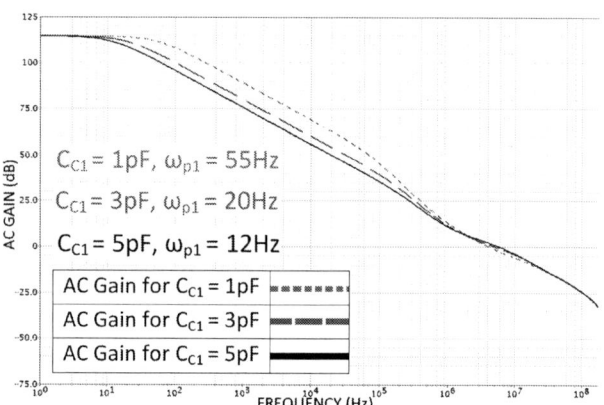

Fig. 6. AC gain plot for varying values of C_{C1}

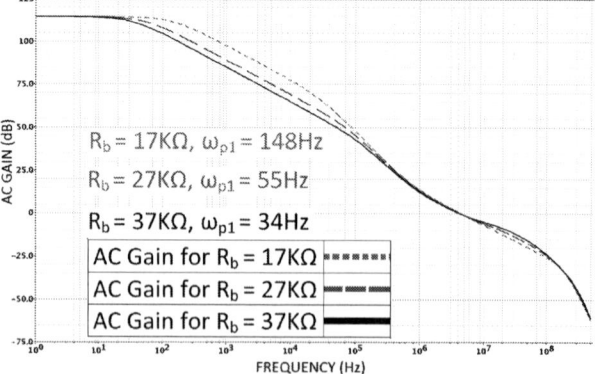

Fig. 7. AC gain plot for varying values of R_b

978-1-5090-0037-1/16 $31.00 © 2016 IEEE

2016 International Conference on VLSI Systems, Architectures, Technology and Applications (VLSI-SATA)

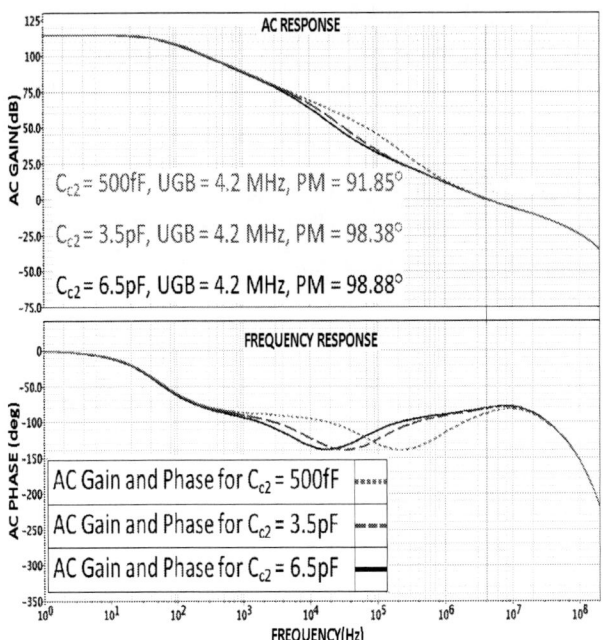

Fig. 8. AC gain and phase plot for varying values of C_{C2}

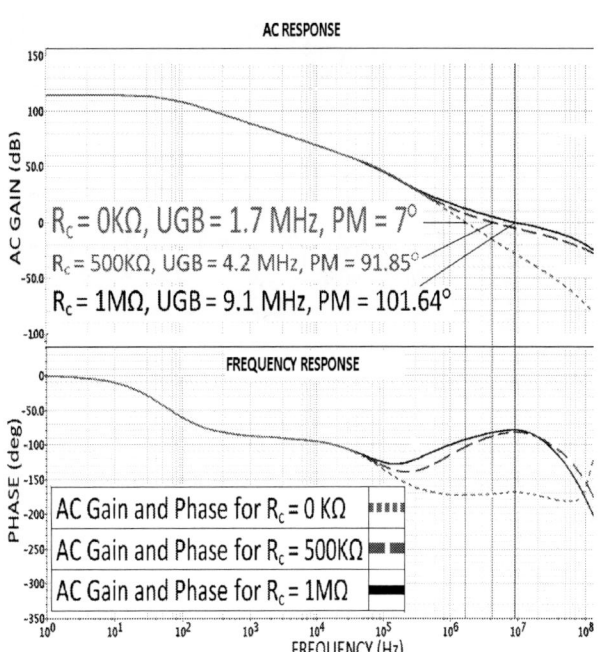

Fig. 9. AC gain and phase plot for varying values of R_C

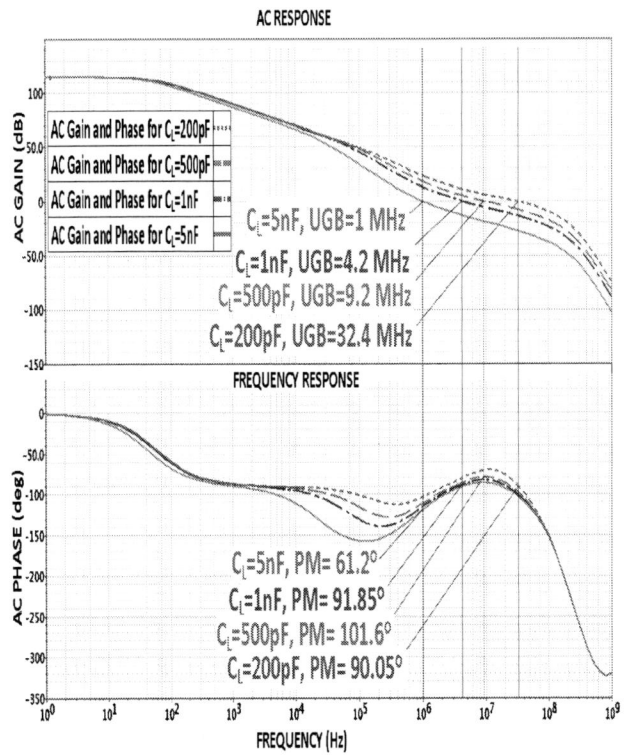

Fig. 10.   AC gain and phase plot for varying values of C_L

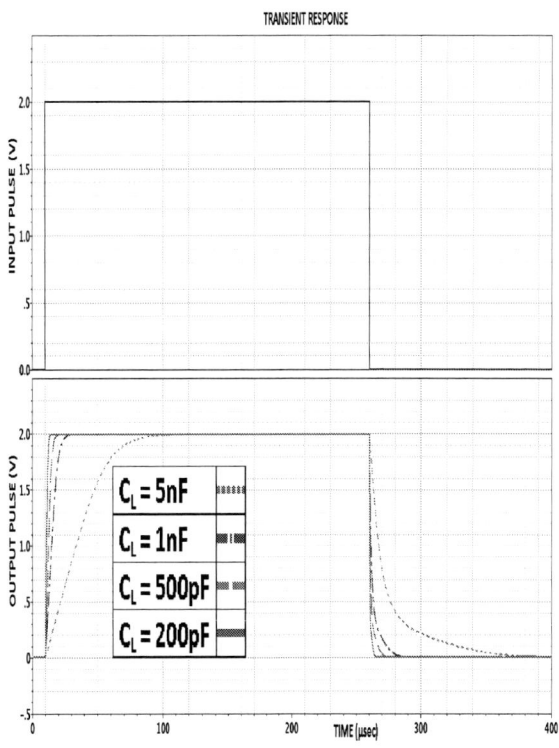

Fig. 11. Transient response for varying values of C_L

978-1-5090-0037-1/16 $31.00 © 2016 IEEE   75

2016 International Conference on VLSI Systems, Architectures, Technology and Applications (VLSI-SATA)

TABLE I. PERFORMANCE PARAMETERS FOR DIFFERENT CAPACITIVE LOADS

Parameter	C_L=5nF			C_L=1nF			C_L=500pF			C_L=200pF		
	TT	FF	SS	TT	FF	SS	TT	FF	SS	TT	FF	SS
Power Supply (V)	2	2	2	2	2	2	2	2	2	2	2	2
Total Bais Current (μA)	50.14	48.12	53	50.14	48.12	53	50.14	48.12	53	50.14	48.12	53
Power Disssipation (μW)	100.28	96.24	106	100.28	96.24	106	100.28	96.24	106	100.28	96.24	106
DC Gain (dB)	114	100	103	114	100	103	114	100	103	114	100	103
UGB (MHz)	1.001	1.031	0.87	4.203	4.238	3.487	9.227	9.322	7.331	32.41	34.59	24.2
Phase Margin	61.2	62.8	56.80	91.85	92.8	88.03	101.6	103.5	98.13	90.05	90.13	94.2
Gain Margin (dB)	43.39	42.36	44.06	28.34	28.3	30.03	22.22	22.19	23.95	14.02	13.94	15.81
SR+/SR- (V/μs)	0.04/ -0.14	0.07/ -0.2	0.03/- .12	0.2/ -0.4	0.24/ -0.92	0.11/ -0.51	0.4/ -.97	0.45/ -1.6	.32/ 0.99	0.75/ -2.2	1.06/ -4.0	0.72/ -2.7

TABLE II. COMPARITION WITH PREVIOUS RNMC TOPOLOGIES

RNMC Class	RNMC Technique	C_{C1}/C_{C2} (pF)	C_L (pF)	DC Gain (dB)	Power (μW)	UGB (MHz)	PM (deg)	$SR_{av.}$ (V/μs)	FOM_S (MHz.pF/ mW)	FOM_L (V/μs.p F/mW)	$IFOM_S$ (MHz.p F/ μA)	$IFOM_L$ (V/μs.p F/ μA)
RNMC with NR	RNMC-FF [7]	6.0/0.5	100	>100	165	2.74	66.1	1.38	1661	836	3321	1673
	RNMC-FNR [7]	5.6/0.5	100	>100	165	3.09	68.0	1.66	1873	1006	3745	2012
	RNMC-FNR2 [7]	5.4/0.5	100	>100	165	3.28	70.1	1.75	1988	1061	3975	2121
RNMC with CB	RNMC-FCB [7]	5.8/0.5	100	>100	165	3.24	69.4	2.04	1965	1236	3927	2472
	RNMC-FCBNR [7]	5.6/0.5	100	>100	165	3.52	67.1	2.16	2134	1309	4267	2618
RNMC with VB	RNMC-FVB [7]	5.4/0.5	100	>100	165	3.96	66.2	2.70	2400	1636	4800	3272
	RNMC-FVBNR [7]	3.5/0.5	100	>100	165	5.92	66.6	4.64	3588	2812	7176	5624
	RNMC-FVBONR [7]	4.2/0.5	100	>100	165	5.36	72.2	3.52	3249	2134	6497	4267
	RNMC-FVBNR2 [7]	2.8/0.5	100	>100	165	7.76	67.5	5.82	4703	3527	9406	7054
-	Proposed work	1/0.5	200	>100	100	32.41	90.05	1.48	**64820**	2950	129640	5900
-	Proposed work	1/0.5	500	>100	100	9.23	101.6	0.69	**46135**	3425	92270	6850
-	Proposed work	1/0.5	1000	>100	100	4.20	91.8	0.30	**42030**	3000	84060	6000
-	Proposed work	1/0.5	5000	>100	100	1.00	61.2	0.09	**50050**	4500	100100	9000

TABLE III. TRANSISTOR RATIOS

Transistor	M1	M2	M3	M4	M5	M6	M7	M8	M9	M10	M11	M12
Ratio(μm/μm)	2.5/0.36	20/0.36	20/0.36	1/0.36	1/0.36	1.25/0.36	1/0.36	1/0.36	3/0.36	3/0.36	1/0.36	1/0.36

B. Discussion

It is observed that the simulated asymptotic and projected phase plots follow the same pattern. It is observed from Fig. 6 and Fig. 7 that the value of dominant pole decreases with increasing C_{C1} and increasing R_b. Fig. 8, Fig. 9 and Fig. 10 suggest that there is a negligible impact of varying R_C, C_{C2} and

C_L on the dominant pole. This validates the expression of ω_{p1} given in (3).As suggested by Fig. 9, R_C introduces a stabilizing zero in the system. The value of R_C alters the position of ω_{z1}. For R_C=0, LHP ω_{z1} is lost and the gain falls at a rate of -40 dB/decade up to UGB. The phase does not rise appreciably at UGB for this configuration, due to the lack of first LHP zero,

978-1-5090-0037-1/16 $31.00 © 2016 IEEE

which results in a low phase margin. For higher values of R_C, LHP ω_{z1} is introduced, which increases the phase at unity gain frequency and considerably improves phase margin. The introduction of ω_{z1} also increases the slope (dB/decade) of AC gain starting at ω_{z1} and hence, the gain falls to unity less steeply (approximately -20 dB/decade from ω_{z1} up to ω_{z2}), resulting in improved UGB. Keeping other factors constant, with larger values of R_C, ω_{z1} shifts towards origin which results in the slope improvement of AC gain at a relatively lower frequency leading to a relatively improved bandwidth.

From Fig. 8 it can be inferred that variation in C_{C2} with other factors constant, does not impact UGB or dominant pole. Hence, from a design perspective, a lower valued C_{C2} could be used while maintaining the constraint of ($C_{C2} \gg C_{p2}$).

Fig. 10 shows the variation in UGB with C_L. This variation occurs due to the impact of C_L on non-dominant poles. It can be inferred from Fig. 10, that larger C_L values shift ω_{p2} towards origin, which results in the slope of AC gain to decrease (approximately to -40 dB/decade from ω_{p2} up to ω_{z1}), at a relatively lower frequency. This results in gain to fall to unity at a relatively lower unity gain frequency.

Fig. 11 shows the output transient response of the amplifier in unity gain feedback for an input pulse of 2V, for varying capacitive loads. It indicates that the amplifier is stable in closed loop feedback for all the capacitive loads ranging from 200pF to 5nF.

Table II compares the figures of merits [7] of previous variants of RNMC with the proposed technique to demonstrate the improvement of its performance compared to those in [7]. V_{DD} for all the previous topologies in [7] is 2V and the performance parameters correspond to 350nm CMOS Technology [7]. These figures of merits provide a basis for comparison of the proposed technique with the previous RNMC variants mentioned, by normalizing their unity gain bandwidth and average slew rate against load capacitance, power and total current. Relatively higher value of a figure of merit implies a relatively better performance. The figures of merits are defined as under:

$$FOM_S = \frac{UGB \times C_L}{Power}, \quad IFOM_S = \frac{UGB \times C_L}{I_{TOTAL}}$$

$$FOM_L = \frac{SR_{av} \times C_L}{Power}, \quad IFOM_L = \frac{SR_{av} \times C_L}{I_{TOTAL}}$$

As indicated by Table II, the proposed compensation technique has significantly higher figures of merits for bandwidth ($IFOM_s$ and FOM_s) compared to the other variants of RNMC mentioned in [7], indicating a significantly improved performance of the amplifier for bandwidth relative the other variants mentioned. The amplifier also has considerably higher $IFOM_L$ and FOM_L compared to other RNMC variants, indicating a good performance for slew rate. The proposed topology also dissipates lesser power compared to the other variants of RNMC mentioned in Table II. Compared to these variants of RNMC [7], which have been tested for a single load of 100pF, the proposed technique can

drive a wide range of loads, ranging from lower capacitive loads (200pF) to large capacitive loads (5nF), while providing a good phase margin. It should also be noted that this range can be increased at both the ends, but at the expense of UGB for heavier loads (greater than 5nF) and phase margin for lower valued loads (lesser than 200pF). Slew rates can be improved for the technique further by increasing the current in the output branch of the proposed amplifier. In [7], the minimum sizes of compensation capacitances correspond to RNMC-FVBNR2. The proposed technique uses lower values of Miller capacitances compared to RNMC-FVBNR2.

IV. CONCLUSION

This paper proposes a hardware optimized frequency compensation technique on a three stage op-amp due to significantly lower values of compensation capacitors used. Lower values of compensation capacitors result in 180% reduction in chip area, in comparison to RNMC-FVBNR2 [7]. This percentage reduction in area due to compensation capacitors is even higher in comparison to other variants of RNMC mentioned in [7]. The impact of varying compensation circuit parameters and load values on unity gain bandwidth and dominant pole is indicated and explained by means of simulation graphs. The capability of the proposed technique to drive a large range of loads makes it suitable for applications where a single IC is required to support a wide range of loads. The performance parameters of the proposed technique are compared with other variants of RNMC, which indicate an overall superior performance of the technique proposed.

ACKNOWLEDGEMENT

We thank the Department of Science and Technology, Govt. of India for their continual support towards research activities at our institute.

REFERENCES

[1] Z. Yan , P.-I. Mak and R. Martins, "Two-stage operational amplifiers: power-and-area-efficient frequency compensation for driving a wide range of capacitive load," *IEEE Circuits Syst. Mag.*, vol. 12, no. 1, pp.26-42, 2011.

[2] Z.S. Yan, "Two-Stage Large Capacitive Load Amplifier with Embedded Capacitor-Multiplier Compensation," in IEEE ISCAS, pp.2481-2484, May 2009.

[3] A. D. Grasso, D. Marano, G. Palumbo and S. Pennisi, "Advances in Reversed Nested Miller Compensation," IEEE Trans. on Circuits and Systems I, vol.54, no.7, pp. 1459-1470, July 2007.

[4] A. D. Grasso, D. Marano, G. Palumbo and S. Pennisi, "Reversed double pole–zero cancellation frequency compensation technique for three-stage amplifiers," in Proc. IEEE PRIME'06, pp. 153–156, June 2006.

[5] A. D. Grasso, G. Palumbo and S. Pennisi, "Active reversed nested Miller compensation for three-stage amplifiers," in Proc. IEEE ISCAS'06, vol. 1, pp. 911–914, May 2006.

[6] R. Mita, G. Palumbo and S. Pennisi, "Design guidelines for reversed nested Miller compensation in three-stage amplifiers," IEEE Trans. Circuits Syst. II, vol. 50, no. 5, pp. 227–233, May 2003.

[7] D. Marano, G. Palumbo and S. Pennisi, "Analytical figure of merit evaluation of RNMC networks for low-power threestage OTAs," in Proc. IEEE ISCAS, pp.777- 780, 2010.

2016 International Conference on VLSI Systems, Architectures, Technology and Applications (VLSI-SATA)

Design and Analysis of Different Low Noise Amplifiers in 2-3GHz

Prameela B
Research Scholar, Division of Electrical Engg.
School Of Engineering, CUSAT
Kerala,India
prameela.ec@adishankara.ac.in

Asha Elizabeth Daniel
Associate Professor, Division of Electrical Engg.
School Of Engineering, CUSAT
Kerala,India
ashapalal@gmail.com

Abstract—**This paper presents the design and simulation of four basic Low Noise Amplifier topologies based on 180nm Silicon technology. A common source stage with inductive degeneration, a resistive feedback, a folded cascode and a cascode stage has been designed, simulated and the performance has been analyzed. The LNA's are designed to be stable in the 2-3GHz. Of the four topologies the cascode stage has a high gain of 19.84dB and a very low Noise Figure (NF) of 0.59dB at 2.5GHz and at a supply voltage of 2.5V.The cascode stage has a P1dB of -17.76dBm and IIP3 of -2.49dBm. The power consumption of the circuit is 18.875mW. The simulations are done in cadence virtuoso SpectreRF. The cascode stage with very low Noise figure and good gain can be used for wireless applications such as Global Positioning System (GPS), Wireless LAN, WCDMA etc.**

Keywords-Low Noise Amplifiers,1dB compression point,Noise Figure,Third-order intercept point

I. INTRODUCTION

The demand of high performance wireless front end system is increasing nowadays. The main focus is on low power mobile devices with low cost. As the Low Noise Amplifier (LNA) is the first stage of an RF receiver it contributes to the noise of the entire receiver. So the study of various LNA topologies is of importance. The main function of the LNA is to amplify the weak signals from the antenna. According to Frii's formula the first stage of the receiver contributes mainly to the overall Noise Figure (NF) of the receiver. So it is important to design LNA's with minimum NF. The main performance parameters of LNAs are gain, Noise Figure (NF), linearity and stability [1]. The forward gain of the LNA is defined by the S parameter, S21. The NF is defined as the ratio of signal to Noise Ratio (SNR) at the input to the SNR at the output. Third-Order intercept point (IP3) and 1dB compression point (P1dB) are the measures of linearity. The stability of an LNA is also important. An LNA is stable if it satisfies the condition Rollet's stability factor, K>1. But always there will be a tradeoff between the design parameters.

LNAs are used in various wireless applications like Wi-Fi, bluetooth, wireless voice, data and video. Many of these applications operate in the 2-3GHz. For designing an LNA there are various topologies and processing technologies are available. Based on our application we can choose a better one. LNA design involves a lot of trade-offs in power,

linearity, gain, frequency and noise. For a better system the gain must be high and noise figure must be low. So to provide a solid structured design method for the development and optimization of LNA for a given set of performance parameters is critical.

Since many of the wireless applications reside in 2-3GHz range, the design of an LNA in this range is of great relevance nowadays. This paper presents the design and simulation of a basic Common source (CS) stage with inductive degeneration, a resistive feedback, a folded cascode and a cascode stage. The designed cascode stage is having a NF which is very less compared to the state of the art LNAs. The cascode LNA can be used in an environment with weak signal strength as it is having a very good gain.

This paper is organized as follows. Section II deals with the basic concepts of different LNA topologies. Section III presents the design considerations. Simulation results are reported in section IV and section V provides the conclusion.

II. EXISTING BASIC LNA TOPOLOGIES

A large variety of LNA topologies are available. Each topology has its own advantages and disadvantages. The popular LNA topologies are the CS with inductive degeneration, the resistive feedback LNA, cascode and folded cascode.

A. Common Source(CS) with inductive degeneration

The inductively degenerated topology [2] is narrowband since the input matching circuit consisting of the source inductors and the gate to source capacitance, Cgs, resonates at a single frequency. The inductively degenerated LNA (L-CSLNA) is the dominating topology for narrowband systems due to its advantages such as low NF, ease of input matching, high gain, and low-power consumption. The input of an L-CSLNA forms a series *RLC* network Fig 1(a).

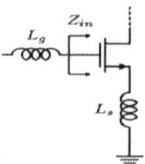

Fig. 1(a) Input stage of an Inductively degenerated CS LNA

978-1-5090-0037-1/16 $31.00 © 2016 IEEE

The input impedance of the stage is given by the equation

$$Z_{in} = s\left(L_s + L_g\right) + \frac{1}{SC_{gs}} + \left(\frac{g_{m1}}{C_{gs}}\right)L_s$$

$$\approx \omega_T L_s (at\ resonance) \qquad (1)$$

The disadvantage of the configuration is the difficulty in simultaneously matching the port impedances for optimum power transfer and minimum noise.

B. Common Source with Resistive feedback

The resistive feedback topology provides wideband input and output matching and small die area because no inductor is required for input matching [3]. But the configuration has poor NF and consumes a large amount of power.

C. Cascode

The most commonly used topology for LNA design is the cascode amplifier with inductive source degeneration [4]. The cascode topology has a higher gain, due to the increase in the output impedance. The cascode transistor suppresses the Miller capacitance of the input device thereby increasing the reverse isolation. The suppression of the parasitic capacitances of the input transistor is suppressed which improves the high frequency operation of the amplifier.

D. Folded Cascode

In a cascode stage, relatively large bias voltage is required due to the stacking architecture of the common-source and common-gate transistors [5]. So for low-voltage applications, a folded topology is used. Due to the absence of stacking gain stages, the operating voltage of the folded cascode LNA can be reduced by one transistor overdrive. The advantages of the cascode stages are good linearity, noise figure, and bias stability. But the low gain is one of the disadvantages of the folded cascode topology.

III. DESIGN CONSIDERATIONS

Various applications in the 2-3GHz have varying design constraints. For GPS application, the gain and NF should be extremely good around 20dB and 1dB respectively. For WCDMA and WLAN the gain and NF should be moderate. Table I shows the design specifications formulated for the design of LNA.

TABLE I DESIGN SPECIFICATIONS

Frequency	2-3GHz
Technology	180nm
Noise Figure (dB)	<3dB
Gain(dB)	>15dB
Supply Voltage	2.5V
Power consumption(mW)	<20mW
Third Order Intercept Point(dBm)	<3dBm

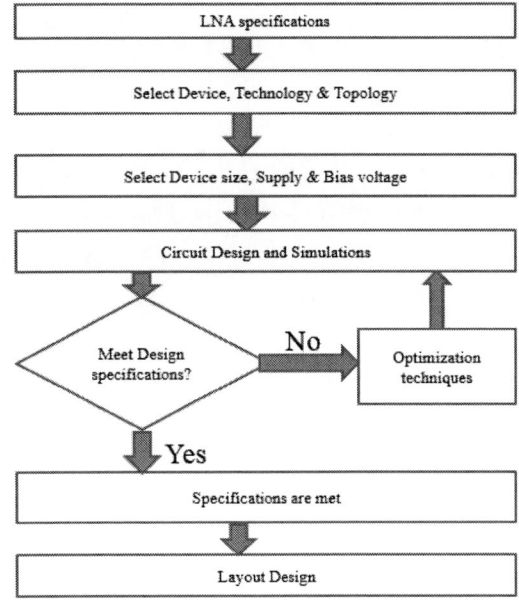

Fig. 1(b) The Basic design flow of an RF LNA

The Fig 1(b) shows the design flow of an RF LNA. After choosing the suitable technology and topology, next step is to choose the size of the input device. The equations from (2) to (7) are used to design the source inductor L_s, gate inductor L_g, drain inductor L_d and width of the input device W[6].

$$L_s = \frac{g_m}{c_{gs}} = \frac{R_s}{\omega_T} \qquad (2)$$

$$\omega_T = 2\pi f_T \qquad (3)$$

$$L_g = \frac{Q_L R_s}{\omega_0} - L_s \qquad (4)$$

$$L_d = \frac{1}{\omega_0^2 C_L} \qquad (5)$$

$$C_{gs} = \frac{1}{\omega_0^2 (L_g + L_S)} \qquad (6)$$

$$W = \frac{3C_{gs}}{2C_{ox}L_{min}} \qquad (7)$$

where g_m is the trans conductance of the device, C_{gs} is the gate source capacitance and R_s is the source resistance which is equal to 50Ω in equation (2). f_T in equation (3) is the unity gain frequency of the MOS transistor. Q_L (equation (4)) is the Q factor of the inductance which is chosen as 2.67. ω_0 is the center frequency which is chosen to be 2.5GHz. In equation (7) C_{ox} is the oxide capacitance and L_{min} is the minimum channel length which is 180nm in this design.

To fix the bias voltage the IV characteristics is plotted where the gate source voltage, V_{gs} is varied from 0.5 to 1V.

2016 International Conference on VLSI Systems, Architectures, Technology and Applications (VLSI-SATA)

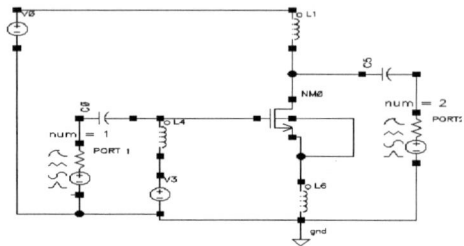

Fig. 2(a) Circuit diagram of an L- CSLNA

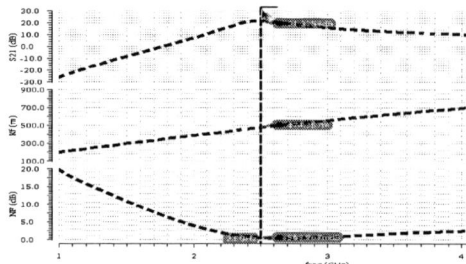

Fig. 2(b) Gain, Stability factor and NF of L-CSLNA

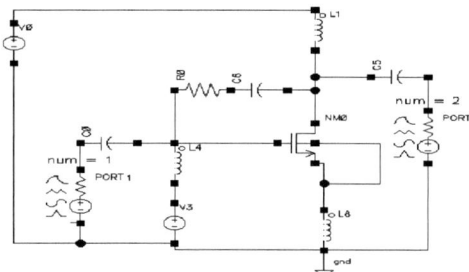

Fig. 3(a) Circuit diagram of resistive feedback LNA

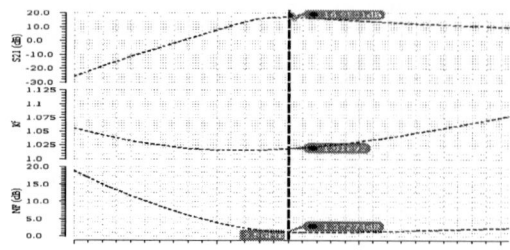

Fig. 3(b) Gain, Stability factor and NF of Resistive feedback

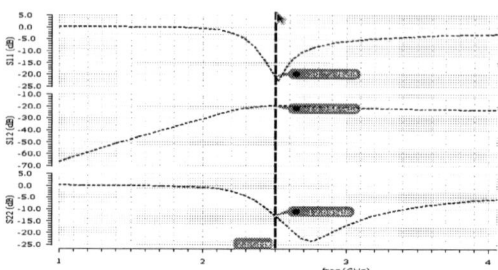

Fig. 3(c) S Parameters of Resistive feedback LNA

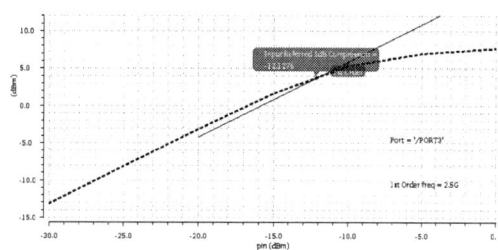

Fig. 3(d) P1dB of Resistive feedback LNA

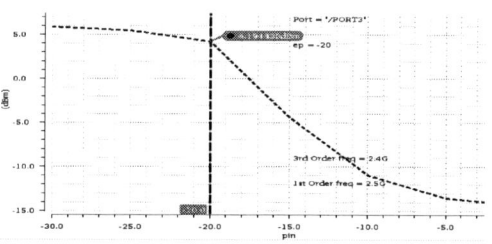

Fig. 3(e) IIP3 of Resistive feedback LNA

IV. SIMULATION RESULTS AND DISCUSSIONS

The performance parameters of the four different LNA topologies are discussed in this section.

A. Inductive degenerated LNA(L-CSLNA)

Fig. 2(a) shows the schematic of an inductively degenerated CS LNA. The performance parameters of the LNA are shown in Fig. 2(b). The stability factor, Kf is less than one. But the LNA is having a gain of 21.3dB and a NF of 0.474dB.

B. Common Source with Resistive feedback

In order to improve the stability of the CS LNA an RC feedback network is designed (Fig. 3(a)). The stability factor, NF and S parameters of the LNA are shown in Fig. 3 (b) and 3(c). Here Kf >1, but compared to the basic CS LNA, the gain is reduced to 16.95dB and the NF increases to 1.27dB at 2.5GHz. The P1dB is -12.26dBm and IIP3 is 0.004dBm as shown in Fig. 3(d) and 3(e). The power consumption of the circuit is 19.17mW.

C. Folded Cascode

A folded cascode topology is designed to further improve the NF. The NMOS and PMOS devices are having the same aspect ratio. To make the circuit stable, an RC network is added in parallel with the inductor, L1 (Fig. 4(a)). The gain of the folded cascode stage is 11.56dB and the NF reduced to 0.702dB at 2.5GHz(Fig. 4(b)). The S parameters of the folded cascode LNA is shown in Fig. 4(c). The P1dB of the circuit is -15.76dBm and IIP3 is -2.75dBm as per Fig 4(d) and 4(e). The power consumption of the circuit is 19.17mW.

978-1-5090-0037-1/16 $31.00 © 2016 IEEE   80

2016 International Conference on VLSI Systems, Architectures, Technology and Applications (VLSI-SATA)

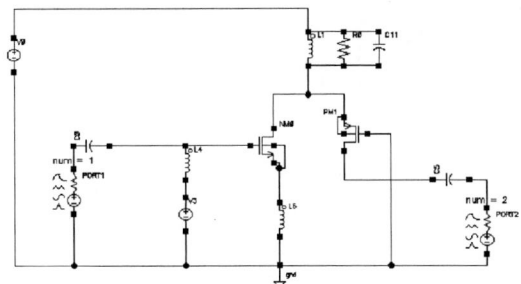

Fig. 4(a) Circuit diagram of folded cascode LNA

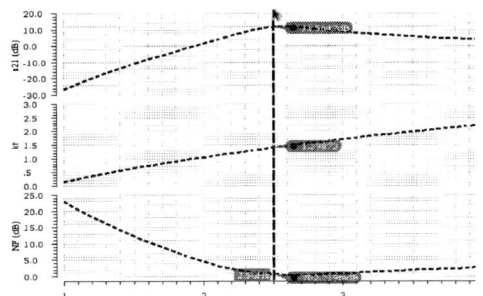

Fig. 4(b) Gain, Stability factor and NF of folded cascode LNA

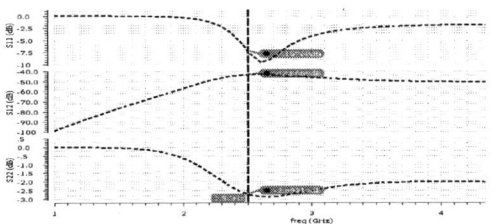

Fig. 4(c) S Parameters of Folded cascode LNA

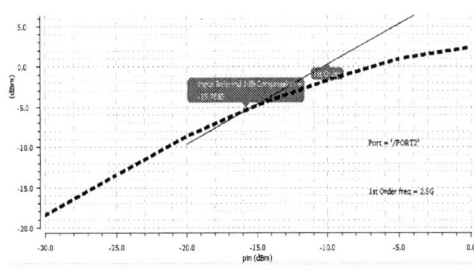

Fig. 4(d) P1dB of folded cascode LNA

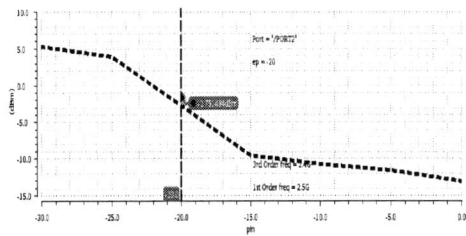

Fig. 4(e) IIP3 of Folded cascode LNA

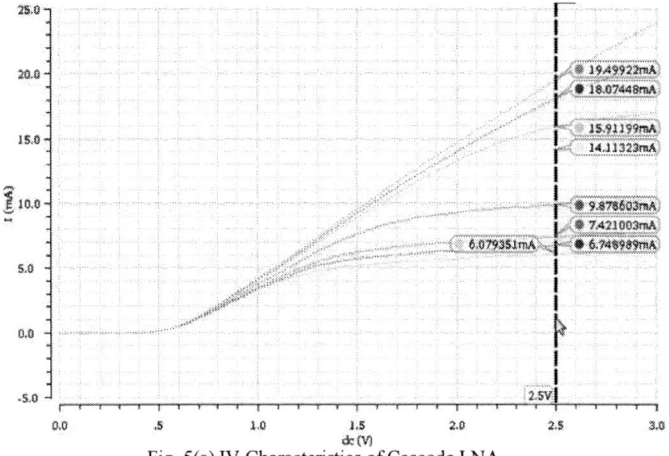

Fig. 5(a) IV Characteristics of Cascode LNA

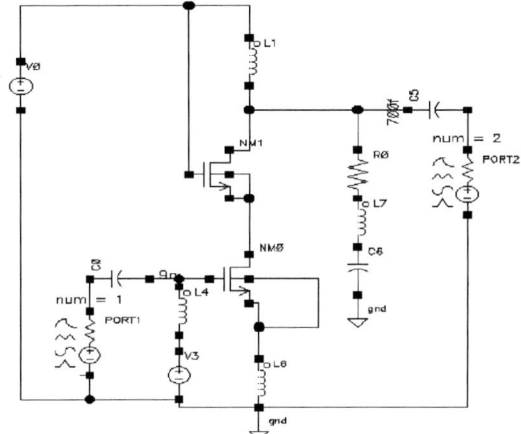

Fig. 5(b) Cascode LNA

D. Cascode LNA

To further reduce the power consumption and improve the NF and gain, a cascode stage is designed. The aspect ratio of the input device is 50μ/180n. The size of the cascode device is chosen to be half of the input device inorder to get better performance. The I-V characteristics of the cascode LNA is shown in Fig 5(a). The gate bias is chosen in such a way that the current is 6.7mA. Initially the circuit is not stable. By adding a series RLC network at the output node the circuit is stabilized as shown in Fig. 5(b). From Fig. 5(c), the gain of the cascode stage is 19.84dB and the NF is reduced to 0.59dB at 2.5GHz. The S Parameters of the cascode LNA is shown in Fig. 5(d). The P1dB of the circuit is -17.76dBm and an IIP3 of -2.49dBm (Fig 5(e), 5(f)). The power consumption of the circuit is 18. 875mW.

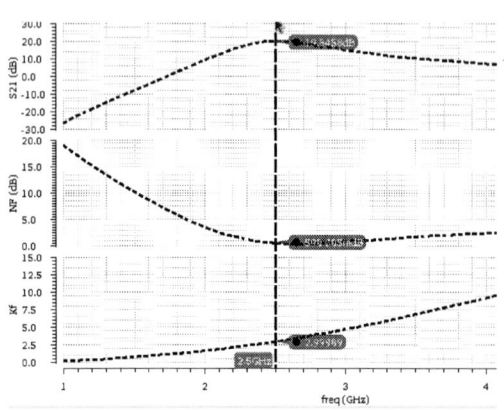

Fig. 5(c) Gain, Stability factor and NF of Cascode LNA

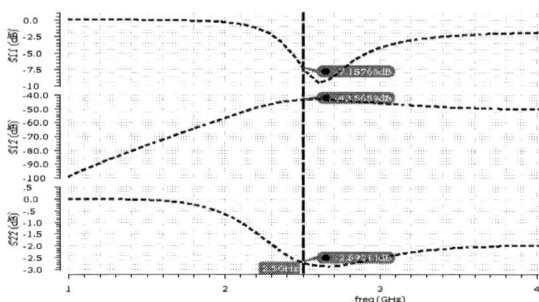

Fig. 5(d) S Parameters of Cascode LNA

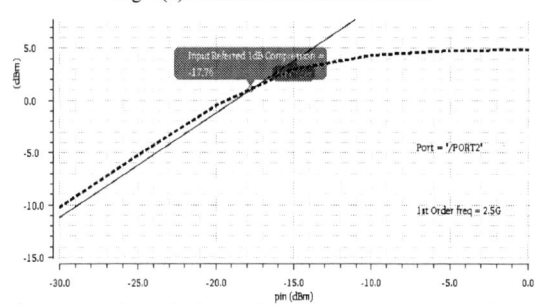

Fig. 5(e) P1dB of Cascode LNA

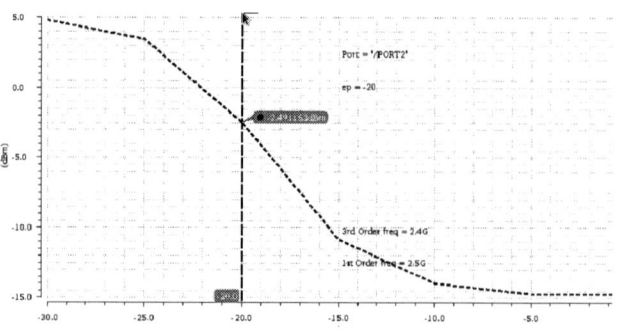

Fig. 5(f) IIP3 of Cascode LNA

V. CONCLUSION

In recent years, the domain of wireless communications has undergone a great evolution, moving rapidly through a series of generations. The design of the receiver with minimum noise is a major design constraint. Hence the design and analysis of the performance parameters of an LNA is of great importance. The design and simulation of basic four types of LNA's are discussed in this paper. Table II summarizes the performance parameters of the four LNA topologies. From the analysis it can be concluded that the cascode stage has a better NF and high gain compared to the other topologies. The gain of the cascode stage is 19.84dB and the NF is reduced to 0.59dB at 2.5GHz. The cascode LNA is stable over the 2-3GHz range and it is having good linearity. The power consumption of the circuit is 18.875mW. From Table III it can be concluded that the designed silicon based cascode topology is having a minimum Noise Figure and high gain compared to the other existing designs in the same frequency range. The S12 of the LNA is also very low. So this LNA can be used for wireless applications such as GPS in the 2-3GHz.

TABLE II. PERFORMANCE PARAMETERS OF THE LNA'S IN THIS WORK AT 2.5GHZ

Topology	VDD(V)	Kf	S21(dB)	NF(dB)	S11(dB)	S22(dB)	S12(dB)	P1dB(dBm)	IIP3(dBm)	Power(mW)
Common Source(CS)	2.5	.474	21.3	.568	-	-	-	-	-	-
CS with Resistive feedback	2.5	1.04	16.95	1.27	-20.7	-12.68	-19.9	-12.26	4.19m	19.17
Folded cascode	2.5	1.39	11.56	0.702	-4.46	-7.8	-26.8	-15.76	-2.75	19.17
Cascode	2.5	2.99	19.84	0.59	-7.17	-2.69	-43.56	-17.76	-2.49	18.875

TABLE III. COMPARISON OF EXISTING LNA'S IN 2-3GHz RANGE

Reference	Frequency (GHz)	Technology	S21(dB)	NF(dB)	P1dB (dBm)	IIP3(dBm)	Power(mW)
[7]	2.4	130nm	24	2	-	-11	2.6
[8]	2- 4.6 GHz	180nm	9.8(max)	2.3-5.2		-7	12.6
[9]	2-4.6 GHz	180nm	9.5(max)	3.5	-6	-0.8	16.5
[10]	2.5-4	90nm	19.6	4	-	-0.8	8
[11]	2.2	350nm	8.6	1.92	-	-2.6	16.2
[12]	2.45	250nm	15.1	2.88	-	2.2	24.3
Cascode(This work)	2.5	180nm	19.84	0.59	-17.76	-2.49	18.875

REFERENCES

[1] B. Razavi, RF Microelectronics, second edition ed. Prentice Hall,2011.

[2] D. K. Shaeffer and T. H. Lee, "A 1.5-V, 1.5-GHz CMOS low noise amplifier," IEEE J. Solid-State Circuits , vol. 32, no. 5, pp. 745–759, May 1997.

[3] F. Bruccoleri, E. A. M. Klumperink, and B. Nauta, "Wide-band CMOS low-noise amplifier exploiting thermal noise canceling," IEEE J. Solid-State Circuits , vol. 39, no. 2, pp. 275–282, Feb. 2004.

[4] Lorenzo, Michael Angelo G., and Maria Theresa G. de Leon. "Comparison of LNA topologies for WiMAX applications in a standard 90-nm CMOS process." In Computer Modelling and Simulation (UKSim), 2010 12th International Conference on, pp. 642-647. IEEE, 2010.

[5] Hsieh, H-H., J-H. Wang, and L-H. Lu. "Gain-enhancement techniques for CMOS folded cascode LNAs at low-voltage operations." Microwave Theory and Techniques, IEEE Transactions on 56.8 (2008): 1807-1816.

[6] T. Lee, The design of CMOS Radio Frequency Circuits, second edition ed.Prentice Hall,2003.

[7] B. J. Sanghoon Joo, Tae-Young Choi, "A 2.4-GHz Resistive Feedback LNA in 0.13 µm CMOS," IEEE journal of solid-state circuits, vol. 44, november 2009.

[8] Kim, C. W., M. S. Kang, P. T. Anh, and S. G. Lee, "An ultra-wideband CMOS low noise amplifier for 3-5 GHz UWB system," IEEE Journal of Solid State Circuits, Vol. 40, No. 2, pp. 544-547, Feb. 2005.

[9] A. Bevilacqu, C. Sandner, A. Gerosa, and A. Neviani, "A fully integrated differential CMOS LNA for 3–5-GHz ultra-wideband wireless receivers," IEEE Microw. Wireless Compon. Lett., vol. 16, no. 3, pp. 134–136, Mar.2006.

[10] S. C. Blaakmeer, E. A. Klumperink, D. M. Leenaerts, and B. Nauta, "A wideband noise-canceling CMOS LNA exploiting a transformer," in Proc. IEEE Radio Frequency Integrated Circuits Symp. , Jun. 2006.

[11] X. Fan, H. Zhang, and E. Sanchez-Sinencio, "A noise reduction and linearity improvement technique for a differential cascode LNA," IEEE J. Solid-State Circuits , vol. 43, no. 3, pp. 588–599, Mar. 2008.

[12] X. Li, T. Brogan, M. Esposito, B. Myers, and K. K. O,"A compariison of CMOS and SiGe LNA's and mixers for wireless LAN application,"in Proc. IEEE Custom Integrated Circuits Conf. (CICC), 2001,pp. 531–534.

2016 International Conference on VLSI Systems, Architectures, Technology and Applications (VLSI-SATA)

Fully-Digital Time based ADC/TDC in $0.18\mu m$ CMOS

Vineet Sharma, Nupur Jain and Biswajit Mishra

VLSI and Embedded Systems Research Group

Dhirubhai Ambani Institute of Information and Communication Technology

Gandhinagar, 382007 Gujarat

Email:{vineet_sharma, nupur_jain, biswajit_mishra}@daiict.ac.in

Abstract—This paper proposes a fully digital sensor interface. For this, an analog to digital converter (ADC) and time to digital converter (TDC) based on a common time based ADC (TAD) architecture has been investigated. It is concluded that the proposed fully digital time-based ADC architecture can also be operated as TDC. The fully digital circuit has a ring delay line (RDL), latch, encoder and a synchronous counter. The circuit is implemented in $0.18\mu m$ digital CMOS, achieving $139\mu V$/LSB (14-bit, 1-MS/s, 1.6 mW) in ADC mode and 227 ps/LSB (V_{IN} = 1.0 V, 14-bit), 94 ps/LSB (V_{IN} = 1.8 V, 14-bit) in TDC mode respectively. In addition to the scalable design, the resolution of both TDC as well as ADC, can be set by a variable input voltage, V_{IN}.

Keywords—Fully-digital time based ADC, TDC, $0.18\mu m$ CMOS.

I. INTRODUCTION

Recently, there has been a lot of interest on Internet of things (IoT) and Information and Communication Technology (ICT) applications specifically in the field of agriculture, environmental monitoring, body sensor networks and wireless sensor networks. The implementation of these applications require the use of sensors that sense and record data. These sensors, typically, sense analog signals and generates analog or digital outputs to communicate to the μcontroller or a DSP processor. Therefore, interfaces are important for either converting the analog data to digital data or for sending the digital output from sensor to the processing unit. Furthermore, a low power implementation of such circuits such as capacitance to time conversion [1], [2] are useful for sensor interface circuits and have been demonstrated. The analog and digital conversions fall under either voltage to digital or time (frequency) varying signals. Therefore, a circuit that can operate both as a ADC and a TDC [3] is useful as it can interface both the voltage and time varying signals.

In this paper, we propose an ADC for voltage signals and TDC for the time/frequency signals on a common Time based Analog to Digital Converter (TAD). A modified ring delay line (RDL), encoder, low power latch, which forms the ADC are custom designed. A novel coupling unit is designed to generate control signals for the RDL. The proposed design is scalable and in this paper the results for 14 bits TDC are reported. A novel coupling unit is designed to generate the necessary control signals and is discussed in section II D. This paper discusses the circuit level changes applied to an existing architecture that results in low power operation of TDC. Furthermore, the paper reports custom cell designs for

delay cells, coupling unit, latch and D flip-flop that can be operated at variable supply voltages. We show the benefits of the design using 0.18 μm physical models using simulation results.

The paper discusses the principle of operation of the TDC in section II. It also describes the function of the elements of TDC and reports detained analysis of every element. The simulation results for ADC and TDC are discussed in section III followed by conclusion in section IV.

II. PRINCIPLE OF OPERATION OF TDC AND THE PROPOSED ADC ARCHITECTURE

Time-based ADC (TAD) performs data conversion through a scheme of quantizing time at predefined amplitude intervals whereas, a conventional ADC quantizes amplitude at predefined time intervals. This quantization of time at predefined amplitude intervals can be used, in future, as time to digital conversion that is useful for frequency interval signals. A TDC in its simplest form can be realized by a counter. For high resolution time to digital conversion (TDC), precise and fine tuned time steps are required which an active delay line is able to provide. The TAD provides the digital equivalent of voltage, which when multiplied by the delay of the circuit would yield the time equivalent of analog voltage. Thus, the TAD architecture can also be realised as a time to digital converter (TDC) and is discussed in detail in subsequent sections. As discussed earlier, the fine tuned delay line is the key to high time resolution TAD, a brief discussion on its principle of operation is presented followed by the proposed architecture of the ADC. In the process of time amplification, a pulse is passed through a chain of delay elements, which forms the delay line. The number of delay elements traversed denotes the distance travelled by the pulse in the time window.

Fig.1 shows a conventional delay line which consists of buffers acting as delay elements. A start pulse is fed as input to the *n*-element delay line. The pulse propogates through a

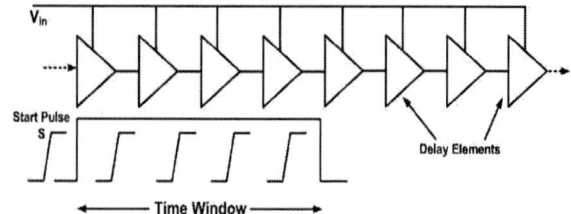

Fig. 1. A Conventional Delay Line

978-1-5090-0037-1/16 $31.00 © 2016 IEEE

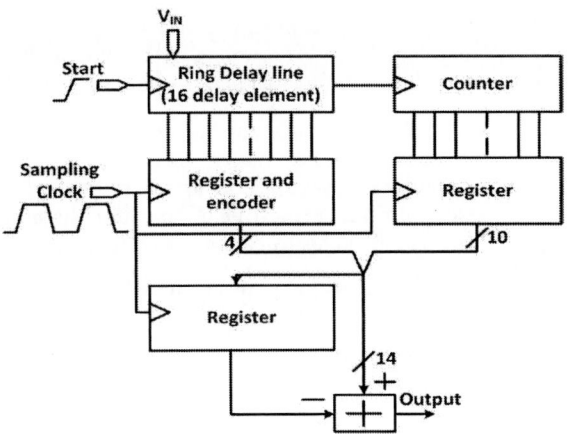

Fig. 2. Architecture of time based ADC

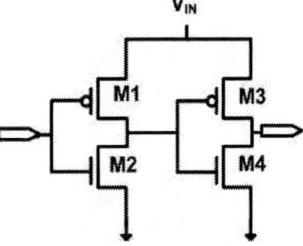

Fig. 3. Delay element with variable supply voltage

segment of delay elements which determines the delay encountered by the pulse within the time window. This delay depends upon the charging and discharging of parasitic capacitances of the delay elements and can be modulated by controlling the amount of the current available for charging (or discharging). Several methods exist in the literature such as [4–6] to get variable delays. In [5], the delay is regulated by changing the current using current mirror, whereas [4] and [6] change the supply voltage and the gate to source voltage (V_{GS}) of nMOS to achieve the same. In this work, we vary the supply voltage to obtain the desired delay. The proposed architecture of time based ADC (TAD) is shown in Fig.2 is adopted from [4], [7] and the circuit level changes made to the design are discussed in this work. A start signal (S) is applied at the input of the ring delay line and the delay is controlled by analog input voltage V_{IN}. The position of S is latched at the rising edge of the sampling clock. A counter is used to count the number of times the start signal S has traversed through the entire delay line. The total number of delay elements traversed by S are determined by the output of both encoder and the counter. At every sampling clock, the current output of the register is subtracted from the value captured at the previous pulse. This result denotes the number of delay elements the start pulse has passed through within the sampling period. The subtraction operation also eliminates the need to reset the circuit after every sample.

It should be noted that, once the supply voltage of the delay element is fixed, the start pulse is able to pass over a definite number of delay elements in a particular interval of time. Therefore, this enables us to use the proposed architecture both as, ADC and TDC. The proposed architecture yields itself to be used as a frequency (or time) meter. In case of TDC, a time varying signal is applied at the clock input of the circuit and the counter output is captured at the two consecutive rising edges of the signal. The former is subtracted from later and the result is multiplied with the delay encountered that gives the time period of the signal.

A. Delay Element

The delay element is the basic unit of the delay line. The delay element shown in Fig. 3, has analog input V_{IN} as the supply voltage. They can be connected in multiple ways as reported in [4] and [6]. These delay elements are connected

in cascade i.e. the output of one stage serves as input to the next stage delay element. Also both, delay as well as output, depends upon V_{IN}, which makes it difficult for the subsequent stages to read the input as the output of one stage has to first latch onto the input of next stage. To ensure proper latching of output, V_{IN} should be held above the logic threshold and was overcome by careful design iterations and following analysis on the saturation current for the delay elements.

According to [8] and [9], the saturation current in a deep-submicron MOS transistor can be approximated as

$$I_{DS} = \frac{W}{2L}\mu_n C_{ox}(V_{GS} - V_t)^\alpha \tag{1}$$

where α denotes velocity saturation index which lies between 1.2-1.4 for 0.18μm CMOS technology. W/L, C_{ox}, μ_n are effective aspect ratio of the transistor, oxide capacitance of the inverter, and electron mobility, respectively. The delay of an inverter is taken as the average of the delay of low to high (T_{PLH}) and high to low (T_{PHL}) transitions. Assuming that the discharge current of the nMOS transistor between voltage V_{DD} to $1/2 V_{DD}$ can be approximated by its saturation current, according to [10] the T_{PHL} (delay of high to low transition) can be written as

$$T_{PHL} \approx \frac{0.5 \cdot C_L V_{DD}}{I_{DS}} = \frac{0.5 \cdot C_L V_{DD}}{\frac{W}{2L}\mu_n C_{ox}(V_{GS} - V_T)^\alpha} = A\frac{C_L V_{DD}}{(V_{DD} - V_T)^\alpha} \tag{2}$$

where C_L denotes load capacitance, A is constant and V_{GS} is replaced by V_{DD} (assuming approx. zero rise time of V_{GS}). Similarly, the T_{PLH} low to high transition can be approximated as

$$T_{PLH} \approx B\frac{C_L V_{DD}}{(V_{DD} - V_T)^\alpha} \tag{3}$$

where B denotes a constant and the other symbols stands for thier usual representation.

B. Analysis of Delay Line

We now present, the delay specification of the gates forming the delay line. Accoring to [11], the NAND gate can be regarded as equivalent NOT gate in terms of functionality. Then the propagation delay of delay line, T_d, can be given as eqn 4, where n and T_{NOT} are the numbers of NOT and NAND gates in the delay line and the propagation delay of NOT gate, respectively.

$$T_d = n.T_{NOT} \tag{4}$$

978-1-5090-0037-1/16 $31.00 © 2016 IEEE   85

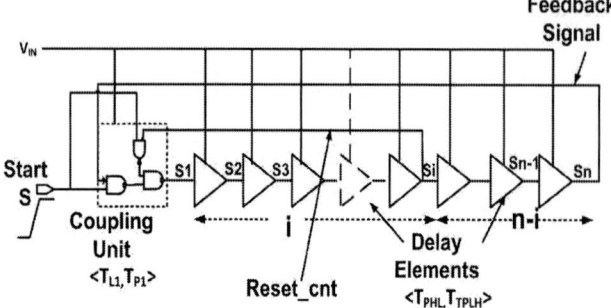

Fig. 4. Ring Delay Line

The delay of a buffer consisting of two NOT gates is given by eqn 5, where k is a constant. The sizing of various transistors employed in the design is done in accordance to eqn 5.

$$T_b = 2.T_d = A\frac{C_L V_{DD}}{(V_{DD} - V_T)^\alpha} + B\frac{C_L V_{DD}}{(V_{DD} - V_T)^\alpha}$$

$$= k\frac{C_L V_{DD}}{(V_{DD} - V_T)^\alpha} \qquad (5)$$

The length of the delay line is primarily determined by the time taken by start pulse to complete a loop in the delay line. Because the output from the last delay element has to be fed as the counter clock, the period of delay line must be greater than the minimum clock period required for counter operation. Here, 16 delay elements satisfy the criteria as discussed above. The delay elements give 4 bits ($= \log_2 16$) output while the rest 10 bits are provided by the counter.

C. Ring Delay Line

In the ring delay line, buffers are connected back to back to form the ring topology and its operation is discussed in the following. This topology makes the start signal pass through delay line several times. A counter keeps track of number of rotations (B_{CNT}) the start pulse has made through the delay line. Therefore, the number of delay element traversed by the start pulse consists of two components: *(i)* complete loops the start pulse has traversed and *(ii)* fraction of the loop traversed by the start pulse during the last loop (B_{DL}). According to [12], it is given by

$$B = n \cdot B_{CNT} + B_{DL} \qquad (6)$$

where n is the number of delay elements in delay line.

A count reset pulse (labelled reset_cnt in Fig. 4) is used to initialize the output of delay element before the start pulse is traversed back to the input of delay line, which otherwise would make the delay line advance into stable state and the counting to stop. This can be explained as shown in Fig. 5 for a 16-element delay line.

1) Start pulse is applied at the input of delay line.
2) This sets the output of delay element after a delay of T_d
3) This happens upto the last delay element, and
4) When the signal is traversed back to the input of delay line, the output of delay elements does not change because the it was already set.
5) So no further count occurs.

To avoid the above explained condition, a precise count reset pulse is generated. The timing behaviour of the count reset pulse is shown in Fig.6 and can be explained as follows:

- When the start pulse passing through the delay line reaches the end, the output of delay elements in the beginning of delay line are already reseted by a count reset pulse generated from a tap within the middle of the delay line chosen carefully and is design specific.

- This count reset pulse avoids the stable condition that could have occurred as in previous case.

- Now as delay line does not advance in stable state, the counting continues.

D. Coupling Unit Design

To generate the count reset, a coupling unit is designed and discussed in the following. Coupling unit is used to feed the start pulse back into the delay line once it reaches the end of the delay line. The schematic of coupling unit is shown in Fig.7.

Based on the Table I,

$$Y = Start(Reset\_cnt\prime + Feedback) \qquad (7)$$

E. Reset pulse in ring delay line

In the loop delay line design, a count reset pulse is used to prohibit the circuit from advancing into the stable condition. Once the circuit proceeds into the stable state it is not able to count futher. The output voltage of the coupling unit must be set to 0 before the start pulse traverse back to the coupling unit to continue counting. For this ,the count reset pulse is generated from the output of a buffer in delay line and fed back to coupling unit which resets its output thereby circumventing the possibility of locking the delay line into the stable state.

While generating the reset signal, the location of buffer is so chosen as to avoid the stable state from occuring. It was analysed that the stable condition can occur in one of the following two cases:

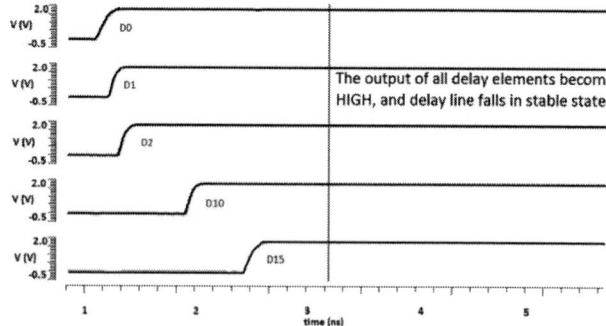

Fig. 5. Ring delay line in stable state

TABLE I. TRUTH TABLE FOR COUPLING UNIT DESIGN

Start signal	Count reset signal	Feedback signal	Output of coupling unit
0	x	x	0
1	0	0	1
1	1	0	1
1	0	1	1
1	1	1	1

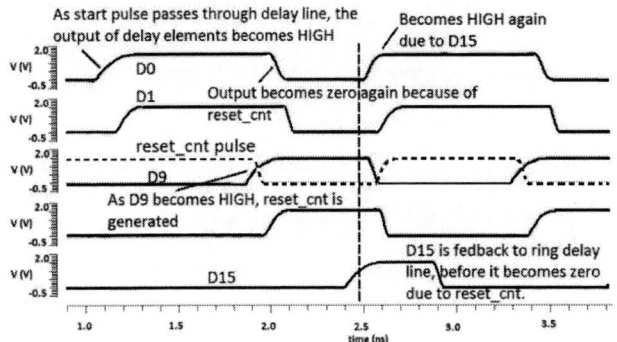

Fig. 6. Ring delay line in unstable state

Case 1: Either start pulse approaches and coincides with the count reset pulse or,

Case 2: Count reset pulse coincides with the start pulse

Let us assume there are *n* delay elements in which *n-1* are of same kind (buffer consisting of two inverters) and a coupling unit. Further, the delays of the buffer and the coupling are assumed as $< t_{PL}, t_{PH} >$ and $< t_{L1}, t_{H1} >$ for passing $< 0, 1 >$, respectively as shown in Fig. 4.

The two cases are now discussed in detail. Case1: Let *x* denote the current position of the start pulse in terms of number of delay elements passed. When the start pulse reaches the ith delay element *i.e.* $x = i$, for the first time. The first reset pulse gets generated and subsequently occurs at the output of the coupling unit after $(t_{INV} + t_{L1})$ duration. However, the number of delay elements passed by the start pulse (*i.e. x*) should be less than $(n-i)$, to avoid the stable state from occuring. Thus *x* can be written as,

$$x = \frac{t_{INV} + t_{L1}}{t_{PH}} \quad \text{and} \quad x \leq n - i \quad (8)$$

Once the condition specified by eqn 8 is met, the count reset pulse is able to loop through the delay line before start pulse completes its second loop thereby, avoiding the stable state. The relation between the delays of buffer and coupling unit is given by the following eqn.

$$(n - i - x)t_{PH} + t_{H1} + (n - 2)t_{PH} \geq (n - 1)t_P L \quad (9)$$

Case 2: To avoid start pulse from being caught by reset pulse. The time taken by the count reset pulse to reach the end of delay line (LHS of the inequality 10) must be greater than the time start pulse takes to tranverse all the elements in the delay line (RHS of the inequality 10).

$$(n - 1)t_{PL} \geq (n - i - x)t_{PH} + t_{H1} \quad (10)$$

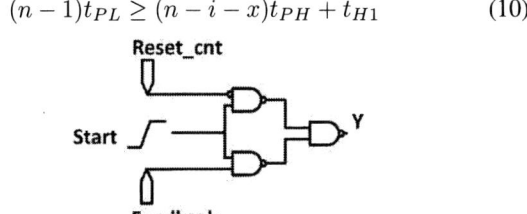

Fig. 7. Coupling Unit

F. Counter

The counter shown in Fig. 2 keeps track of the number of times the start pulse has crossed the entire ring delay line. The counter module contains a 10-bit synchronous counter designed using D flip flops [13] shown in Fig. 8.

G. Encoder Unit

Encoder circuit is designated to record the position of pulse in the delay line on the arrival of sampling pulse. The state of the delay line is captured by a register on the arrival of sampling pulse and the output of the register is fed to encoder circuit. The design of encoder circuit is shown in Fig.9. Two pulses, start and reset, are circulating in the delay line. The start pulse changes the output of delay element from logic '0' to logic '1', and is encoded by the circuit shown in Fig.9. The working of encoder unit is divided in two phases and is discussed now.

Phase 1 — — > START pulse in D0-D14
In this case, one output of the encoder is necessarily at logic 1. Therefore, the location of start pulse is determined by unique pair of logic '0' followed by logic '1'. The output of AND gate which is directly fed by this logic '1' will be high and it will be encoded in 4-bits by 16:4 encoder.

Phase 2 — — > START pulse is at D15.
In this case, the output of encoder will be 0 and the counter will be incremented. Here an increment by 1 is equivalent to 16 delay suffered.

III. SIMULATION RESULTS

A. Analog-to-Digital Converter

In this section, we discuss and quantify the results of the TAD architecture both as ADC and TDC. The delays provided by the buffer and coupling unit are plotted as a function of supply voltage (varied from 1.0-1.8V), V_{IN} and are shown in Fig.10. The delay of the buffer is observed as linearly varying for V_{IN} between 1.6-1.8V.

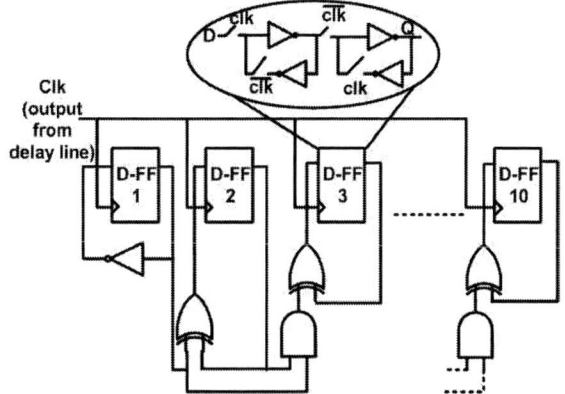

Fig. 8. 10 bit synchronous counter

TABLE II. SIZING OF WIDTH IN DELAY ELEMENT

MOSFET	Width
M1	$480nm$
M2	$480nm$
M3	$480nm$
M4	$720nm$

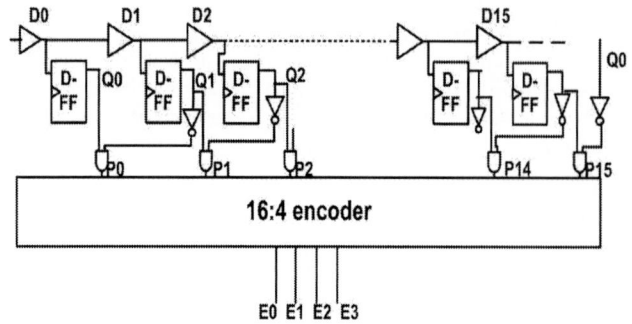

Fig. 9. Encoder Unit

The sizing of the transistors shown in Fig. 3 is shown in Table II. Fig. 11 represents of the time period of the delay line plotted as a function of input voltage. The time period varies inversely with the input voltage.

TABLE III. ADC COMPARISON

	This work		[7]	
Conversion rate	1MS/s	20 MS/s	100kS/s	20MS/s
Input range	200 mV	200 mV	800 mV	800 mV
Resolution	139μV/LSB	2.81 mV/LSB	15μV/LSB	3.1 mV/LSB
	(10.5-b)	(6.15-b)	(15.7-b)	(8.1-b)
Power consumed	1.6 mW	1.65 mW	1.3mW	1.7 mW

Table III shows the results depicting characteristics of ADC. The ADC characteristics at sampling frequency $F_s =$ 1MHz ($25°C$) are shown in Fig.12. The resolution of the ADC is computed by taking the ratio of the change in voltage and the change in count over the voltage change. For sampling frequency of 1MHz, the count values are 9819 and 11251 for V_{IN} 1.6V and 1.8V, respectively. Therefore, the change in counts is 1433 (= 11251-9819) for the voltage range of 200mV (= 1.8-1.6). Thus, the resolution is estimated as $139\mu V/LSB$ (= 200m/1433) which amounts to 10.5-bit(= $\log_2(139\mu)$).

The power consumption is computed using Cadence Virtuoso and is estimated as 1.6 mW for supply voltage of 1.8V. The ADC characteristics at $F_s = 20MHz$ ($25°C$) are shown

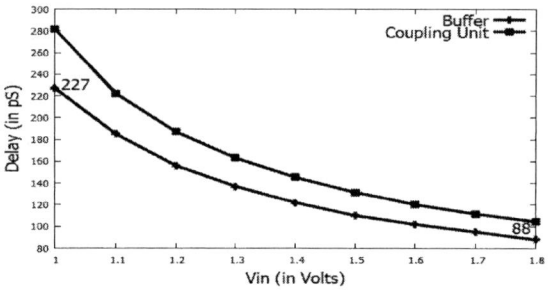

Fig. 10. Variation of delay of buffer and coupling unit with the input voltage

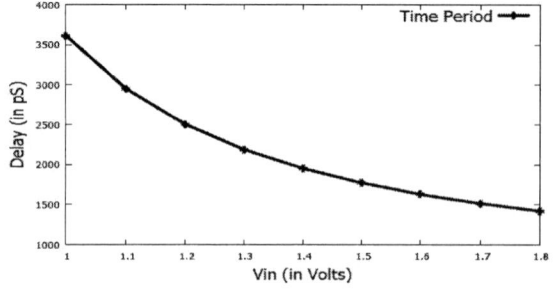

Fig. 11. Variation of time period of delay line with supply voltage

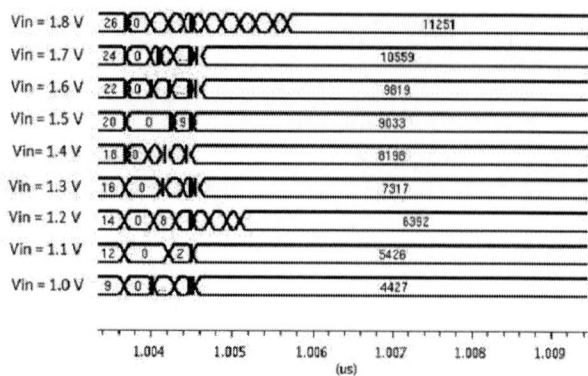

Fig. 12. Simulation results of ADC implemented with physical library cells at sampling rate 1 MHz with V_{IN} varied from 1 V to 1.8 V.

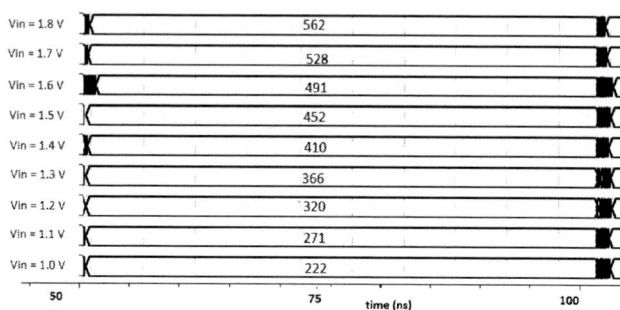

Fig. 13. Simulation results of ADC implemented with physical library cells at sampling rate 20 MHz with V_{IN} varied from 1 V to 1.8 V.

in Fig. 13. Tthe resolution is approximated as 2.81 mV/LSB (=200m/71). The power consumption is 1.65 mW for 1.8V supply voltage.

B. Time-to-Digital Converter

The TAD architecture is also demonstrated to be used as time to digital converter (TDC). The TDC resolution is controlled by a settable V_{IN}. Its characteristics are now discussed. The value of counts (number of delay elements) against the measured time period for two different V_{IN} are shown in Fig.14. For the measurement of 50 ns time interval, the power consumed for V_{IN} at 1.0V and 1.8V are 0.73 mW and 1.65 mW, respectively. Results of TDC are shown in Table IV.

The maximum possible time period that can be measured by the proposed architecture is approximated as the product of the combined output of RDL and counter, and the delay of the delay element. The output can go to a maximum of 16384 (= 2^{14}) and delay for different values of V_{IN} is shown in Fig.10.

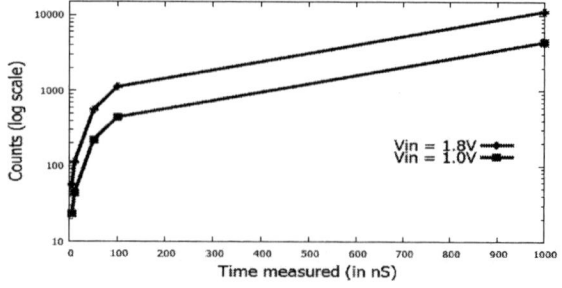

Fig. 14. Variation of counts with respect to time measured at V_{IN} 1.0 V and 1.8V.

TABLE IV.   TDC COMPARISON

	This work		[7]	
V_{IN} (in Volt)	1.0	1.8	1.0	1.8
Measurable range	5ns-3.71μs	5ns-1.44μs	20ns-12.3ms	20ns-4.2ms
Time resolution	227ps	88ps	368ps	126ps
Output bits	14	14	25	25
Power consumed	0.73 mW	1.6mW	1.3mW	1.7mW

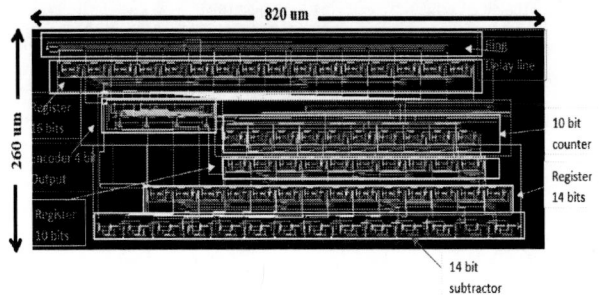

Fig. 15.   Layout of the proposed TAD Architecture

Therefore, the maximum measurable time period is estimated as 3.71μs (= 16384×227 ps) and 1.44μs (= 16384×88 ps) for V_{IN} of 1.0 V and 1.8 V, respectively. The minimum measurable time period depends delay of the subtractor and forms the lower bound for the time measured by the design. Therefore, the measurable range for TDC at V_{IN} level = 1.0 V is between 5ns - 3.69μs while at V_{IN} level=1.8 V, it is 5 ns - 1.45μs.

This work is compared with a similar work presented in [7]. The lower bound on the time measured by the TDC is reported as 50ns in the latter, whereas the proposed design can measure a time period as low as 5ns. This is because the proposed design has employed the subtractor of width 14-b as opposed to 25-b used in [7]. This work offers an improved time resolution at both 1V and 1.8 V as compared to [7] because of the faster performance of the RDL. The TDC design discussed in this work has 14-b output as compared to the 25-b output reported in [7], but the proposed design remains scalable and the number of bits can be increased by entending the size of the counter employed in the design. By doing so, the measurable time range of the proposed TDC increases but at the expense of increase in power and area of the design. The variations due to process parameters (V_{DD}, V_{th}, temperature) at low voltages influence the pulse width that can vary considerably [14]. The focus of this work is the implementation and circuit level design improvement and hence the simulation at different process corners is left as future work. However, it was observed in the post layout simulation that the deviation of frequency and power is within the 5% bound. To mitigate such variations, selecting appropriate delay cells control logic is required and is left as future work. The novel autocalibration method discussed in [2] is suitable for the proposed work and is left as future work.

The layout of the proposed design was made using Cadence Virtuoso Assura Extraction tool and is shown in Fig.15. The DRC (Design Rule Check) and LVS (Layout Versus Schematic) were complied in the design. The layout has a dimension of 820μm × 260 μm. It has all the optimized circuit blocks of the proposed design.

IV. CONCLUSION

This work presents a time-based ADC architecture for sensor interface. The proposed circuit is implemented using 0.18μm CMOS and has an added benefit of lower power and higher resolution as compared to existing designs, [4] and [7]. This architecture is implemented in the digital circuit for sensor application. The same circuit can be used as TDC for measuring time. This architecture is compact, scalable and offers a high resolution. The simulation results show that ADC with 14-bit output has a 139μV/LSB resolution at 1MS/s, and TDC has a measurable range between 5ns -1.45μs at 1.8V supply voltage. The resolution of ADC/TDC can be controlled by altering F_s and V_{IN}.

V. ACKNOWLEDGEMENT

Authors thank the Gujarat Council of Science and Technology (DST), Gandhinagar for financial support under grant reference GUJCOST/MRP/2014-15/2582.

REFERENCES

[1] M. Shulaker, J. Van Rethy, G. Hills, H. Chen, G. Gielen, H. Wong, and S. Mitra, "Experimental demonstration of a fully digital capacitive sensor interface built entirely using carbon-nanotube FETs," in *IEEE International Solid-State Circuits Conference Digest of Technical Papers (ISSCC)*, Feb 2013, pp. 112-113.

[2] A. Savaliya and B. Mishra, "A 0.3V, 12nW, 47fJ/conv, fully digital capacitive sensor interface in 0.18m CMOS," in *International Conference on VLSI Systems, Architecture, Technology and Applications*, Jan 2015, pp. 1-6.

[3] D. Wentzloff and Y. Park, "A cyclic vernier TDC for adplls synthesized from a standard cell library," *IEEE transactions Circuits and Systems I: Regular Papers*, vol. 58, no. 7, pp. 1511-1517, July 2011.

[4] T. Watanabe, T. Mizuno, and Y. Makino, "An all-digital analog-to-digital converter with 12μV/LSB using moving-average filtering", *IEEE Journal of Solid-State Circuits*, vol. 38, no. 1, pp. 120-125, Jan 2003.

[5] Y. Tousi,G. Li, A. Hassibi, and E. Afshari, "A 1mW 4b 1Gs/s delay-line based analog -to-digital converter," *IEEE Internationsl Symosium on Circuits and Systems*, May 2009, pp. 1121-1124.

[6] H. Pan and A. Abidi, "Signal folding in A/D converters", *IEEE Transactions on Circuits and Systems I: Regular Papers*, vol. 51, no. 1, pp. 3-14, Jan 2004.

[7] T. Watanabe and T. Terasawa, "An all-digital ADC/TDC for sensor interface with TAD architecture in 0.18μm digital CMOS", in *16th IEEE International Conference on Electronics, Circuits, and Systems*, 2009, Dec 2009, pp. 219-222.

[8] J. Rabaey, *Low Power Design Essentials*. Springer Science & Business Media, 2009.

[9] T. Sakurai and A. Newton, "Alpha-power law MOSFET model and its applications to CMOS inverter delay and other formulas", *IEEE Journal of Solid-State Circuits*, vol. 25, no. 2, pp. 584-594, Apr 1990.

[10] Y. Leblebici, *CMOS Digital Integrated Circuits: Analysis and Design*, McGraw-Hill College, 2009.

[11] C. Chen and H. Chen, "A Low-Cost CMOS Smart Temperature Sensor Using a Thermal-Sensing and Pulse-Shrinking Delay Line", *IEEE Sensors Journal*, vol. 14, no. 1, pp. 278-284, Jan 2014.

[12] S. Henzler, *Time-to-digital converters*. Springer Science & Business Media, 2010. vol. 29.

[13] W. Harris *et al.*, *Cmos VLSI Design: A Circuits And Systems Perspective, 3/E*, Pearson Education India, 2006.

[14] B. Mishra, B. Al-Hashimi, and M. Zwolinski, "Variation resilient adaptive controller for subthreshold circuits", in *Design, Automation Test in Europe Conference Exhibition*, Apr 2009, pp. 142-147.

A Portable Platform to Estimate Power Consumption of Software Modules

Abhishek Bhardwaj
Department of Computer Science
PDPM IIITDM-Jabalpur, Madhya Pradesh
Jabalpur, India
E-mail: abhishekbhardwaj@iiitdmj.ac.in

Saket Saurav
Department of Computer Science
PDPM IIITDM-Jabalpur, Madhya Pradesh
Jabalpur, India
E-mail: saket@iiitdmj.ac.in

Abstract— Researchers perform various hardware and software tests in order to measure power consumption of mobile devices, network modules and embedded system solutions. In this paper, we discuss a portable platform to measure changes in power consumption. We run several tests on this platform to compute CPU utilization by the test programs and the operating system and relate it with execution time and memory utilization. Using the data obtained, we use regression analysis to relate power consumption with CPU utilization.

Keywords—Green-Computing, RaspberryPi, CPU utilization, Ardiuno.

I. Introduction

As the computation is being ubiquitous, there is a growing concern of the power consumed by the computing devices. Obvious concerns are for the computing solutions requiring battery power, and therefore energy efficiency is the major concern for the designers as well as manufacturers of such systems. In this study we analyze the various peripherals that affect power consumption of embedded computers and understand the relationship between power consumption and CPU utilization. We have used Raspberry Pi an embedded computer with a processor of ARM-7 family .That can resemble the CPUs of many mobile phones and network devices like routers and switches. By conducting experiments on our setup, we could comment on the power consumption of these modules.

Green computing [1] a branch in itself is gaining popularity because of its trends in the direction of environmentally sustainable computing. Our study will not only help green researchers but also the developers to design and develop hardware as well as software keeping power consumption in their minds. Till now while developing mobile applications and embedded solutions we focus on its time complexities and functionalities only , introducing power efficient algorithms will not only increase the life of devices but saves our non-renewable resources as well.

II. Related Work

There is altogether a community of Green Miners[2] which has developed an android based environment for testing mobile phones and its peripherals. They conducted test on Galaxy Nexus with Android OS 4.2.2. Their work has been gaining popularity in mobile industry as their frame work is easy to use and is portable. Gupta [3] has also worked on interpreting energy consumption by various components of a Window Phone.

PowerTOP [4] is another Linux tool to diagnose issues with power consumption and power management. Since it is software it doesn't compute the power consumed by peripheral that gets switched on during execution of software programs and function calls. For example , an application requiring high running frequency switched on the oscillator attached to the microprocessors or microcontrollers during run time. Similarly, if parallel algorithm based software is being executed, selected processors get activated whereas serial application can runs on a single core . Hence the power execution is affected by the procedures we follow.

Green Miner[2] , reported that Reddit reader app on a mobile phone had abnormally high power usage even compared to web browsers. After clarifying with the developers they analyzed the code and concluded that the code used the GlobalLayoutListener function excessively leading CPU to overload and use higher current thus the redundant code was removed. This example clearly demonstrates the usability of our scope of study in near future.

In our work we have presented a much simpler application using python that log, plot and report which can be accessed easily, more we help present a way of testing network modules and embedded computers. Our hardware is easy to arrange and is cost effective.

III. Platform

We have developed hardware and software platform to test the power consumption with logging and reporting capabilities of Raspberry Pi. It comprise of a self-prepared power distribution board and an Ardiuno Uno to log and report.

A. Apparatus

Fig 1 shows that Ardiuno Uno, power board and Raspberry Pi are placed on an insulated plank and screwed up. Raspberry Pi is connected to laptop via Ethernet RJ45 port to initiate test and applications. Ardiuno Uno is also connected to the laptop for logging values serially. Raspberry Pi is chosen because it is a low power ARM CPU that can represent major mobile industry as well as network devices. It comprises Broadcom BCM2835 chip of ARM-7 family. It required 5v supply and a SD card for file system which in my case is SanDisk class 4, 4 GB of capacity. With 512 MB of SDRAM it is capable of running wide variety of applications.

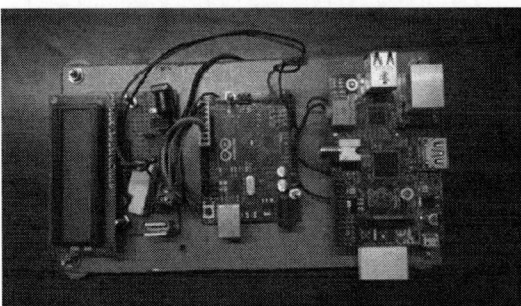

Figure 1:Testing Apparatus

B. Methodology

We have used the basic embedded technology to log and report data. As shown in the flow chart after setting apparatus right we initiate the script and log the data on the computer. We logged in Raspberry Pi via SSH so as to execute script. Fig 2 shows the control flow that our test follows during test.

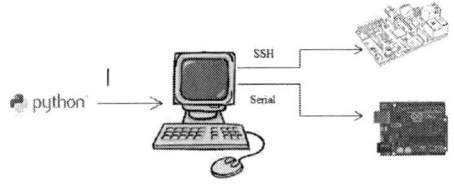

Figure 2 : Connection Flow Chart

C. Instrumentation

All our experiments our being done on Ardiuno Uno with Atmega-328 microcontroller with analog sampling rate of 10000 ADC input per second. Our python application is communicates with microcontroller through serial communication at 9600 baud rate.

D. Current Calibration and Power Calculation

We used the traditional method for computing current by measuring the potential drop across shunt .We have log the values of current using Ardiuno Uno via analog pins. The Ardiuno Uno interfaces with python based program which plots graph through serial communication. This method requires lot of tuning and calibration for accuracy. Hence in we have used the recommended Ardiuno IDE software which contains map function in their libraries to map 8 bit ADC signal to 0-255 values. We have performed the experiment 20 times and observed 20 voltage values across shunt with their corresponding analog value and hence obtained the following relationship:

1ADC unit = 4.072 mV

We also discussed on the placement of shunt resistors. Two arrangements are possible for placing the shunt as depicted in fig-3. After going through current measuring procedures, we concluded that placing shut toward ground is well suited when we need to share the ground with measuring device (i.e. Ardiuno Uno in this case).

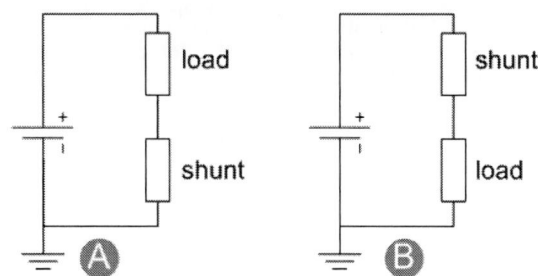

Figure 3 : Placement of Shunt Resistance

After obtaining voltages from Ardiuno , we calculate current i.e. I = Vs/Rs (Vs is voltage across shunt and Rs is shunt resistance). Now this current is used to calculate the power consumed by Raspberry Pi i.e. P = I2 Rr (Where Rr is the resistance of Raspberry Pi).Since we are adding a shunt resistance Rs current should be written as I = V / (Rr + Rs)

V is the total voltage applied. This tell us that I is function of Rs as well so we need to judicially select the value of Rs such that Rs << Rr so that I = V / (Rr + Rs) ≈ V / I = V / Rr

As shown in fig-3 we made a small circuit using a ceramic resistance of 1ohm with a positive error of .4 ohms as shunt resistor. USB power cable is inserted in USB port and a DC power adapter is also plugged in the DC port which can be seen in the fig-1. We have use 7805 IC as 5V voltage regulator to provide Raspberry Pi with ample of current at a constant voltage. The data through serial channel was recorded at 9600 baud rate.

E. Selecting Operating System

Initially we experimented all the stress testing software and programs on Raspbian[5] which is Debian ported for Raspberry Pi to achieve a reference of power Raspberry Pi consumes for basic processing and computations. In the latter part of the study we ported other variants of Linux and Raspberry Pi and ran several mobile applications to comment on their power usage and to proof read the internal power management system of the operating system.

F. Visualization

We have developed a program using python to communicate serially with the Ardiuno Uno to receive value and plot the graph for every observation. We just have to enter the number of observation as input and the real time plotting takes place. For making this application we have used three major python libraries NumPy[6], Matplotlib[7], PySerial[8] in this program

IV. REFERENCING OF RASPBERRY PI ON RASPBIAN

Initially to obtain references of power consumption by Raspberry Pi we observe the current required for basic applications like connecting the LAN wire, pushing data through LAN, Video operations via RCA or HDMI and running some programs as well.

A. Loading Raspbian

After loading Raspbian , as we boot Raspberry without any peripherals connected we have observed a uniform disturbed curve along the time access. We can also observe in the Fig-4 that there are few spikes that demonstrate the of the processor and other associated components for which require some extra current. It can also be a possibility that because of several processes being triggered and terminated during booting lead to the unpredictable disturbances are seen in the following fig. We have found that to start Raspberry Pi with Raspbian over it takes around 2.235 Watt on an average. Table-I shows the status of other peripherals during the test such as video output not connected or the lan not plugged in etc.

TABLE I.   PERIPHERAL STATUS DURING TEST

PERIPHERALS	STATUS
Video Output	Not Connected
Lan Connectivity	Not Connected
USB	Not Connected
Mouse and Key Board	Not Connected
CPU clock	692 MHz

B. Connecting Mouse and Keyboard

Zebronics Doublet keyboard and mouse increase the current consumptions by 21.4%.This means these input devices on an average increases the power consumption by 25.49% while in use. This data we can used to estimate the

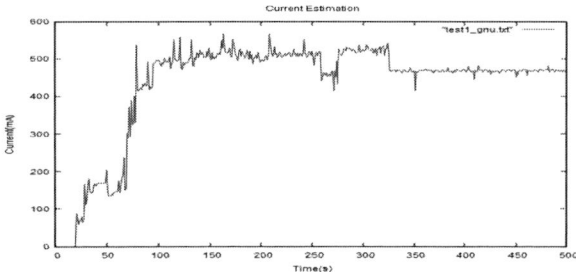

Figure 4: Current with Mouse and Keyboard

power consumed by a USB keyboard and a mouse which is around 0.5Watt. Now, a day's USB hosted devices are coming into market to provide multiple interface options so to reduce the pains on battery we require to analyzes and choose a power efficient device.

TABLE II.   PERIPHERALS DURING KEYBOARD AND MOUSE TESTING

PERIPHERALS	STATUS
Video Output	Not Connected
Lan Connectivity	Not Connected
USB	Not Connected
Mouse and Key Board	Connected
CPU clock	701 MHz

C. Connecting LAN

LAN connectivity for mobile computers in embedded systems and network modules is an essential feature that is required to share, communicate and control information via Ethernet .So as we plugged in LAN cable it increases the current consumption by 5.48% on Raspberry Pi whereas it reached to 11.54% when we started downloading some files using sudo apt-get update. (Fig-5 shows the comparison between rises in current). This peripheral increase the usability of mobile computers many folds .We have found that LAN communication required around 0.275Watt.

TABLE III.   PERIPHERALS DURING LAN TEST(ACTIVE AND IDLE)

PERIPHERALS	STATUS
Video Output	Not Connected
Lan Connectivity	Connected (Active/Idle)
USB	Not Connected
Mouse and Key Board	Connected
CPU clock	684 MHz

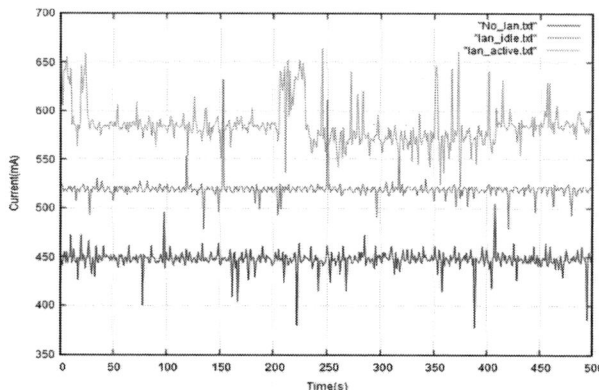

Figure 5: Current Comparison of Local Area Network in Active and Idle Modes

D. Taking Video Output

There are two ways to extract video output from Raspberry Pi i.e. RCA or HDMI. Although RCA lags in its picture quality but it provides an astonishing results in terms of current. We used Intex USB TV-tuner[9] to observe the RCA. It reduced the current consumption by 40%. Its coaxial cable provides a small potential which drives few components and it end up with the reduction. Whereas HDMI increases the power consumption by many folds due to its high resolution and quality. It consumes around 0.576 Watt that is around 25.72% of the total power consumptions of Raspberry Pi. This experiment clearly interprets the power consumption of high definition screens of our mobile computers.

978-1-5090-0037-1/16 $31.00 © 2016 IEEE

TABLE IV. PERIPHERALS STATUS DURING RCA TESTING

PERIPHERALS	STATUS
Video Output	RCA
Lan Connectivity	Connected and Active
USB	Not Connected
Mouse and Key Board	Not Connected
CPU clock	678 MHz

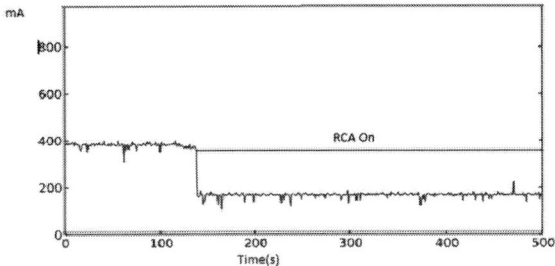

Figure 6: Drop of Current after turning RCA on

V. TEST DEFINITIONS

A. CPU Utilization and its Power Study

To test and load CPU, Memory and drive we have used two software "stress [10]" and "cpulimit [11]". Stress is a single file called stress.c whose internal organization is in essence a loop that forks worker processes and then waits for them to either complete normally or exit with an error. Whereas Cpulimit is a simple program which attempts to limit the CPU usage of a process.

1) Memory Utilization

This test was conducted to observe the power requirements of 512MB SDRAM of Raspberry Pi during stressful computations. To generate load on RAM , Stress was used with the following command.$ stress -m 4 --vm-bytes 110M [12]

This command loads 4 processes of 110MB each in the RAM. This test increased the average power consumption by 10.2% .This increase also in cooperates the increased in power consumption of CPU as increase RAM utilization ultimately results in increase of computation.

TABLE V. PERIPHERALS STATUS DURING MEMORY, DISK AND CPU UTILIZATION TEST

PERIPHERALS	STATUS
Video Output	Not Connected
Lan Connectivity	Connected and Active
USB	Not Connected
Mouse and Key Board	Connected
CPU clock	678 - 706 MHz

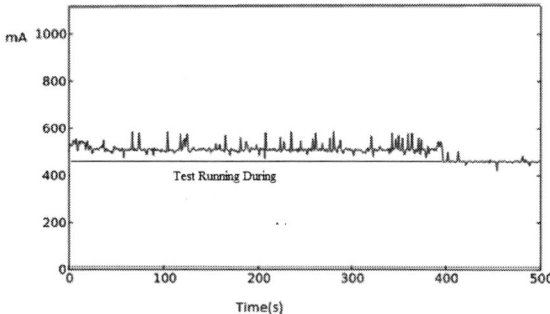

Figure 7 : Power Consumption by RAM during Stress

2) Disk Utilization

Unlike memory utilization disk utilization will provide more reliable results as disk stress test doesn't require many computations. Hence it prevents the overhead of power consumed by CPU. We can generate 2 process reading and writing 512MB of files to and fro into Hard disk (in our case SD Card). $ stress -d 2 --hdd-bytes 512M [10]

After conducting this test we observed that reading or writing a disk is quite more feasible that loading processes in RAM. It consumed around 0.18 Watt to read and write 512MB file from the SD card.

3) CPU Utilization

The most important analysis of our study is CPU utilization and its corresponding power consumption. CPU is the major component that consumes most of the power during computation, in which a significant amount has been lost as heat. It is difficult to identify the amount of power our CPU consumes for a certain process but here in this study we execute some scripts to help us plot CPU utilization (%) versus power consumption to derive a mathematical model for a particular hardware to compute accurate power consumption during run time. We added the following command in our script which restrict the CPU usage for stress to a certain percentage X.$stress -c 1 & cpulimit -p $(pidof -o $! stress) -l X

We executed this script, it stalled the CPU on a certain CPU usage and then increments it by 1%. Fig 8 shows power consumption increases with increment in CPU utilization. We repeatedly experiment 10 times and then took the mean to obtain final CPU utilization at a certain percentageWe analyzed this data using regression[13] model of Microsoft Excel to derive mathematical equation to compute power consumption on Raspberry Pi.

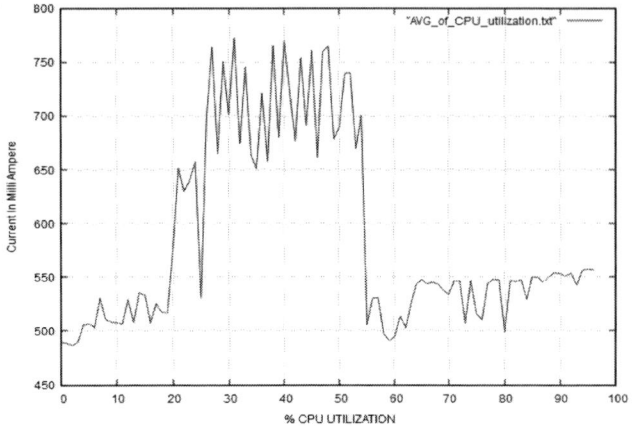

Figure 8 : CPU Utilization vs. Power Consumption

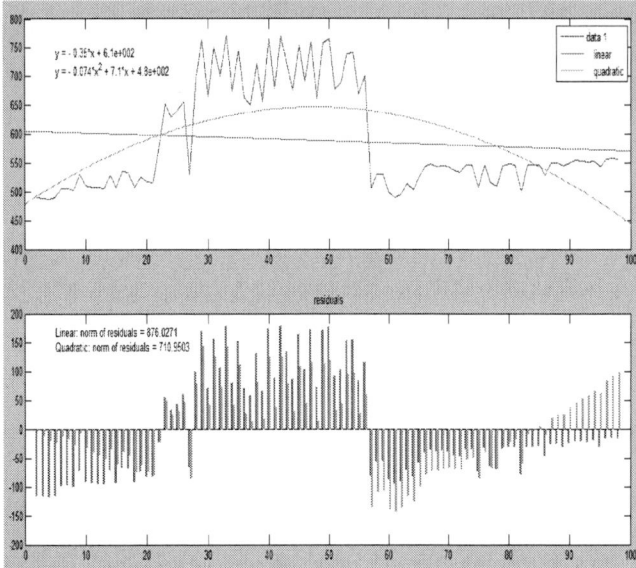

Figure 9: Matlab Based Linear Regression Model to Compute Power

We used Matlab curve fitting toolbox to attain conclusion from our experiments with the help of mathematical modeling. Using the toolbox we generated linear estimation of our model which is as follows.

$$y = -0.348x + 605.3 \qquad (1)$$

Where y is the Power consumed at x percentage of CPU utilization.

Regression is an approach for modeling the relationship between a scalar dependent variable y and one or more explanatory variables denoted by x [14] Using the toolbox we also estimated the error from norm of residual describe the data fit in statistical model[15] . If norm of residual tends towards 0, more is the degree of association between X and Y less is the error. Since the value of norm of residual in our case is 393.63, it determines a good accuracy of our hypothesis to compute power consumption based on CPU utilization. We also analyzed the other regression models to

compare the accuracy as shown in the table VI to compute power consumption .It concludes that polynomial model of degree 10 is the best in computing power from our hypothesis as it has value of norm of residual is minimum 393.63.On further increasing the degree of polynomial model the value remains nearly same. So the best possible mathematical relation to compute power for certain x CPU utilization is as follows.

$$y=(-9.3034e\text{-}014)x^{\wedge}10+(4.612e\text{-}011)x^{\wedge}9+(-9.7553e\text{-}009)x^{\wedge}8+(1.1513e\text{-}006)x^{\wedge}7+(-8.3323e\text{-}005)x^{\wedge}6+(0.0038337)x^{\wedge}5+(-0.11235)\ x^{\wedge}4+(2.0243)\ x^{\wedge}3+(-20.361)\ x^{\wedge}2+(98.658)x+(335.23) \qquad (2)$$

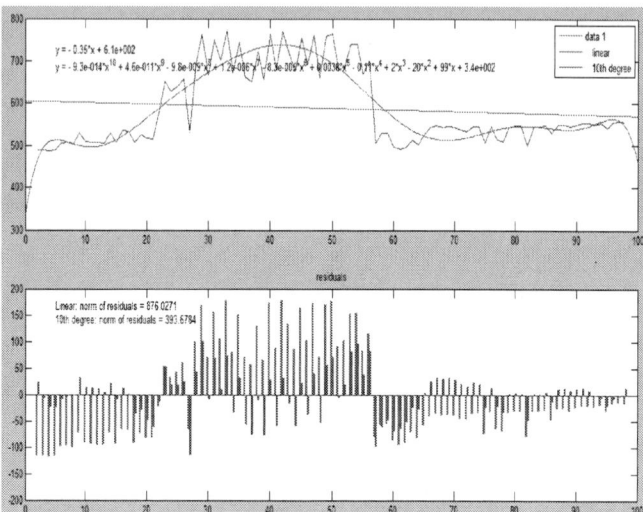

Figure 10: Polynomial with Degree 10 covering all data points

TABLE VI.   REGRESSION MODEL WITH THEIR RELATION AND R² VALUE

Regression Type	Mathematical Relation	Residual
Linear	$y = -0.348x + 605.3$	876.02
Polynomial degree-2	y=-0.074123x^2+7.0642x+478.45	710.95
Polynomial Degree-3	y=0.0024218x^3+-0.43739 x^2+(21.811)x+346.58	580.04
Polynomial Degree-10	y=(-9.3034e-014)x^10+(4.612e-011)x^9+(-9.7553e-009)x^8+(1.1513e-006)x^7+(-8.3323e-005)x^6+(0.0038337)x^5+(-0.11235) x^4+(2.0243) x^3+(-20.361) x^2+(98.658)x+(335.23)	393.68

VI. RESULTS AND CONCLUSIONS

Our test successfully estimated the power consumption by the embedded computer i.e. Raspberry Pi and its peripherals in an intuitive way. In our study , we defined a platform to record power consumption and devised its mathematical model using regression to compute power taking CPU utilization as its metric.

In conclusion , we present the possible area where our platform can be implemented. This platform will help software developers to develop code, operating systems, compilers, graphical and mobile applications with a power efficient approach. It enables us to monitor and compare the power consumed by different software modules during run time.

With the introduction of artificial intelligent Natural Language Processing in mobile applications like Google talk, Siri, Cortana ,audio and video processing power consumption has increases exponentially ,so while developing modules of such power consuming application an extensive testing needs to be done to check power usage. For this purpose our platform will be very handy to use. Its portability to connect with any device like mobile phones, network routers or tablets makes it an effective framework.

VII. REFERENCES

[1] Hindle, "Green mining: Investigating power consumption across versions," in Proceedings, ICSE: NIER Track. IEEE Computer Society, 2012, http://url.ca/84vh4Green Miner

[2] Greens 2014 :GreenMiner: A Hardware Based Mining Software Repositories Software Energy Consumption Framework Abram Hindle, Alex Wilson, Kent Rasmussen, Jed Barlow, Joshua Campbell and Stephen Romansky (University of Alberta, Canada)

[3] http://2014.msrconf.org/program.php

[4] Gupta, T. Zimmermann, C. Bird, N. Naggapan, T. Bhat, and S. Emran. Energy Consumption in Windows Phone. Technical Report MSR-TR-2011-106, Microsoft Research, 2011.

[5] PowerTOP: https://01.org/powertop

[6] Raspberry Pi Foundation Raspbian :a Debian for Raspberry Pi http://www.raspbian.org/

[7] Sphinx .NumPy: http://www.numpy.org 2005

[8] John D. Hunter.Matplotlib:http://matplotlib.org/,Oct 2002

[9] Chris Liechti.PySerial a serial communication python library.http://pyserial.sourceforge.net, Oct2001.

[10] Intex: IT TV 150 FM(USB STICK)

[11] Stress : http://www.stresslinux.org

[12] AngeloMarletta.Cpulimit http://github.com/opsengine ,March 2012

[13] TestSyntex : http://www.linux.com/learn/tutorials/613523:stresslinux-torture-tests-your-hardware

[14] Linear Regression Analysis By George A. F. Seber, Alan J. Lee

[15] R-Squared Measures for Count Data Regression Models With Applications to Health Care Utilization A. Colin Cameron Dept. of Economics University of California Davis CA 95616-8578 USA

[16] http://www.phoronix.com/scan.php?page=article&item=878, Oct2007.

[17] N. Amsel and B. Tomlinson, "Green tracker: a tool for estimating the energy consumption of software," in Proceedings of the 28thof the International Conference of Extended Abstracts on Human factors in Computing Systems, ser. CHI EA '10. New York, NY,USA: ACM, 2010, pp. 3337–3342

[18] MOZISSLE:http://moze.free.fr/blog/index.php?post/2013/02/Underclock-and-Overclock

[19] M. Larabel. Ubuntu's power consumption tested.

[20] POSIX systems.Stress : simple workload generator.

2016 International Conference on VLSI Systems, Architectures, Technology and Applications (VLSI-SATA)

Towards Formal Verification of Adaptive Cruise Controller using SpaceEx

Ambuj Mishra
IIIT Bangalore
ambuj.mishra@iiitb.org

Subir K Roy
IIIT Bangalore
subir@iiitb.ac.in

Abstract—A formal mathematical model of an Adaptive Cruise Controller (ACC)in SpaceEx is presented with a view to formally verify it to ensure its safety critical behavior. SpaceEx (an academic open source tool) is a hybrid systems modeling and verification platform which employs efficient implementation of reachability and safety verification algorithms which are scalable under certain assumptions, to circumvent the difficult problem of formal verification of hybrid systems. In this paper, application of SpaceEx in the comprehensive verification of an Adaptive Cruise Controller for automobiles is presented.

Keywords: Hybrid Systems, Formal Verification, Hybrid Automata

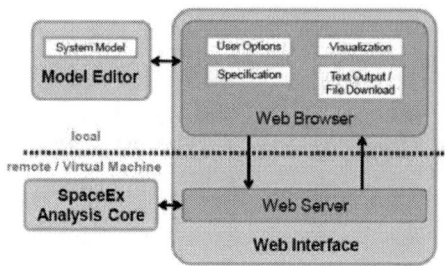

Figure 1. Software Architecture of SpaceEx Platform(source[6])

I. INTRODUCTION

Formal methods have been successfully deployed in the design and verification of digital systems. Hybrid systems are systems with behaviour containing both discrete, as well as, continuous time nature. The ever increasing complexity of real life engineering systems demands design of such systems which ensure safety criticality and fail safe behaviour vis-a-vis human lives. Continuous-time behaviour are modeled by differential equations, while discrete-event dynamics are modeled by automata. Validating the behaviour of such systems with the two different diverse dynamics, interacting and executing concurrently is known to be very hard.

Numerical simulation to validate continuous time domain behaviour while inexpensive, is unable to cover corner case behaviours. To cover these will require the generation of large number of test scenarios. Automated methods based on formal analysis is the best possible way to tackle this problem [1].For the verification of hybrid systems, several tools have been proposed. Some of these tools include d/dt [2], PHAVer [3], Checkmate [4], C2E2 [5], SpaceEx [6]. Commonly used methodologies for representing hybrid systems include timed automata, linear hybrid automata and polyhedral invariant hybrid automata [7].

In this paper we propose a mathematical model for ACC. It is first verified in a simplified setting where the leading vehicle is assumed to be stationary. This simplifies the problem so as to bring it under the capacity of SpaceEx to enable its comprehensive verification over a pre-specified Initial Continuous Set (ICS). Later, we relax this condition by introducing a non-stationary leading vehicle in the ACC model. This renders the system verification task difficult and can stretch the capability of any hybrid systems formal verification tool, to perform

a complete analysis over the same ICS. In such situations we show methods to infer correct behaviour by selectively carrying out verification with respect to a few corner points in the specified ICS, which are taken to completion by SpaceEx.

This paper is organized as follows: In Section II we give a brief overview of the hybrid system verification platform, SpaceEx. Section III gives a brief introduction to adaptive cruise controllers. Section IV describes the proposed model for ACC in SpaceEx for the simplified setting. In Section V, we present details about the verification results obtained from the Simplified ACC system, while in Section VI we discuss the more complex ACC model and show how it can be verified given the limitations of SpaceEx. Finally in Section VII, we briefly summarize the conclusion and gives some pointer towards future work and research directions.

II. SPACEEX

SpaceEx is a platform for modeling hybrid systems with a view to verifying them formally by computing the set of reachable states from the state transition graph model of the hybrid system. It has features to act as a development platform for implementing, exploring and developing newer formal verification algorithms. SpaceEx has three major components

1- **Model Editor**-It is a graphical editor to create complex hybrid system models using nested components. These models are created in the "sx" format supported by SpaceEx.

2- **Analysis Core**- It is a command line program which analyses the model. It takes the sx model file, a configuration file which contains information, such as, ICS, scenario, number of iterations and other options - based on these it carries out the analysis. Output plots are generated after the analysis core has completed its execution.

978-1-5090-0037-1/16 $31.00 © 2016 IEEE

2016 International Conference on VLSI Systems, Architectures, Technology and Applications (VLSI-SATA)

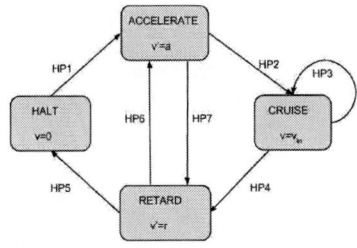

Figure 2. State Transition Graph for ACC

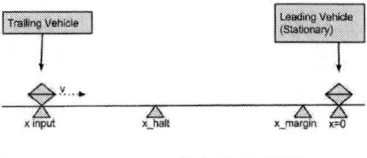

Figure 3. Relative Position of Inputs and Constraints in Simplified ACC Model

3-Web interface- It is a browser based interface to enable a user to specify initial states and other analysis parameters easily. It accesses the analysis core via a web server.

Each verification algorithm implemented in SpaceEx's Analysis Core comes with its own set of representation, applicable to its own class of models and produces different kinds of outputs. In SpaceEx such groupings, or bundles, are called as a scenario [8]. SpaceEx currently supports the following scenarios:

• "The PHAVer scenario implements the basic algorithm from the tool PHAVer. It is applicable to hybrid systems modeled as Linear Hybrid Automata, with piecewise constant bounds on the derivatives."

• "The LGG Support Function scenario implements a variant of the Le Guernic-Girard (LGG) algorithm [9]. It is applicable to hybrid systems with piecewise affine dynamics with non-deterministic inputs."

• "The STC algorithm [10] is an enhancement of the LGG algorithm that produces fewer convex sets for a given accuracy and computes more precise images of discrete transitions."

The above three scenarios were employed to formally verify the proposed ACC model. We present below results for the STC scenario as this gave the most accurate results.

III. ADAPTIVE CRUISE CONTROLLER

Adaptive Cruise Controller (ACC) in a vehicle is an automated highway speed control and monitoring system, enabling vehicles to go through various states of engine control with reference to their respective velocities and accelerations. Behaviour of an ACC system typically consists of the HALT, ACCELERATE, CRUISE and RETARD states (Figure 2). The variable x represents the relative distance between the trailing vehicle and the leading vehicle, while v represents the velocity of the trailing vehicle. The values of these variables at any given time govern the assignment of different dynamics corresponding to different states, besides governing the transition between states in the trailing vehicle. *HP* signifies a hyperplane which separates two states and describes the conditions under which transition takes place.

To construct the hybrid system mathematical model for the ACC system, its dynamic behaviour in each state is defined in terms of differential equations over it system variables as follows : **Halt** $x' = 0, v' = 0$; **Accelerate** $x' = v, v' = a,$; **Cruise** $x' = v, v' = 0$; **Retard** $x' = v, v' = r,$;

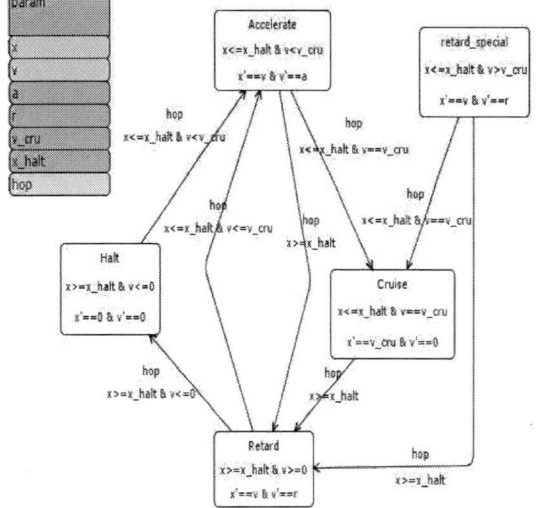

Figure 4. SpaceEx Transition Graph for the proposed Model

IV. ANALYSIS OF SIMPLIFIED MODEL OF ACC IN SPACEEX

We defer our analysis of ACC in SpaceEx for the more general and interesting setting involving the relative motion of two vehicles, viz., the leading vehicle and the trailing vehicle (having the ACC system) to a latter section. Instead, we first analyze the simpler setting, in which we assume the leading vehicle to be stationary(v=0), i.e. acting like a wall - for safety critical behaviour the trailing vehicle needs to stop, while maintaining a minimum specified margin of separation, irrespective of the initial conditions and the constraints assumed for the trailing vehicle, before touching the wall. The position of the stationary leading vehicle's location given by x=0, serves as a reference for all computations needed to implement the ACC behaviour (Figure 3). Figure 4 describes the hybrid automata of the ACC system for the simplified setting as modeled in SpaceEx. Exact interpretations of the different variables are as follows : x = relative distance between the two vehicle; v = velocity of the trailing vehicle; a = acceleration; r = retardation; v_cru = cruise velocity; x_halt = relative distance value at which the retardation action must start;

V. VERIFICATION OF SIMPLIFIED ACC USING SPACEEX

Given an initial separation and an initial velocity of the trailing vehicle, verification of the ACC system essentially

978-1-5090-0037-1/16 $31.00 © 2016 IEEE   97

2016 International Conference on VLSI Systems, Architectures, Technology and Applications (VLSI-SATA)

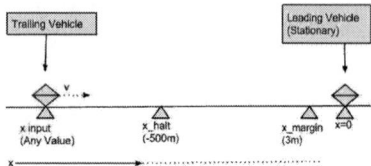

Figure 5. Relative Position of Inputs and Constraints in Simplified ACC Model for chosen practical values

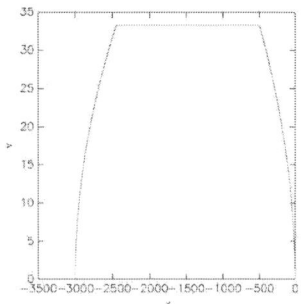

Figure 6. Result with IC- x=-3000m, v=0m/s

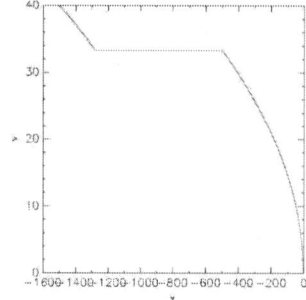

Figure 7. Result with IC- x=-1500m, v=40m/s

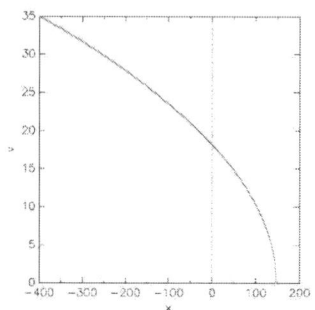

Figure 8. Result with IC- x=-400m, v=35m/s

consists of determining whether it can collide with the leading vehicle. The corresponding non-collision (or safety) condition states that the trailing vehicle should stop before, or at the margin, which is decided a priori (3m in this case). The specified margin is obtained through calculations carried out with respect to the worst possible scenario that ACC is designed to handle, which is the trailing vehicle attaining cruise velocity when the separation distance is 500 meters, for the given rate of deceleration. Different initial conditions are specified with respect to the variables x and v in the simplified ACC model to validate its behaviour. The graphical plots generated by SpaceEx for each set of different initial conditions are described below.

Figure 5 is similar to Figure 3 and is shown along with the initial values and the constraints (or parameters) assumed for the different variables specified in the ACC model. The values shown in the figure are as follows : x_halt = -500 m; v_cru = 33.3 m/s; a = 1 m/s^2; r = -1.12 m/s^2;

In all the plots shown from Figure 6 onwards, corresponding to different initial conditions and constraints, the horizontal axis represents the distance x between the trailing vehicle and the stationary leading vehicle, while the vertical axis represents the speed v of the trailing vehicle with respect to the stationary leading vehicle.

Figure 6, depicts the graphical plot when the input values are kept as x= -3000 m and v= 0 m/s. For these initial values the trailing vehicle starts in the acceleration state, goes to the cruise state after it attains cruise velocity; when it is at a distance of 500m from the leading vehicle it enters the retardation state and finally stops after transiting to the halt state, maintaining the specified safety margin with respect to the leading vehicle; thereby, avoiding any collision.

Figure 7, shows the critical case in which the vehicle starts

with a velocity greater than v_cru and an initial seperation of 1500m. As can be seen from the graph ACC initiates the retardation action immediately to enable the vehicle to attain cruise velocity before the seperation is less than 500m to ensure that no collision takes place.

Figure 8, illustrates the "Failure Case" in which the vehicle starts with a velocity greater than v_cru and an initial seperation of 400m. As can be seen from the graph ACC initiates the retardation action immediately, however with the initial seperation being only 400m it is unable to decelerate the vehicle to attain a velocity much less than the cruise velocity and with an initial seperation less than 500m it cannot stop collision. As expected the trailing vehicle collides with the leading vehicle for the given constant retardation rate (-1.12m/s^2). This is shown in the plot as crossing the vertical dotted line at the position given by x=0. Any trajectory which crosses this vertical line is a failure trajectory which results in a collision.

In Figure 9, we specify the ICS as -2000≤x≤-1000 and 10≤v≤20. This plot clearly shows the collection of trajectories for each and every point in the ICS and none of the trajectories are seen to cross the vertical line located at x=0, thereby implying that collision will never occur between the two vehicles for the given ICS.

Thus, by changing the ICS we can analyze the control algorithm modeled in the ACC system for its safety behaviour. Figure 10 depicts the plot for a less constrained ICS where we can specify the initial velocity to be larger than the cruise velocity. The ICS is specified as -2000≤x≤-400 and 0≤v≤35.

978-1-5090-0037-1/16 $31.00 © 2016 IEEE   98

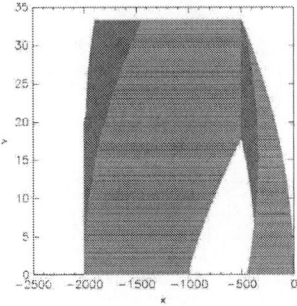

Figure 9. Result with ICS- -2000m≤ x ≤-1000m and 10m/s≤ v ≤20m/s

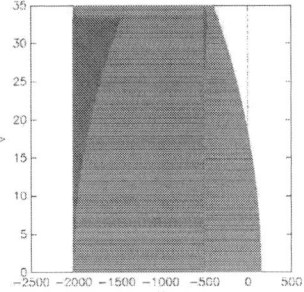

Figure 10. Result with ICS- -500m≤ x ≤0m and 33.3m/s≤ v ≤35m/s

As shown in the plot for initial velocities higher than the cruise velocity (i.e. 33.3≤v≤35) and a set of initial separation between the two vehicles (500≤x≤0) and the given rate of retardation, the control algorithm will not allow for a fail-safe behaviour implying that there are initial conditions which can lead to collisions for the given ACC system parameters. Thus, SpaceEx indirectly can enable us to arrive at different system parameter values (such as higher retardation rates) which can render the ACC system to have a fail-safe control behaviour. In Figure 10 the coloured strip of lines after zero indicates the collection of trajectories corresponding to failure cases which belong to -500≤x≤0 and 33.3≤v≤35 set of initial conditions.

In Figure 4, it can be clearly observed that, since we are using x as the relative distance between the leading and trailing vehicle, the leading stationary vehicle (previously referred as wall too), will not make any considerable difference in the system even if it was not-stationary. The transition from retard state to accelerate state ensures proper functioning in a situation which can only occur if the leading vehicle is non-stationary. Focusing on this point to further check the feasibility of such system we introduce in the next section the ACC system for the more complex setting involving two non-stationary vehicles.

VI. VERIFICATION OF FULL ACC

While a complete analysis has been possible using SpaceEx for the simplified ACC model, we need to strengthen the model to account for more realistic and practical situations, by relaxing our requirement that the leading vehicle is stationary.

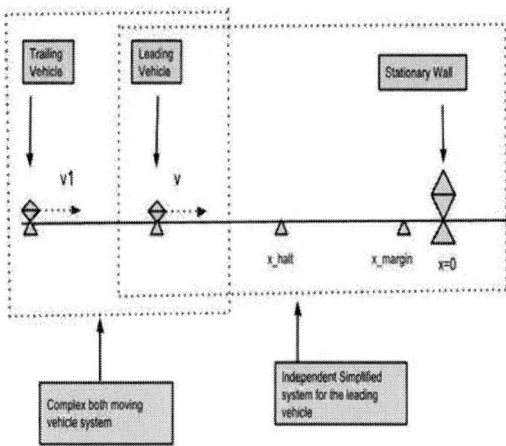

Figure 11. Relative Position of Inputs and Constraints in the Full ACC Model

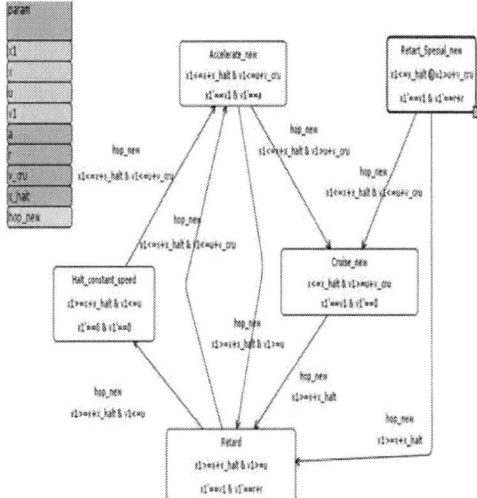

Figure 12. ACC full model SpaceEx Transition Graph

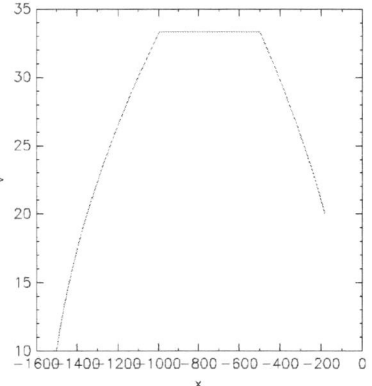

Figure 13. Corner Case-1: x vs v

2016 International Conference on VLSI Systems, Architectures, Technology and Applications (VLSI-SATA)

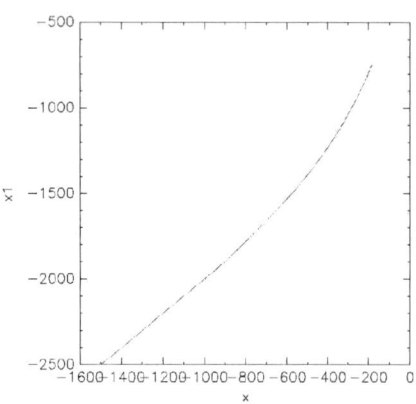

Figure 14. Corner Case-1: x vs x1

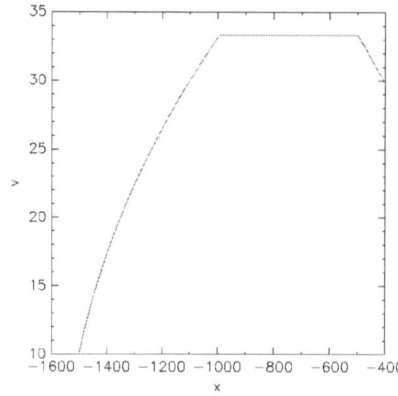

Figure 17. Corner Case-3: x vs v

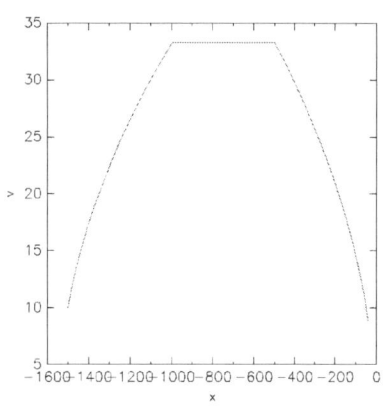

Figure 15. Corner Case-2: x vs v

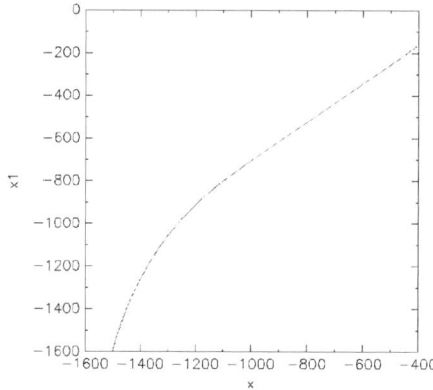

Figure 18. Corner Case-3: x vs x1

In the full model we allow the leading vehicle to be non-stationary. We can view the more general setting as consisting of two separate simplified models, one with respect to the leading vehicle and the other with respect to the trailing vehicle, with a fictitious wall placed in front of both the vehicles at an infinite distance. The objective of the full ACC model is to ensure that the trailing vehicle is always behind the leading vehicle and the distance of separation between them is always more than a pre-specified margin. This is achieved by modifying the simpler ACC model applicable to the trailing vehicle suitably by changing the governing equations with respect to the leading vehicle.

The modified SpaceEx model representing the Full ACC system is shown in Figure 12. The relative position of the trailing vehicle, the leading vehicle and the wall along with the symbolic inputs and constraints needed in the Full ACC Model is shown in Figure 11. To validate the control behaviour of the Full ACC Model we need to run both the models, viz. the modified ACC model for the trailing vehicle and

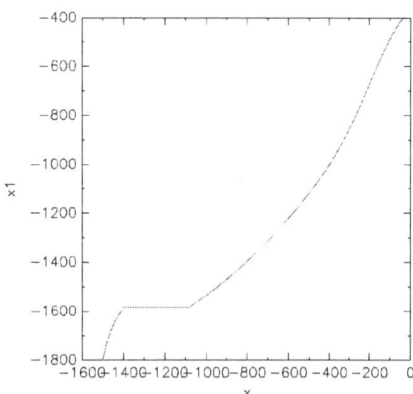

Figure 16. Corner Case-2: x vs x1

978-1-5090-0037-1/16 $31.00 © 2016 IEEE

the simple ACC model for the leading vehicle concurrently in SpaceEx. The following semantic was imposed on the overall behaviour of the two non-stationary vehicle system - the position and velocity values obtained from the leading vehicle are given as inputs to the modified ACC model for the trailing vehicle to compute its separation from the leading vehicle and use this to govern its position and velocity, by transiting to the different states, viz., accelerate, retard, cruise and halt appropriately. Imposing this semantics in SpaceEx turned out to be problematic when a full verification was attempted with an entire ICS.

It also turned out to be problematic for many individual points representing different initial conditions. However, the trajectories were completed for many initial conditions. We report our results and the analysis based on these completed trajectories. On further investigating the inadequacies in the SpaceEx platform, we found that in the STC scenario, the derivatives of x and v cannot be defined in both the simplified ACC_template for the leading vehicle and the modified simple ACC_template for the trailing vehicle. The composition operator in SpaceEx merges these in a way that the result corresponds to neither dynamics.

With respect to Figure 11 the meaning of different variables are as given below: $x(x1)$ = position of the leading(trailing) vehicle; $v(v1)$ = velocity of the leading(trailing) vehicle; For the system to be successful i.e. avoid collision x-x1$\geq$0 i.e. x-x1 should be positive all the time. From Figure 13 onwards, we present the different trajectories or plots resulting from the full ACC model for different points in the ICS.

Corner Case-1: x=-1500m, v=10m/s, x1=-2500m, v1=10m/s

Figure 13 and 14 are the graphs obtained after running the system with the conditions given under Corner Case-1. Figure 13 shows that the leading vehicle and wall system function properly (given the tool limitation), till x=-200m. In Figure 14 we plot the distance of the leading vehicle with respect to the wall versus the distance of the trailing vehicle with respect to the wall. It can be seen in Figure 14 that for the entire plot *x-x1* is always positive, thereby, proving that no collision will take place for the given initial conditions, respectively, for the two simplified ACC models.

Corner Case-2: x=-1500m, v=10m/s, x1=-1800m, v1=40m/s

Figure 15 and 16, give the results under corner case 2, in which the value of position of trailing vehicle is reduced to -1800m. From Figure 16 it is seen that no collision can take place for this set of initial conditions.

Corner Case-3(The failure Case): x=-1500m, v=10m/s, x1=-1600m, v1=50m/s

Figure 17 and 18, give the result under corner case 3, in which the value of position of trailing vehicle is reduced to -1600 and its velocity is increased to 50m/s. Since the initial seperation between the two vehicle is only 100m and the relative velocity is higher for the trailing vehicle with its retardation rate being constant, collision occurs as is evident from the trajectory in Figure 18 showing that *x-x1* becomes

negative after some time elapse, indicating collision takes place.

VII. CONCLUSION AND FUTURE WORK

In this paper we proposed a new approach to the verification of a complex hybrid system such as the Adaptive Cruise Controller. The approach also illustrates the use of SpaceEx to formally validate the fail-safe behavior of the ACC system. This work also points out some of the limitations in SpaceEx. Future work will be in the direction of making the implementation of PhaVer and STC scenarios more robust as they are currently inadequate in being able to handle compositions of different sub-system models from a causality perspective.

REFERENCES

[1] Lata K., Roy S.K. and Jamadagni H.S, "Towards formal verification of analog mixed signal designs using SPICE circuit simulation traces", Quality Electronic Design, 2009. ASQED 2009. 1st Asia Symposium on, July 2009.

[2] Eugene A, Thao D, and Maler, O, "The d/dt Tool for Verification of Hybrid Systems", CAV'02 - Computer Aided Verification, 2002, Copenhagen, Denmark, July 2002, 365-370, LNCS 2404. http://www-verimag.imag.fr/ tdang/ddt.html .

[3] Goran Frehse, "PHAVer algorithmic verification of hybrid systems past HyTech", STTT, 10(3) 263235, 1994.

[4] CMU CheckMate website, http://users.ece.cmu.edu/ krogh/checkmate .

[5] University of Illinois C2E2 website, http://publish.illinois.edu/c2e2-tool.

[6] Verimag SpaceEx website, http://spaceex.imag.fr .

[7] Alur R., Kanna S., La Torre S, "Polyhedral flows in hybrid automata", Hybrid Systems: Computation and Control, pp 5-18, April 1999.

[8] Goran Frehse, "An Introduction to SpaceEx v0.8",2010.

[9] Colas Le Guernic and Antoine Girard, "Reachability analysis of linear systems using support functions. Nonlinear Analysis: Hybrid Systems", 4(2):250 262, 2010. IFAC World Congress 2008.

[10] Goran Frehse, "A Brief Experimental Comparison of the STC and LGG Analysis Algorithms in SpaceEx", 2012.

2016 International Conference on VLSI Systems, Architectures, Technology and Applications (VLSI-SATA)

Efficient Network on Chip (NoC) using heterogeneous circuit switched routers

Anuja Naik
Dept. of Electronics and communication
Amrita Vishwavidyapeetham, Bangalore, India.
naikanuja@gmail.com

Dr. Tirumale K. Ramesh
Advanced Computing Consultant, Virginia, USA
(Formerly with Computer Science and Engineering, Vishwa
Vidyapeetham, Bangalore, India)

Abstract—**Network-on-Chip (NoC) architecture in recent years has been considered as the overwhelming communication solution to provide scalability in multi core systems over traditional bus-based communication architecture. There is an increased use of multi-core with NoC in embedded systems solutions. Energy efficiency in the Network-on-Chip (NoC) is one of the key challenges as these embedded systems are typically battery-powered. Router architecture impacts the performance of NoC. With increased interest towards circuit-switched routers, in this paper, we have proposed an area and energy efficient 5-port, 4 Lane circuit switched router using a CLOS network which presents the advantages of area and energy efficiency. To further improve energy efficiency of NoC, we use a hybrid architecture by mixing buffered and *bufferless* routers. Our results shows that by using CLOS switch network, we can gain 32% reduction in area and 26% reduction in power compared to Crossbar switch of the same size. Our comparison of using an 8×8 mesh heterogeneous router topology shows a reduction of 3% to 18% in silicon area and 10% to 15% in total power using *bufferless* routers compared to a fully buffered configuration.**

Keywords- NoC, circuit switching, clos, crossbar, heterogeneous, bufferless routers

I. INTRODUCTION

With increase in Very Large Scale Integration (VLSI) density, it is now possible to integrate general purpose processors, memory blocks, application specific intellectual property blocks (IP), digital signal processor (DSP), Graphic processor unit (GPU) and mixed signal functions on a single system-on-chip (SoC). Low power high performance embedded computing systems are major example of the use of system on chip (SoC). With more applications that require battery powered embedded system units, the energy and area efficiency of the SoC is a very important factor.

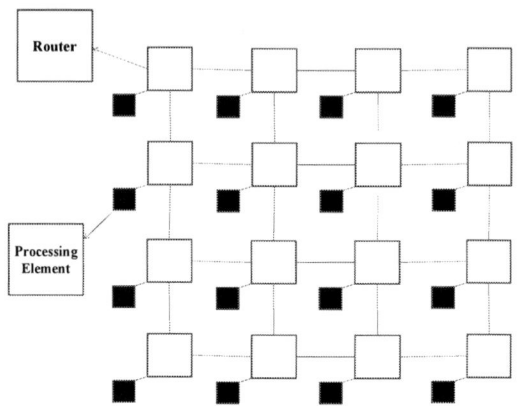

Figure 1. Mesh topology of Network-on-Chip (NoC)

Initially SoC was made up of large number of intellectual property (IP) cores and memory blocks connected to each other using single or multiple layers of shared buses for communication providing dedicated point-to-point connections in SoC. They offer some advantages like simple topology and low area and cost [1] but prevent the system to meet desired performance in terms of bandwidth, latency (due to (non-predictable wire delays), non-scalability and power to mention a few [2]. As today's application requires higher degree of parallelism, the communication becomes a major issue within a chip. Network-on-chip (NoC) though not considered as a new on-chip communication solution today, it continues to attract as a viable communication solution. With comparatively larger area than that of bus based system, the advantage of NoC is that it can provide fault tolerance, shorter switch-to-switch links, energy efficiency reliability, scalability and reusability [2]. Router is a major component of NoC. There are generally two types of routers: *circuit switching* router and *packet switching* router. In the circuit switching router, a path from source to destination is established before the transmission of data and it cannot be allocated to other resources till the desired data transmission is completed [3]. A disadvantage of circuit switched router is that there is a bit delay in the router due to connection time. However, there are many advantages of this router which will be briefly discussed in following sections. In comparison, packet switching does not require setting up a path before data transmission. It depends on buffering schemes and flow control schemes to

978-1-5090-0037-1/16 $31.00 © 2016 IEEE

ensure data transfer completion where data is transmitted in packets through network independently and they are combined together at destination. Packet switching NoCs are not new and have been studied intensively over the years by many researchers.

In this paper, we focus on *circuit switching* NoC as we believe that it offers many advantages compared to packet switched NoC. Many earlier router solution based on Crossbar switch fabric [6] requires more energy. Some earlier work [4] suggested that CLOS networks are known to perform well in terms of area and power compared to *Crossbar* network. So, our first attempt is to demonstrate energy efficiency advantage by combining circuit switched router in CLOS network topology. In addition, we also included a mix of buffered and buffer less routers termed as a *heterogeneous routers* in a single topology to gain higher energy efficiency.

Our results show that this combined configuration provides significant reduction in silicon area and power. The rest of the paper is organized as follows: in section II, we provide some analysis on the related work. Section III provides overview of NoC and *circuit switched* router and CLOS switching fabric. In section IV, we provide our heterogeneous router approach mixing *buffered* and *bufferless* routers. In section V, we presented the implementation and discussed our results. Finally in section VI, we provide the conclusions and future research work.

II. RELATED WORK

Packet switching used for majority of the current router implementations. But, there are many benefits of circuit switching routers. Authors in [3] as noted earlier have investigated the merits of circuit switching in NoC. Packet switching requires significant buffering adding to silicon area and power. Some of the merits of circuit switching consist of easier implementation, minimal amount of control, energy efficiency, higher throughput, no buffering schemes etc. Authors in [3] concluded that the average transmission time for circuit switching NoC is slightly higher than packet switching NoC due to added circuit set up time prior to data transmission. With proper optimization, this time can be reduced. For comparatively smaller packet size, it is noticed that circuit switching NoC gives higher performance.

Crossbar versus CLOS

Authors in paper [4] presented a survey on various single stage (Crossbar) and multi stage (CLOS and Benes networks) non-blocking interconnection networks. It is noticeable that multistage networks Benes and Clos network uses smaller sized Crossbar switches. Benes network has lower transistor count and lower area but the delay is longer compared to Crossbar. Further, authors in [4] proposed CLOS network with larger sized switches. Their study show lower timing delay, smaller area and power consumption compared to Crossbar in large scale network.

Circuit Switching versus Packet Switching

Authors in [5] conducted an evaluation and analysis of circuit switched NoC and packet switched NoC. They first considered the effect of packet size on performance of NoC. It is noticeable that circuit switching is relatively independent of packet size. The authors in [5] concluded that for a larger packet size, it is preferable to use circuit switched NoC. This is primarily due to congestion penalty in packet switched NoC for large packet sizes authors in [5] also conclude that circuit switched NoC can operate at higher frequency than packet switched router. The authors further stated that the cross-over point above which the latency of circuit switched router is lower.

Energy Efficiency in Circuit Switched Router

Authors in [6] proposed an energy efficient reconfigurable circuit switched router design using the lane division multiplexing scheme (LDMS). By exploiting lane division multiplexing, the authors reported a reduction of overall energy consumption by factor of 3.5 compared to packet switching. By using lane division multiplexing scheme, they were able to remove *blocking* of a connection between two router ports and able to move multiple data simultaneously. In general, Crossbar switch consumes maximum power and area compared to other components. To improve efficiency of circuit switched router in terms of throughput, authors in [7] proposed an implementation of a *pipelined based circuit switched router* by adding pipeline registers in various stages. Due to reduction in critical path, it offers higher throughput but, it increases the area and power.

III. NOC ROUTER ARCHITECTURE

As a result of our understanding of many combined advantages as we discussed in the section II, we propose a *circuit switched* router using a CLOS network to gain area and energy efficiency. The following subsections describe some salient features of our architecture.

A. Circuit switching

Switching describes how a data is transmitted from source to destination node. A fixed amount of time is blocked for all intermediate nodes until the data originated from the source reaches its destination. This is similar to a telephone switching system used in earlier days. The payload is sent only after the entire path has been reserved. The path remains connected and available during the entire communication. Once the data transmission is completed, the path is released and used by another data. This method however will increase the delay, but once the path is decided, it can give guaranteed throughput and quality of service (QoS). The method particularly very useful for streaming based applications where the circuit set up time tends to be much larger than the data transmission time. Authors in [6] conducted extensive research on the suitable applications for circuit switching like HiperLAN/2, UMTS(Universal Mobile Telecommunications System) and DRM (Digital Radio Monodial) and observed that the amount of expected guaranteed throughput traffic is much higher in these applications.

B. Switching fabric inside router

For any type of router, the switching fabric defines how inputs are connected to the outputs. A switching network is said to be strictly *non-blocking* if it can handle all circuit request from input to output simultaneously without the rearrangement of already set up circuits. Crossbar network and CLOS network is an example of *non-blocking* scheme. A switching network is said to be 'rearrangeably non-blocking' if it can connect any unconnected input to any unconnected output by re-arranging existing connections. Benes network [8] is an example of this type.

An $m \times n$ Crossbar switch shown in Figure 2 connects m input channels to n outputs directly without any intermediate stage. It is a single stage network using either multiplexers or tri-state gates or wired OR gates. To construct a $m \times n$ Crossbar, we need n number of $m \times 1$ multiplexers. One multiplexer for one output [9]. One major drawback of cross bar is scalability. It is economical only for relatively smaller network size. As the network size increases, the cost for $n \times n$ grows as $O(n^2)$. The number of cross points increases drastically.

To make switching fabric of a router more cost effective, we propose a CLOS network based solution that consists of multiple small sized Crossbar switches. It is a 3-stage network characterized by (m, n, r) where m is the number of middle stage switches, n is the number of input ports on each input switches and r is the number of input and output switches. Figure 3 shows a CLOS network.

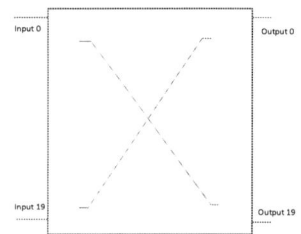

Figure 2.   Crossbar network of size 20×20

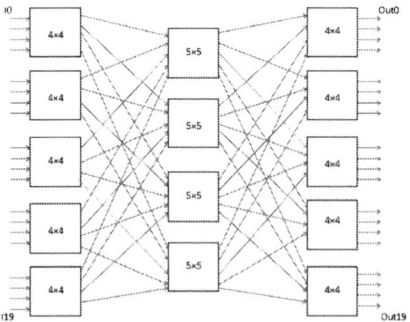

Figure 3. CLOS network of the size 20×20

The first (input) stage in CLOS network consists of total 5 Crossbar switches of the size 4×4. The 5 switches will provide total 20 input channels. The second stage of CLOS consists of total 4 Crossbar switches of the size 5×5. The third (output) stage is same as the first stage. It has 5 Crossbar switches of the size 4 × 4. The third stage will provide required 20 output channels. In summary, a 3-stage CLOS switch will have 20 inputs and 20 outputs.

IV. PROPOSED HERTEROGENOUS ROUTER APPROACH

A. Circuit switched router

We propose the concept of lane division multiplexing (LDM) suggested by earlier work [6]. Using LDM, a single port is segmented into smaller sets of bus which can be used by different data streams simultaneously. Our implementation terms a router as *R(5,4)* consist of 5 ports where each port is divided into 4 lanes of equal size in one direction. For example, the router shown in Figure 4 has eight lanes per port with four incoming and four outgoing lanes. We recognize switching network inside a router consumes major silicon area. To reduce this silicon area and power dissipation, we propose using multistage CLOS network where a single CLOS switch is made up of multiple small Crossbar switches.

To simplify our design, we allocated each small Crossbar of the first stage to different ports – North, South, East, West and Tile. Figure 5 shows an example where each Crossbar can carry simultaneously 4 inputs from one direction and send them to 4 different directions except the one coming from the tile. We assume that data will not backtrack to the same router from where it arrived from.

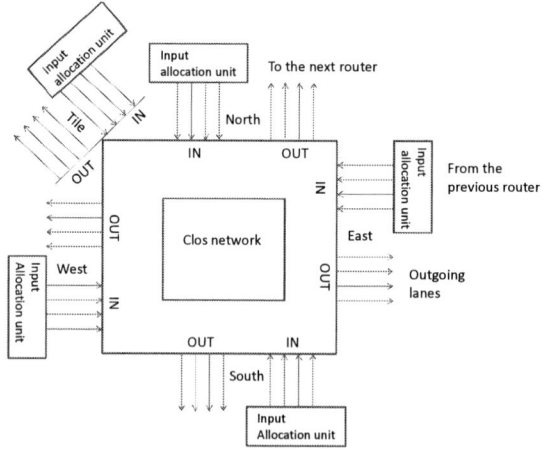

Figure 4. R(5,4) Router Configuration

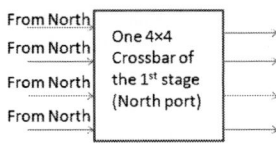

Figure 5. A single unit of small Crossbar from Multistage CLOS network for illustration purpose

Input allocation unit:

Input allocation unit checks the incoming four data arriving to a router port and allocate appropriate lane to each of them. The allocation depends upon the destined direction. (i.e. data destined to go in the South will be allocated to that particular lane). The unit stores incoming data in the temporary buffers till the routing decision takes place. The allocation algorithm checks for the destination address of each input and then send it to appropriate lane and set the flag "high" for that channel. A higher flag for the lane suggests that the particular lane reserved to transmit the message cannot be used by any other data until the previous transmission is completed. Flags for all the lanes cleared every time before initiating a data transfer. Once all four input channels of a single port are allocated, data can travel to the desired output ports simultaneously.

B. Heterogeneous NoC

Many researchers consider heterogeneous NoC mean differently. For example, authors in [10] implemented integrating *bufferless* switches with pipeline channels to construct on-chip network that behaves as packet switching if message has only one packet and as circuit switching for multiple packets. With a 2-depth channels, the authors in [10] stated improvement in area and power but at the cost of slight reduction in the throughput. To reduce the silicon area, many *bufferless* router schemes have been proposed [11],[12] Removing buffers from the router entirely will reduce the cost of it but may reduce the performance reduce due to deflection and re-transmitting of packets. In paper [13], authors proposed mix of *buffered* and *bufferless* router.

To further improve gain energy efficiency, in our *circuit switched* implementation, we included a mix of *buffered* and *bufferless* router. To make room to use the heterogeneous router concept in a circuit switched router which normally does not have input and output First in First out (FIFO) data storage buffers, we store incoming data temporarily before routing decision takes place. We have arbitrarily removed some input data buffers and replaced with *bufferless* routers. As we cannot store data, we send them to the nearest buffered router and initiate further transmission of the data arriving to its neighbor. However, we noticed an increase the latency as expected because of minimum extra mandatory *hop* needed and maximum more than one *hop*. But, using heterogeneous router reduced the silicon area for the overall network. We have implemented three different schemes of *buffered* and *bufferless circuit switched* routers in mesh topology and compared with all buffered routers.

V. IMPLEMENTATION AND RESULTS

We designed *circuit switched* routers using CLOS network as switching fabric using Verilog and implemented using 180 nm technology library of Cadence RTL compiler for the synthesis. Using those routers we have constructed an 8 × 8 mesh topology for all *buffered* routers shown in Figure 6. We made the mix of buffered and *bufferless* routers in 8×8 mesh topology as shown in the Figure 7. Figure 6 and Figure 7 are module level view in which each square represents a single router configuration as shown in Figure 4. Topology A represents a case of 8 *bufferless* routers. Topology B represents a case with 16 *bufferless* routers and topology C represents a case with 32 *bufferless* routers.

The result of comparison between CLOS switch and Crossbar switch of the same size are shown in Table I. From the results we conclude that CLOS network of the same size gives 32% reduction in area and 26.6% reduction in dynamic power as compared to a cross bar network of the same size. Table II shows the comparison of circuit switch router implementation using both Crossbar and CLOS network for the same size.

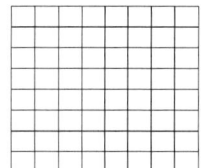

Figure 6. An 8×8 Mesh topology using all the buffered routers

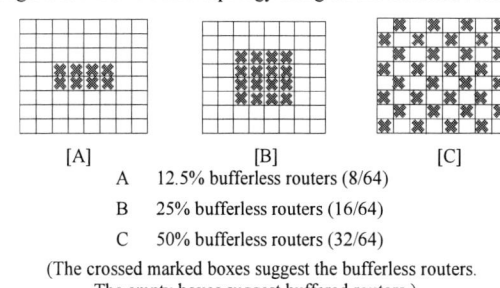

[A]	[B]	[C]

A   12.5% bufferless routers (8/64)

B   25% bufferless routers (16/64)

C   50% bufferless routers (32/64)

(The crossed marked boxes suggest the bufferless routers. The empty boxes suggest buffered routers.)

Figure 7. Various plans for heterogeneous NoC.

TABLE I.
COMPARISON RESULTS OF CROSSBAR AND CLOS SWITCH

Size 20×20	Total Area	Total Power
Crossbar switch	26279 µm²	1099.94 µw
CLOS Switch	17696 µm²	806.54 µw

TABLE II.
COMPARISON RESULTS OF CS ROUTER USING CLOS AND CROSSBAR

Circuit Switching	Total Area	Total Power
CS Router (Crossbar)	38906 μm^2	9275.6 μw
CS Router (CLOS)	30540 μm^2	8027.7 μw

Our results shown in Table II indicate about 20 % reduction in area and about 13 % reduction in power consumption. Using CLOS switch in the routers, the results of mesh topology is shown in Table III. The comparison of all buffered router topology and mixed topologies suggests that removing some gives definite advantage in silicon area and dynamic power. But, it affects the latency because of the added mandatory extra *hop* for each message from *bufferless* router. The latency results are shown Table III. We have used random traffic pattern to calculate latency. We have used the same source and destination for a data transmission in all the types of topologies to compare the time it takes to reach destination.

Table IV shows that by reducing the number of buffered routers in topology gives gradual decrease in total power and power dissipation.

TABLE III.
LATENCY RESULTS OF VARIOUS MESH TOPOLOGY CASES

8×8 mesh topology	Average latency
All buffered.	925 ns
A	1295 ns
B	1514 ns
C	1795 ns

TABLE IV.
AREA AND POWER RESULTS OF VARIOUS MESH TOPOLOGY CASES

8×8 mesh topology	Area	Total Power
All buffered	1805211 μm^2	440.54 mw
A	1744264 μm^2	393.11 mw
B	1622485 μm^2	346.24 mw
C	1474653 μm^2	245.71 mw

The *bufferless* router represents a module which cannot take routing decision on its own, it passes the data immediately to the nearest buffered routers. We can easily predict that as the number of *bufferless* routers increase in any topology, the routing performance degrades even if area and power improves greatly. In the topology C shown in Figure 7, the number of *buffeless* routers is more. It is best in terms of area and power performance. Also the proximity to buffered routers is very less but there will be always more than one

closest buffered router for each *bufferless* router. So it can pass data go to any one of them. Hence the chances of data taking longer route to reach destination is extremely high in this case.

Topology B in Figure 7 gives good performance in area and power with slight effect on latency compared to fully buffered topology.

VI. CONCLUSION AND FUTURE RESEARCH

We proposed a circuit switching based NoC as compared to traditional packet switching based NoC due to some significant advantages of guaranteed throughput. We showed an implementation of an efficient 5-port circuit switched router within a multi-stage CLOS network. Our results shows that the use of this CLOS switch network produces 32% reduction in area and 26% reduction in power compared to Crossbar switch of the same size. The entire Circuit switched CLOS network NoC using an 8×8 mesh heterogeneous router topology shows a reduction of 3% to 18% in silicon area and 10% to 15% reduction in total power using *bufferless* routers compared to a fully buffered configuration.

As single chip energy efficient heterogeneous computing is drawing wide attention today, we believe that developing energy efficient NoC is very important. Further research includes addressing a general graph theoretical model for optimum configuration for buffered and *bufferless* router placements for mesh and other regular topologies. For future work, our objective is to extend the router architecture for NoC communication for single chip heterogeneous computing solutions assess overall energy efficiency.

REFERENCES

[1] R. Kamal and N. Yadav, "NoC and bus architecture: a comparison", in International Journal of Engineering Science and Technology, Vol. 4, No 04, April 2012.

[2] É. Cota, De Morai Amory, and S. Lubaszewski M, "Reliability, Availability and Serviceability of Networks-on-Chip", Springer Science + Business Media, LLC 2012.

[3] N. Chin-Ee and N. Soin, "A study on circuit switching merits in the design of Network-on-Chip", in International conference on Computer and communication engineering, Kuala Lumpur, Malasia, May 2010.

[4] Y. Jiang and M. Yang," On circuit design of on-chip-non-blocking interconnection networks", in System-on-Chip conference (SOCC), 27th IEEE international, September 2014.

[5] S. Liu, A. Jantesch, and Z. Lu, "Analysis and evalution of circuit switched NoC and Packet switched NoC", in the 16th Euromicro conference on Digital System Design, September 2013.

[6] P. T. Wolkotte, Gerard J.M. Smit, G. K. Rauwerda, and L. T. Smit, "An energy efficient reconfigurable circuit-switched Network-on-Chip", in the 19th IEEE International Parallel and Distributed Processing Symposium, April 2005.

[7] U. Saravanakumar, R. Rangarajan, and K. Rajasekar, "Hardware implementation of pipeline based router design for on-chip network", in ICTACT Journal on Communication Technology 3

(4), 646 – 650, December 2012.

[8] C. Hau Hoo and A. Kumar, "An area efficient partially reconfigurable crossbar switch with low reconfiguration delay", in Field Programmable Logic and Applications (FPL), 22nd International Conference, IEEE, August 2012.

[9] William James Dally and Brian Towels, "Principles and practices of Interconnection networks", Morgan Kuffman Publishers, 2004.

[10] H. Luo, S. Wei, D. Chen, and D. Guo, "Hybrid Circuit switched NoC for low cost on-chip communication", Proceedings of IEEE International Conference on Anti-Counterfeiting, Security and Identification, August 2012.

[11] T. Moscibroda and O. Mutlu, "A case for bufferless routing in on-chip networks", in International Symposium on Computer Architecture-36, June 2009.

[12] M. Hayenga, N. Jerger, and M. Lipasti, "Scrab: A single cycle adaptive routing and bufferless network", in Microarchitecture MICRO – 42, November 2009.

[13] H. Zhao, M. Kandemir, W. Ding, and M. J. Irwin, "Exploring Heterogeneous NoC design space", in proceedings of the international conference on Computer Aided-design, IEEE, November 2011.

A Circuit Technique for Leakage Power reduction in CMOS VLSI Circuits

Venkata Ramakrishna Nandyala* and Kamala Kanta Mahapatra
Electronics & Communication Engineering Department
National Institute of Technology, Rourkela
Rourkela-769008, Odisha, India
*e-mail: 512ec107@nitrkl.ac.in

Abstract— Scaling of CMOS technology improved the speed nevertheless the leakage currents are leftover as an adverse effect. The problem has taken a serious turn as the scaling extends into ultra-deep-submicron (UDSM) region. These unsolicited leakage currents should be minimized for the smooth functioning of the circuit. Designing of such leakage free nanoscale CMOS circuits turns to be a challenging task. In this work, we address the issue of leakage power that arises with the device channel length scaling to sub-100nm. We present a circuit technique to mitigate the leakage currents of MOSFET through controlling the voltage at the source terminal of the MOSFET. CMOS inverter designed using the proposed technique results in 98% and 30% improvement in static and total power dissipation respectively compared with its conventional design. The simulation results of NAND and NOR gates designed using the same technique indicates 15.89% and 18.83% improvement in the total power compared with their corresponding conventional designs. 11-stage CMOS ring oscillator designed using the proposed technique is analyzed, and corresponding simulation results are reported. Comparison of the proposed circuits in terms of power dissipation and delay with two existing techniques is presented. The circuits designed using the proposed technique results in good Power-Delay Product (PDP).

Keywords—CMOS; UDSM; leakage power; CMOS inverter; low power dissipation.

I. INTRODUCTION

With the rapid technological growth in the semiconductor industry, the high computational and even complex applications are being implemented in a small size VLSI chip with the use of Complementary Metal Oxide Semiconductor (CMOS) technology. Fortunately, the growth in the semiconductor technology is capable of providing required feature size. With the utilization of each new technology node, the speed of the Integrated Circuit (IC) has increased by 30% roughly [1]. The Requirement of high density chips and high speed systems made MOS devices to scale to smaller dimensions that increased the current drive (g_m) capability. These smaller transistors and shorter interconnects results in less capacitance and altogether increased the speed of the integrated circuit [1]. Nonetheless, the extent of scaling is constrained by physical limitations such as short-channel effects. The main consequences are the leakage currents contributing to massive static power dissipation. The leakage current increases with the scaling of device channel length.

With the overall effect, power dissipation has become the critical issue in the design of microelectronic circuits. Enormous efforts have been paid [2][3][4] and need to do more to mitigate these leakage currents. Few of the efforts for reducing leakage currents include techniques, MTCMOS Power Gating, Super Cutoff CMOS Circuit, Forced Transistor Stacking and Sleepy Stack. Multi threshold CMOS (MTCMOS) inserts extra transistor(s) either PMOS/NOMS or both called sleep transistor(s) into the design. During the normal mode of operation, these transistors set to "*on*" state without disturbing the functionality of the circuit. During the standby mode, these transistors switched to "*off*" state to isolate the power supply from the circuit. The isolated supply voltage causes the leakage currents to minimize [2]. Nevertheless, this technique increases the dynamic power dissipation of the circuit. Super Cutoff CMOS, an alternate to MTCMOS uses low threshold voltage sleep transistors instead of high threshold voltage transistors. It turns off sleep transistors with a small negative voltage to reduce the subthreshold leakage current leaving a difficulty of designing of the controller circuit to generate negative gate voltage for the sleep transistors [2].

Stacking is another technique that uses series connected transistors to reduce subthreshold leakage currents. If the natural stacking of transistors does not exist in the circuit, then one can achieve stacking effect by replacing a single transistor with two transistors in the design called forced stacking [2]. This technique works efficiently when more than one transistor in the stack of series connected transistors are in "*off*" state [3]. Due to extra transistors, this technique results in delay penalty. Sleepy stack is another method that can reduce this delay by adding additional transistor in parallel to the additional stack transistor. However, sleepy stack approach results in higher dynamic power consumption, and it needs an extra complex sleep signal circuitry [2]. LECTOR [5] is a circuit technique for achieving low power dissipation. It inserts two extra transistors in the circuit for achieving low leakage design. Complex circuit techniques have to be developed to get power savings as technology scaling itself doesn't provide enough savings [6].

In this paper, we present a circuit technique to mitigate the leakage currents of sub 100-nm CMOS VLSI circuits by using two extra transistors.

978-1-5090-0037-1/16 $31.00 © 2016 IEEE

The rest of the paper is organized as follows. Section II describes leakage power dissipation in CMOS VLSI circuits and its importance. Section III describes our proposed technique. Section IV describes the results followed by the conclusion in Section V.

II. LEAKAGE POWER DISSIPATION

Leakage or static power dissipation occurs due to the presence of various leakage currents such as subthreshold leakage, reverse-bias source/drain junction leakages, gate oxide tunneling leakage, Gate Induced Drain Leakage (GIDL) [3][4]. Subthreshold leakage current is the dominant leakage current in sub-100nm circuits [2]. Even when the gate voltage is less than the threshold voltage of the device, the current flows between the drain and source due to the diffusion of minority carriers. The subthreshold leakage current, I_{sub} can be expressed as [7]

$$ I_{sub} = I_{D0} \times e^{\frac{V_G}{mV_T}} \times (e^{-\frac{V_S}{V_T}} - e^{-\frac{V_D}{V_T}}) \qquad (1) $$

Subthreshold leakage current is due to the diffusion of minority carriers in the channel of MOS transistor [4]. This current depends on temperature, size of the device, supply voltage and process parameters [4]. For the desired speed of operation, the supply voltage and threshold voltage should be minimized as the CMOS technology subjected to scales down. However, subthreshold current increases exponentially with the decrease in the threshold voltage.

The power dissipation of Intel process technologies are shown in Fig. 1 [8]. In 1μm technology the leakage power is only about 0.01% of the active power. But as the technology scales down to 100 nm, the leakage power contribution raised and became 10 % of the active power dissipation. It means that leakage currents are increasing with the scaling of channel length. This has motivated many researchers to work on it, and many innovations have been proposed to circumvent the leakage power problem.

III. PROPOSED TECHNIQUE

The basic idea behind our proposed approach is to raise the voltage at the source terminal of the MOS transistor to reduce the leakage currents so as to minimize the static power dissipation. It is known that the subthreshold current increases exponentially with the decrease in the threshold voltage. And it is observed that the threshold voltage of a short channel MOSFET reduces with increase in the drain to source voltage (V_{DS}) [3]. It implies more subthreshold current occurs at higher drain to source voltage. With reference to the subthreshold current equation (1), the subthreshold current can be minimized significantly by decreasing drain to source voltage (V_{DS}). Considering these two facts, subthreshold current can be minimized by reducing the drain to source voltage. Drain to source voltage can be reduced by raising the voltage at the source terminal of the MOSFET [9] called source biasing. For better understanding of the approach, we consider CMOS inverter circuit. The corresponding circuit designed using the proposed approach is shown in Fig. 2 along with the conventional circuit.

Here, we have used two extra transistors labeled PM2 (PMOS) and NM2 (NMOS) for the purpose of raising the voltage at the source terminal of the MOSFET. Transistor PM2 is configured to work in "*cut off*" mode (source and gate connected) while the transistor NM2 follows the circuit input conditions. The transistor PM2 is connected between the output node of the circuit and the source terminal of the transistor NM1. The purpose of the transistor PM2 is to supply the leakage currents to the source terminal of the upper NMOS transistor (NM1) to charge the node (source terminal of NM1). During the logic '0' condition at the input, the two NMOS transistors NM1 and NM2 turned *off* and a logic '1' appears at the output node. As the extra PMOS transistor PM2 set to "*cut off*" state, it supplies leakage currents and establishes a certain voltage at the source terminal of the upper NMOS transistor NM2 depending on the amount of leakage current provided by the transistor PM2.

The increased source voltage (V_S) of the upper NMOS transistor (NM1) reduces the drain-to-source voltage (V_{DS}) of the transistor NM1 and then the subthreshold leakage current. The dependency of node voltage on the width of PMOS transistor PM2 is depicted in Fig. 3.

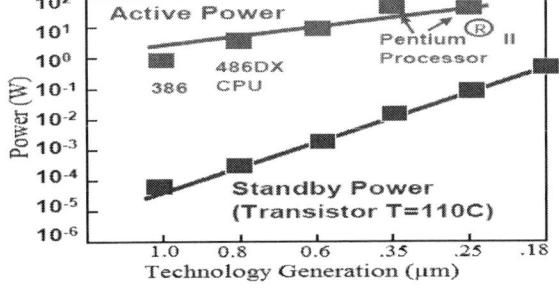

Fig. 1. Power consumption of Intel technologies [8].

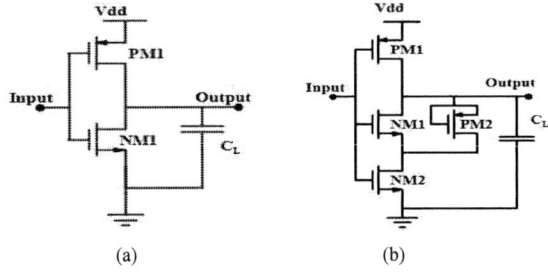

Fig. 2. Circuit schematic of CMOS inverter designed using: (a) Conventional technique, (b) proposed technique.

978-1-5090-0037-1/16 $31.00 © 2016 IEEE

2016 International Conference on VLSI Systems, Architectures, Technology and Applications (VLSI-SATA)

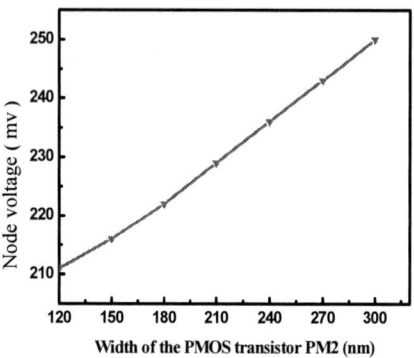

Fig. 3. Dependency of node voltage (source terminal of NM1) on the width of the transistor PM2.

The source voltage of the transistor NM2 increases with the width of PM2 as the amount of leakage in PM2 increases with its width. Exclusive charging of source terminal of the lower NMOS transistor (NM1) in the proposed inverter circuit keeps V_{GS} more negative. Voltage transfer characteristic (VTC) for proposed CMOS inverter circuit along with conventional circuits are shown in Fig. 4. The proposed CMOS inverter circuit exhibits good characteristic curve than the circuit designed using LECTOR technique, and it is comparable to the conventional circuit. We have designed NAND and NOR gates using the proposed approach. Using the proposed CMOS inverter, we have designed an 11-stage CMOS ring oscillator. Fig. 5 shows the circuit schematic of NAND gates designed using conventional, LECTOR and proposed techniques. It also shows the schematic of CMOS inverter designed using LECTOR technique. We have discussed the results of all these circuits in the following section.

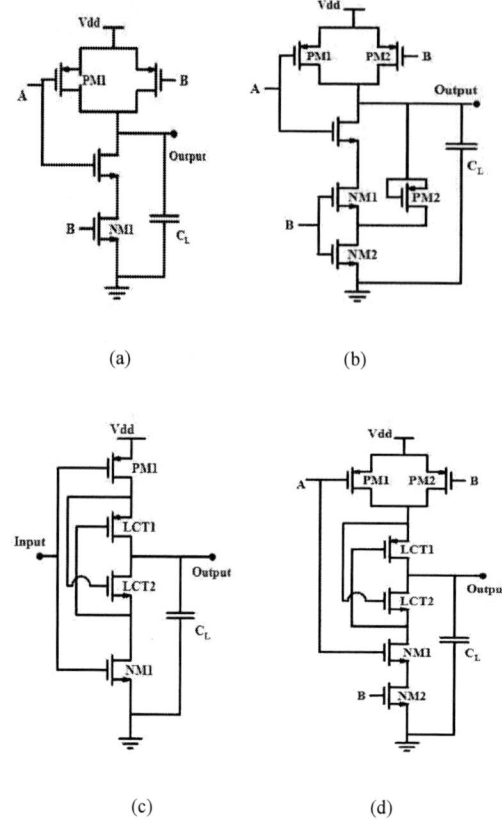

(a)   (b)

(c)   (d)

Fig. 5. Circuit schematic of: (a) conventional NAND gate, (b) proposed NAND gate, (c) LECTOR[5] CMOS inverter, (d)LECTOR [5] NAND gate.

IV. RESULTS AND DISCUSSIONS

In this section, we have made an elaborate discussion on the simulation results of the proposed and conventional designs obtained using 90nm technology file. Simulations are carried out in Cadence Spectre simulation tool. Fig. 6 shows the output waveforms of CMOS inverter circuits designed using conventional, LECTOR [5] and proposed techniques. It is well observed that our proposed circuit results in full swing output voltage similar to the case of conventional circuit. But the output of the circuit designed using the LECTOR technique [5] does not reach to full swing voltage. The reason could be as at least one among two extra transistors inserted for leakage control set to be "*cut off*" state at any time, the output voltage may not get full swing. Among two additional transistors, one is inserted in pull-up path and the other is in the pull-down path. In the case of the proposed design, the working mode of the extra transistor (NM2) inserted in the pull down path is similar to the transistor NM1. There is no additional transistor inserted in the pull up path of the proposed circuit. So the circuit can attain full value (logic "1") at the output.

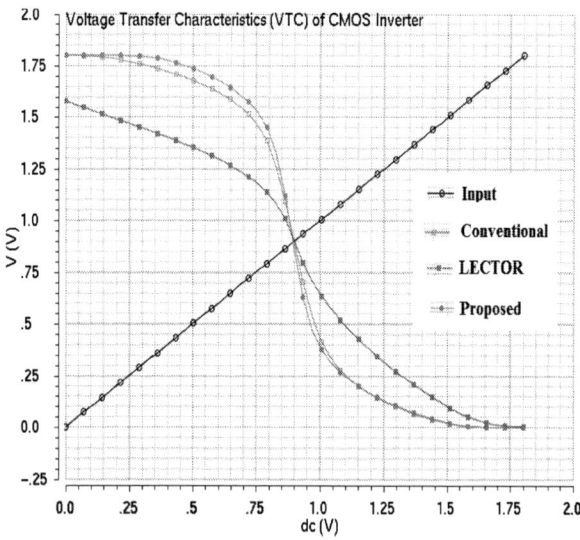

Fig. 4. Voltage Transfer characteristc (VTC) of CMOS Inverter circuit.

978-1-5090-0037-1/16 $31.00 © 2016 IEEE   110

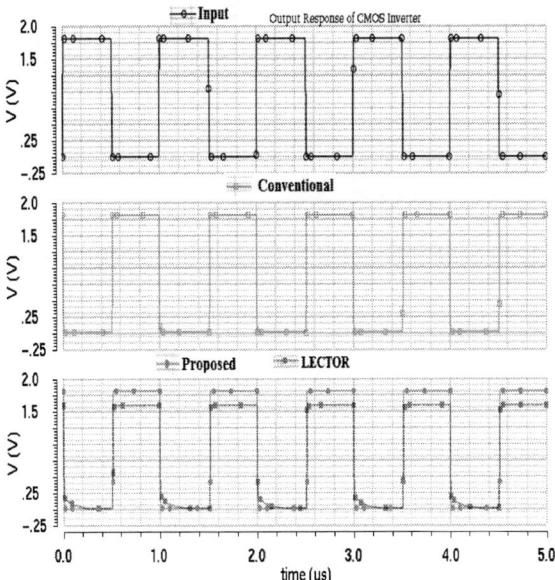

Fig. 6. Output waveforms of CMOS inverter circuit.

Table I shows the static power consumption of the CMOS inverter circuit. Proposed CMOS inverter circuit dissipates less power compared to the other two circuits. The total power dissipated by CMOS inverter, 2- input NAND and NOR gates and 11-stage ring oscillator circuits is presented in Table II. For a considered simulation setup, all the circuits dissipated less power compared to conventional circuits. CMOS inverter and 11-stag ring oscillator designed using proposed approach dissipate less power compared to the corresponding circuits designed using LECTOR technique. Static power consumed by NAND and NOR gates designed using conventional, LECTOR and proposed circuit techniques is summarized in Table III. We have listed the leakage power dissipated during each possible input vector. In the table, 'A' and 'B' represents the inputs to the gate.

Power-Delay-Product (PDP) is the important parameter to assess the quality and performance of logic circuits. It represents the energy consumed for switching event. It is preferred to have the least value of PDP for the logic circuits. We measured PDP for the all three designs (conventional, LECTOR and proposed) and summarized in Table IV. CMOS inverter circuit along with the two logic gates NAND and NOR designed with proposed technique results in less PDP compared with the conventional design. Proposed CMOS inverter exhibits less PDP compared to the CMOS inverter circuit designed through LECTOR approach. PDP plays a significant role in low power designs [10]. On overall comparison, our proposed technique results in full swing operation along with good PDP. All the results of the proposed circuits are obtained by choosing the width of the extra PMOS transistor PM2 equal to 120nm. Extra PMOS transistor PM2 is the transistor connected between the output and the node.

TABLE I.   STATIC POWER DISSIPATION OF CMOS INVERTER

Circuit/ Design	Conventional (w)	LECTOR (w)	Proposed (w)	% Improvement comparison with	
				Conventional	LECTOR
CMOS Inverter	489.2 e^{-9}	253.9 e^{-9}	8.9 e^{-9}	98.2 e^{-9}	96.5 e^{-9}

TABLE II.   TOTAL POWER DISSIPATION

Circuit/Technique	Conventional (μw)	LECTOR (μw)	Proposed (μw)	Improvement (%) Comparison with	
				conventional	LECTOR
Inverter	1.35	0.96	0.94	30.7	2.81
NAND	0.89	0.66	0.75	15.9	-13.1
NOR	1.13	0.76	0.92	18.8	-20.3
11-stage RO	296.67	318.9	285.1	3.9	10.6

TABLE III.   STATIC POWER DISSIPATION OF NAND AND NOR GATES

Input vector		Conventional (nw)		LECTOR (nw)		Proposed (nw)	
A	B	NAND	NOR	NAND	NOR	NAND	NOR
0	0	9.67	1008	9.53	508.47	9.69	528.89
0	1	569.6	47.5	285.69	35.82	572.77	47.49
1	0	336.3	16.12	278.21	14.31	7.97	16.12
1	1	9.96	0.07	6.08	0.07	9.96	0.07

TABLE IV.   POWER-DELAY PRODUCT(PDP)

Design	Conventional (J)	LECTOR (J)	Proposed (J)
Inverter	854.66 e^{-18}	707.31 e^{-18}	621.73 e^{-18}
NAND	537.84 e^{-18}	451.76 e^{-18}	447.97 e^{-18}
NOR	457.32 e^{-18}	358.48 e^{-18}	377.21 e^{-18}

V. CONCLUSION

In this paper, a low leakage power circuit technique that controls the source voltage of the MOSFET through the leakage currents of another MOSFET is presented. Maximum improvement in the power dissipation depends upon the amount of leakage currents that are supplied to the source terminal of the subjected MOSFET. CMOS inverter circuit designed using the proposed technique results in full swing operation along with better power improvement. 11-stage ring oscillator designed using proposed technique attained 3.9 % and 10 % improvement in the total power consumption compared with the conventional and LECTOR techniques respectively. Power-delay product of the proposed circuits improved well compared with conventional designs. The proposed CMOS inverter can be used in the design of inverter based designs like SRAM, CMOS ring oscillator for low power operation.

REFERENCES

[1] Chenming C. Hu, Modern Semiconductor Devices for Integrated Circuits, 1st ed. New Jersey: Prentice Hall, 2010.

[2] B.S. Deepaksubramanyan and Adrian Nunez, "Analysis of subthreshold leakage reduction in CMOS digital circuits," in Proc.13th NASA VLSI Symposium, POST FALLS, IDAHO, USA, June 2007, pp. 1-8.

[3] Kaushik Roy, Saibal Mukhopadhyay and Hamid Mahmoodi-Meimand, "Leakage current mechanisms and leakage reduction techniques in deep-submicrometer CMOS circuits," Proc. IEEE, vol. 91, no.2, pp. 305-327, February 2003.

[4] Ndubuisi Ekekwe and Ralph Etienne-Cummings, "Power dissipation sources and possible control techniques in ultra deep submicron CMOS technologies," Microelectronics Journal, vol. 37, pp.851-860, September 2006.

[5] Narender Hanchate, and Nagarajan Ranganathan, "LECTOR: a technique for leakage reduction in CMOS circuits," IEEE Trans. Very Large Scale Integration (VLSI) Systems, vol. 12, no. 2, pp. 196-205, February 2004.

[6] Marc Belleville, Olivier Thomas, Alexandre Valentian, Fabien Clermidy, "Designing digital circuits with nano-scale devices: challenges and opportunities," Solid-State Electronics, vol. 84, pp. 38-45, June 2013.

[7] Alice Wang, B. H. Calhoun, and A. P. Chandrakasan, Sub-threshold design for ultra low-power system, 1st ed. New York: Springer Science + Business Media, LLC, 2006.

[8] Scott Thompson, Paul Packan and Mark Bohr, "MOS scaling: Transistor challenges for the 21st century," Intel Technology Journal, 1998.

[9] Venkata Ramakrishna Nandyala, MunshiNurul Islam and Kamala KantaMahapatra, "A useful integration of extra nodes and the leakage current in digital circuits," in Proc. IEEE Conf. European Modelling Symposium (EMS), November 2013, pp. 685-690.

[10] Vahid Foroutan, MohammadReza Taheri, Keivan Navi, and Arash Azizi Mazreah, "Design of two low-power full adder cells using GDI structure and hybrid CMOS logic style," INTEGRATION, the VLSI journal, vol. 47, pp. 48-61, January 2014.

A Gain Enhanced Low Voltage Bulk Driven Pseudo-Differential OTA design in CMOS

Antaryami Panigrahi

Dept. of Electronics and Communication Engineering
CIT-Kokrajhar, Kokrajhar, Assam, INDIA
E-mail:- a.panigrahi@cit.ac.in

Abhipsa Parhi

Dept. of Electrical Engineering
Bineswar Brahma Engineering College, Kokrajhar, Assam, INDIA
E-mail:- abhipsa.padhi01@gmail.com

Abstract—a 0.5V low voltage bulk driven pseudo-differential OTA is presented here. Cross coupling technique both at the gate and bulk level is used to increase the effective output resistance of the OTA core. The effective input transconductance is also improved to achieve the high gain. Theoretical analysis for the operation of the OTA is described and simulation is performed to confirm the operation. The simulation results show open loop gain to be 42 dB and UGB of 2.3 MHz and Phase Margin of 85°. The input referred noise is $3.3\mu V/\sqrt{Hz}$, Slew Rate 4.64V/uSec for load of 1pF and 10kΩ. Simulated transient response shows, the OTA achieving full swing of $200mV_{p-p}$. The circuit is designed using 250nm twin-well CMOS and simulated using T-Spice and BSIM 3v3 model. The power dissipation of the proposed OTA is 6.4μWatts.

Keywords—Bulk driven Analog circuit ; CMOS; Low-voltage Ciruit design ;OTA.

I. INTRODUCTION

The demand for low power portable and battery operated systems has veered analog circuit design towards smart use of transistors in weak inversion or in boundary of weak and moderate inversion, losing significant band width and matching properties [1], [2], [3]. This technique has been used for many applications [4], [5], [6], where low power operation is prime importance. Sometimes weak inversion mode has been successfully exploited for emulating trans-linear property of BJT devices in digital CMOS technology for performing arithmetical operations. Another smart approach has been, using the bulk terminal to forward bias the source-substrate junction (assuming no parasitic-bipolar– action), losing the DC gain, increasing the input referred noise (since g_{mb} is 4 to 5 times smaller than g_m). But this approach allows the devices to operate at much lower $|V_{DS,sat}|$, so allowing the circuit to use the least supply voltage. There are many works reported using this approach for various applications [7], [8], [9]. At the circuit level, reducing the no of stacked transistors between the supply rails allows the circuits to operate at much lower supply voltage. Designing the OTA at such lower voltage dropping the tail current source opens the gate for pseudo-differential

technique [10],[15], which helps in getting full swing at such low supply voltage. Many adaption of pseudo-differential OTA has been reported in [10], [11], [12] [13], [14]. Getting higher gain has been a daunting task for bulk driven pseudo-differential OTA because of low bulk-transconductance. So as to increase the gain of the conventional OTA core two basic approaches has been: (1) To increase the effective bulk transconductance by some technique without adding much noise and power consumption. Or (2) To increase the output resistance without sacrificing swing. Approach (1) has been used in some works to enhance effective bulk-transconductance [11], [12], [18] with the help additional devices used for applying the inputs. And approach (2) has been used with the help of self-cascode [13] to decrease the effective g_{ds} loosing significant swing, unless the devices operated in moderate inversion or the boundary of weak or moderate inversion. In this work a new technique of enhancing the gain is presented where, bulk terminals are effectively used to increase the input transconductance without using more stages and another technique of cross-coupling both at the gate level and bulk level are used to enhance the DC gain. Full operation of the circuit is explained in section II. Bulk driven circuit used here is inherently susceptible to threshold voltage variations, so the mismatch analysis is explained and expression for input referred offset voltage is derived in Section III in addition to the noise contributed by the circuit. The issue of linearity is a major area of concern for continuous-time applications, which is also discussed theoretically and simulation of THD is also discussed. The circuit is simulated and results are shown and discussed in Section IV.

II. THE PROPOSED PSEUDO-DIFFERENTIAL OTA

A. Operation of OTA :

The input transconductance (g_{mb}) is obtained from the two transistors M_1 and M_2 whose bulk terminals are connected to the input. The partial positive feedback is formed with the help of two cross coupled pairs: $M_{3,A}$-$M_{3,B}$ which produces a negative g_{mb}. And the cross coupled pair formed by $M_{4,A}$-

$M_{4,B}$ yields a negative g_m. Addition of these g_m and g_{mb} into the equation 1 helps in matching with the other terms in

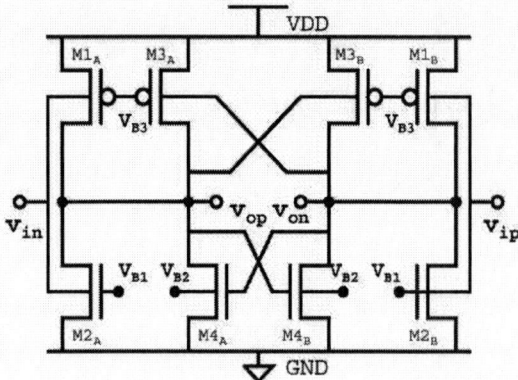

Figure 1 Proposed Pseudo-differential OTA

the denominator, thereby making the denominator as small as possible, causing a large gain without any additional stages. Gate terminals of both the input transistors $M_{2,A}$ & $M_{2,B}$ are tied to the ground to make sure that transistors have enough V_{SG} so as to operate in deep strong inversion. Unlike that in [7], where partial positive feedback is used only to introduce a negative g_{mb} which is formed by the cross coupled pair of $M_{3,A}$ & $M_{3,B}$ yielding a differential gain of 24dB, this work has an differential gain of 36dB. $M_{3,A}$ & $M_{3,B}$ with zero gate bias improves the output resistance of the Pseudo-Differential OTA. The gates of the NMOS transistors are biased such that all the transistors operate in strong inversion and in the saturation mode. The NMOS devices are given a forward substrate bias of 250mV decreasing the threshold voltage to mere 150mV, such that the devices operate in strong inversion and still remains in saturation. Now with the help of small signal model drawn in Fig. 2, the differential gain for the half circuit of OTA can be calculated as;

$$A_{diff} = \frac{(g_{mb1} + g_{mb2})}{g_{ds1} + g_{ds2} + g_{ds3} + g_{ds4} + g_{ds5} - g_{mb3} - g_{m4}} \quad (1)$$

Where g_{mb3} and g_{m5} are the transconductance due to the cross coupled transistors and $g_{ds,i}$ are the output conductances (i=1, 2...5) to get a positive DC gain, the denominator has been made quite small and positive (order of 11×10^{-6}) with proper biasing as well as the aspect ratios of the transistors. The amount of negative conductance added to the denominator terms can be controlled with the help of substrate bias voltages to vary the respective V_{TH} values of the cross-coupled pair, in other way to vary the gate conductance and bulk-conductance. Since the dependence of g_m and g_{mb} on the substrate bias are related as;

$$g_m = \mu_n . C_{ox} . \frac{W}{L} \left[V_{GS} - V_{T0} - \gamma \left(\sqrt{2\phi_F + |V_{SB}|} - \sqrt{2\phi_F} \right) \right]$$

$$(2)$$

where V_{T0}, γ are the terms as expressed in [14]. And g_{mb} is written as in [16]; $g_{mb} = g_m . \frac{\gamma}{\sqrt{2\phi_F + |V_{SB}|}}$ $\quad (3)$

assuming the devices in Saturation. The common mode voltage is 0.25V to maximize the swing. The input bulk-

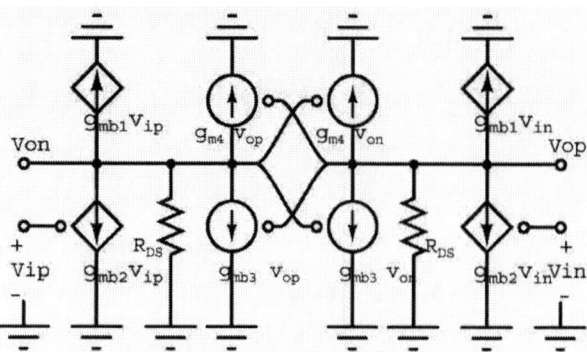

Figure 2 Small signal Model of the Proposed OTA

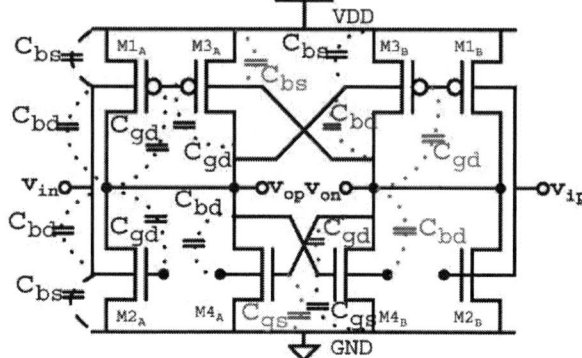

Figure 3 Parasitic Capacitances of the OTA

transconductance is of the order of 100uAmp/Volts. The differential gain obtained from the proposed OTA core is 36 dB. The expression for common-mode gain can be written as;

$$A_{cm} = \frac{(g_{mb1} + g_{mb2})}{g_{ds1} + g_{ds2} + g_{ds3} + g_{ds4} + g_{ds5} + g_{m4}} \quad (4)$$

The g_{m4} term is larger than g_{mb1} making the common mode gain quite small i.e. -50dB. Unlike that of [7], where a local common mode feedback with help of resistor is used to sense the common mode level and to bring down the common mode gain, we get lot more common gain without using any extra circuitry for common mode feedback. In the process area can be saved and thermal noise contribution due to the resistors can be avoided.

B. *The frequency response of the OTA*

The parasitic capacitances for the half circuit that contribute to the overall frequency response is shown in Fig. 3. The total input capacitance (shown in dashed line) for the first stage comprised of transistor pair $M1_A$ & $M2_A$ is given by [17];

$$C_{in} \approx C_{bs1.2} + C_{bsub} + C_{bd1.2} \left(1 + g_{mb1,2} . (r_{ds1} || r_{ds2}) \right)$$

$$(5)$$

978-1-5090-0037-1/16 $31.00 © 2016 IEEE

The output capacitances for the first stage (shown in dotted lines in Fig. 3) consists of the gate-drain capacitance of $M3_A$, bulk-drain capacitance of $M4_A$ & gate-source capacitance of $M4_B$, bulk-source capacitance of $M3_B$ is given by;

$$C_1 \approx C_{gd3.a} + C_{bd4.a} + C_{gs4.b} + C_{bs3.b} + C_{bd3.b}(1 + g_{mb3.b}r_{ds3.b}) + C_{gd4.b}(1 + g_{m4.b}r_{ds4.b}) \quad (6)$$

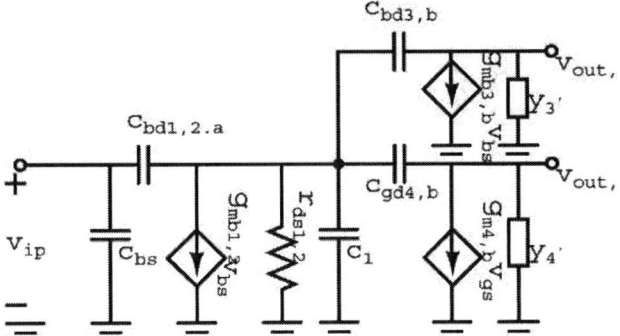

Figure 4 AC equivalent for half circuit of OTA

The admittances Y_3' and Y_4' shown in the Fig 4 can be written as;

$$Y_3' = g_{ds3.b} + s(C_{gd3.b} + C_{bd3.b}) \text{ and}$$

$$Y_4' = g_{ds4.b} + s(C_{gd4.b} + C_{bd4.b}) \quad (7)$$

where $C_{bd3.b}$ and $C_{gd4.b}$ are the capacitances coupling the first stage and cross coupled stage consisting of $M3_B$ and $M4_B$ (shown in blue dotted line in Fig 3). Now the exact transfer function (without using Miller's equivalent for $C_{bd3.b}$ and $C_{gd4.b}$) for the half circuit can be written as;

$$\frac{v_{out,p}}{v_{in}}$$
$$\frac{[(g_{mb3.b}+g_{m4.b})-s(C_{bd3.b}+C_{gd4.b})].(g_{mb1,2}-sC_{bd1,2})}{\left\{\begin{array}{c}[Y_3'+Y_4'+s(C_{bd3.b}+C_{gd4.b})].[g_{ds1,2}+s(C_{bd1,2}+C_1+C_{bd3.b}+C_{gd4.b})]+\\ [(g_{mb3b}+g_{m4.b})-s(C_{bd3.b}+C_{gd4.b})].s(C_{bd3.b}+C_{gd4.b})\end{array}\right\}}$$

$$(8)$$

The approximated location of poles and zeros can be found out by replacing the capacitances in the direct path between the input node and output node with its Miller's equivalent. The dominant pole can be written as;

$$p_1 = \frac{1}{(r_{ds1}||r_{ds2}).C_1}; \quad (9)$$

And the second pole and third pole due to cross coupled pair are given by; $p_2 = \frac{1}{r_{ds3}.(C_{gd3.b}+C_{bd3.b})}$;

$$p_3 = \frac{1}{r_{ds4}.(C_{gd4.b}+C_{bd4.b})} \quad (10)$$

The zeros are $z_1 = \frac{g_{mb1,2}}{C_{bd1,2}}$ and $z_2 = \frac{(g_{mb3.b}+g_{m4.b})}{(C_{bd3.b}+C_{gd4.b})}$. So the cross coupled pair yields two poles and an RHP zero. Since the equivalent capacitance at the output of the first stage is quite larger, making p_1 as the dominant pole. The poles p_2 &

p_3 in Eq. (10) fall at relatively high frequency. The pole zero map is shown in Fig. 5.

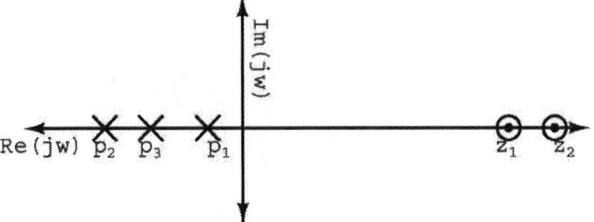

Figure 5 Pole Zero Map of the half circuit

III. Mismatch and Noise Analysis of the OTA

A. Analysis of Mismatch

For two identically designed devices of a differential pair biased with identical V_{GS}, there can be two main sources of mismatch in the drain currents i.e. V_{TH} and β. The mismatch in drain currents can cause the DC offset [17]. For bulk driven transistors of a pseudo-differential pair, the extra mismatch caused due to the mismatch in the substrate bias voltages and also due to difference in bulk-effect coefficients results in an extra degradation of the matching of the transistors which is described below. The following well known expression [16] shows the dependence of substrate bias effects for V_{TH};

$$V_{TH} = V_{T0} + \gamma\left(\sqrt{2\phi_F + |V_{SB}|} - \sqrt{2\phi_F}\right) \quad (11)$$

shows the dependence of V_{TH} and substrate bias V_{SB}. Two transistors biased with V_{SB2} and V_{SB1}, (Assuming Gate terminal is at same bias for both the devices, V_{GB} remaining same) the deviation in the threshold voltage i.e. $V_{TH1} - V_{TH2}$ can be derived by expanding the Eq. (3) using Taylor's Series around a DC operating point V_{SB};

$$\Delta V_{TH} = \Delta V_{T0} + \Delta\gamma\sqrt{2\phi_F} + \Delta\gamma\sqrt{2\phi_F + |V_{SB}|} + \frac{1}{\sqrt{2\phi_F+|V_{SB}|}}(\gamma_1.V_{SB1} + \gamma_2.V_{SB2} - \Delta\gamma.V_{SB}) + \cdots$$

$$(12)$$

Where $\Delta V_{TH} = V_{TH1} - V_{TH2}$; ΔV_{T0} is the deviation in threshold voltage at zero substrate bias; ϕ_F is Fermi potential as expressed in [15] and $\Delta\gamma$ is deviation in body coefficient. The above expression shows the extra mismatch terms due to substrate bias and body coefficients. The matched pairs of the proposed OTA from Fig. 1 are: $M_{1A,B}$, $M_{2A,B}$, $M_{3A,B}$, $M_{4A,B}$, $M_{5A,B}$. The input referred offset voltage due to mismatch in the current for each matched pair of transistors can be calculated as in [16]. The variance of the input referred offset due to $M_{1A,B}$ can be written as;

$$\sigma_{v_{in1}}^2 = \left[\frac{\sigma(\Delta\beta1)}{\beta1}\right]^2 . \frac{I_{D1}^2}{g_{mb1,2}^2} + \sigma^2(\Delta V_T); \text{ For } M_{2A,B}$$

$$\sigma_{v_{in2}}^2 = \left[\frac{\sigma(\Delta\beta2)}{\beta2}\right]^2 . \frac{I_{D2}^2}{g_{mb1,2}^2} + \frac{g_{mb2}^2}{g_{mb1,2}^2} . \sigma^2(\Delta V_T); \text{ For } M_{3A,B}$$

$$\sigma_{v_{in3}}^2 = \left[\frac{\sigma(\Delta\beta3)}{\beta3}\right]^2 . \frac{I_{D3}^2}{g_{mb1,2}^2} + \frac{g_{mb3}^2}{g_{mb1,2}^2} . \sigma^2(\Delta V_T); \text{ For } M_{4A,B}$$

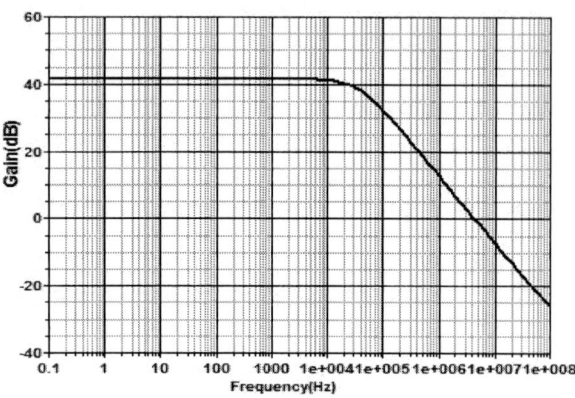

Figure 6 Gain Plot

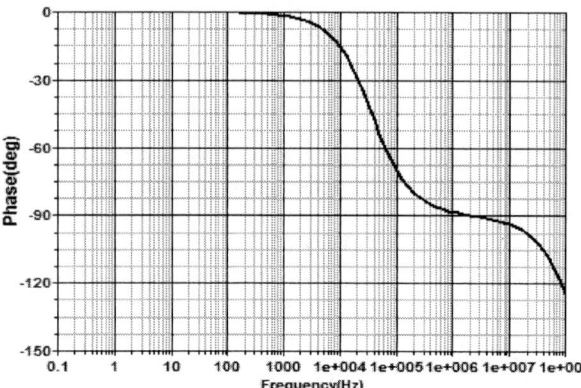

Figure 7 Phase Response of the OTA

$$\sigma_{v_{in4}}^2 = \left[\frac{\sigma(\Delta\beta 4)}{\beta 4}\right]^2 \cdot \frac{I_{D4}^2}{g_{mb1,2}^2} + \frac{g_{mb4}^2}{g_{mb1,2}^2} \cdot \sigma^2(\Delta V_T) \qquad (13)$$

Where $\beta_i = \mu_n \cdot C_{ox} \cdot \left(\frac{W}{L}\right)_i$ and $\frac{\sigma(\Delta\beta)}{\beta} = \frac{A_\beta^2}{W.L}$; $\sigma^2(\Delta V_T) = \frac{A_{VT}^2}{W.L}$. A_{VT} and A_β are the process constants as given in [17]. So variance of total offset voltage due to all matched pair can be expressed as [17];

$$\sigma_{v_{ios}}^2 = \sigma_{v_{in1}}^2 + \sigma_{v_{in2}}^2 + \sigma_{v_{in3}}^2 + \sigma_{v_{in4}}^2 \qquad (14)$$

To reduce the offset, it is always better to choose larger device sizes, but for bulk driven approach of design, the effect of input capacitance ($C_{bulk-Source}$, $C_{bulk-drain}$ and $C_{well-substrate}$) also starts to play causing fall of UGB [17]. So aspect ratios are chosen to have an optimum gain and input referred noise. It can be inferred from eqn. (13), input referred offset voltage can be decreased by decreasing the input bulk- transconductance of the OTA i.e. g_{mb1} due to the $M_{1A,B}$. The ratio of g_m/g_{mb} being less than unity causes severe problems in input referred offset at such low supply voltages [9], [13].

B. Analysis of Noise

Assuming the noise sources in each half of the OTA are uncorrelated their effects on the output can considered individually. Two main sources of noise that can affect the performance of the OTA are:

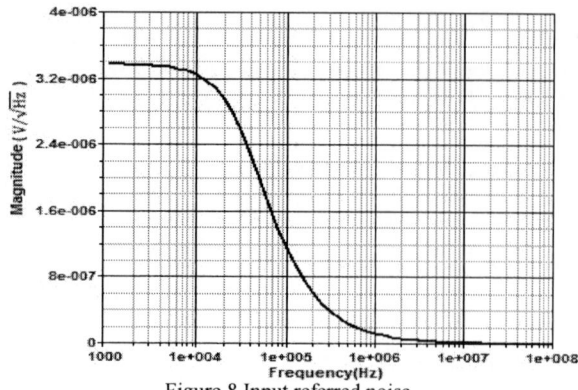

Figure 8 Input referred noise

flicker Noise and thermal noise. The expression for Flicker Noise can be written as;

$$\overline{v_{n(1/f)}^2} \approx \frac{.K_f}{g_{mb1,2}^2 \cdot C_{ox} \cdot f}\left[\frac{g_{m1}^2}{(WL)_1} + \frac{g_{m2}^2}{(WL)_2} + \frac{g_{m3}^2}{(WL)_3} + \frac{g_{m4}^2}{(WL)_4}\right] \qquad (15)$$

Where g_{m1} is the gate transconductance of the i'th transistor (i=1,2,3,4) on the and $g_{mb1,2} = g_{mb1} + g_{mb2}$. Ignoring the thermal noise component due to the Gate resistance and Bulk resistance, the approximated thermal noise expression can be written as;

$$\overline{v_{n(Th)}^2} \approx \frac{4.K.T.\gamma}{3.g_{mb1,2}^2}[g_{m1} + g_{m2} + g_{m3} + g_{m4}] \qquad (16)$$

The above expression shows that the bulk input devices are more susceptible to the flicker noise component owing to the lower value of g_{mb} than gate input devices and thermal noise component due to larger value of g_m/g_{mb} ratio.

IV. RESULTS AND DISCUSSIONS

To verify the circuit performance, 250nm twin well CMOS technology is used and to simulate the proposed circuit, BSIM 3v3 model is used in TSPICE. The CMOS process has V_{THN} and V_{THP} = 0.56 Volt. Substrate bias voltages are carefully used to bring down the threshold voltages of the devices so as to operate the circuit at 0.5V supply. The aspect ratios of the devices are shown in Table I. In this work, the aspect ratios of all transistors are chosen to optimize both gain and input referred noise. An AC signal of magnitude 1V, 10 kHz is applied to the circuit with load capacitance of 10pF and resistance of 10kΩ. The simulated gain and phase response of the circuit is shown in Fig. 5 and 6 respectively. It can be seen that the DC gain is approximately 36dB, UGB is around 2.3 MHz. The phase margin is around 90 degree from Fig. 6. Noise is simulated and input referred noise spectral density is shown in Fig.7. The transient response of the circuit is obtained by simulating the circuit with a common mode 0.25V and sinusoidal signal of 10 mVp-p,10kHz at its input . As it can be seen from Fig. 8, that output common mode voltage is maintained around 0.25V(the dotted line in Fig.7) and output swings between 140 mV to 360 mV is shown in Fig. 6. Single ended output is shown in Fig. 7 to show the DC common mode voltage at the outputs. To simulate for the slew rate of the OTA, a pulse of time period 100nSec

978-1-5090-0037-1/16 $31.00 © 2016 IEEE

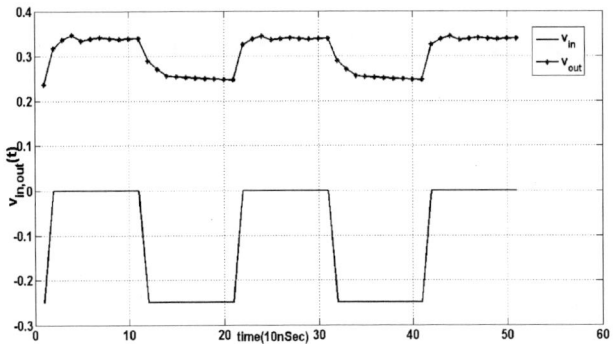

Figure 8 Transient responses for a pulse input (Single ended)

(having 5nSec rise and fall time) is applied to the OTA, the transient response. The slew rate from the response can be found to be 4.64V/µSec with same load conditions (10pF, 10kΩ). Total power dissipation by the circuit 6.4µW. The summaries of simulated results are shown in the Table-II. Input referred noise for this OTA is on higher side, as the flicker noise dominates at lower frequencies and the ratio of g_m/g_{mb} is also lower.

TABLE I   DEVICE SIZES

Devices	Aspect Ratio
$MP_{1,A-B}$	65µm/0.250 µm
$MP_{2,A-B}$	65 µm/0.250 µm
$MP_{3,A-B}$	35 µm/0.250 µm
$MN_{4,A-B}$	34 µm/0.250 µm

TABLE II SUMMARY OF SIMULATED PERFORMANCES OF THE OTA

Parameters	Simulated values
Power (uW)	6.4
DC gain(dB) @ C_L=10PF	42
DC gain(dB) @ C_L=1PF	42
UGB(MHz) @ C_L=10PF	2.3
UGB(MHz) @ C_L=1PF	8.8
Phase Margin(deg) @ C_L=10PF	90
Phase Margin(deg) @ C_L=1PF	88
Input-Referred Noise($\mu V/\sqrt{Hz}$) at 10kHz	3.3
CMRR(dB) at UGB	-86 dB
Slew Rate(V/µSec)	4.64

V. CONCLUSION

A new technique of implementing bulk driven pseudo-differential OTA at 0.5V is presented. The proposed technique is analyzed thoroughly mathematically and the simulated results confirm the operation of the OTA having a DC gain of 42dB from a single stage OTA. Due to the increase in the input transconductance, thermal noise contributed by this OTA is lesser, the problem of flicker noise still remains with bulk driven circuits. This technique not only increases the gain, but decreases the input referred noise density. The circuit is quite simple and has less no of transistors, so less silicon area requirement and less power dissipation.

REFERENCES

[1] E. A .Vittoz, "Weak Inversion for Ultra Low-Power and Very Low-Voltage Circuits, " plenary talk, IEEE Asian Solid-State Circuits Conference November 16-18, 2009 / Taipei, Taiwan, pp.129-132

[2] M.J. Chen, J. S. Ho, and T. H. Huang, " Dependence of Current Match on Back-Gate Bias in Weakly Inverted MOS Transistors and Its Modeling," IEEE Journal of Solid-State Circuits, Vol. 31, No. 2, Feb. 1996, pp. 259-262

[3] D. Gangopadhyay, E. G. Allstot, etal. "Compressed Sensing Analog Front-End for Bio-Sensor Applications," IEEE Journal of Solid-State Circuits, Vol. 49, No. 2, Feb. 2014, pp 426-438

[4] C. Sawigun, and W. A. Serdijn, "Analysis and Design of a Low-Voltage, Low-Power, High-Precision, Class-AB Current-Mode Subthreshold CMOS Sample and Hold Circuit," IEEE Transactions On Circuits And Systems -I, Vol. 58, No. 7, Jul. 2011 pp.1615-1626

[5] A. Tajalli, E. J. Brauer, etal., "Subthreshold Source-Coupled Logic Circuits for Ultra-Low-Power Applications," IEEE Journal of Solid-State Circuits, Vol. 43, No. 7, Jul 2008 pp. 1699-1710

[6] A. Panigrahi, P. K. Paul "A novel bulk-input low voltage and low power four quadrant analog multiplier in weak inversion," Springer, Jornal of Analog Integrated Circuits and Signal Processing, Vol. 75, No. 2, Sept. 2013, pp. 237-243

[7] S. Chatterjee, Y. Tsividis, and P. Kinget, "0.5-V Analog Circuit Techniques and Their Application in OTA and Filter Design," IEEE Journal Of Solid-State Circuits, Vol. 40, No. 12, Dec. 200, pp. 2373-2383.

[8] J. Rosenfeld, M. Kozak, and G. Friedman, "A Bulk driven CMOS OTA with 68 dB DC gain" Proc. of 11th IEEE International Conference on Electronics, Circuits and Systems, 2004, pp. 5-8

[9] C. L. Chien, C. C. Hung & C. W. Chen, "A Pseudo-Differential OTA with Linearity Improving by HD3 Feedforward" IEEE Asian Solid-State Circuits Conference November 16-18, 2009 / Taipei, Taiwan

[10] F. Bahamani, E. Sinencio, "A new highly linear Pseudo-differential Transconductance", Proc. of the 30th European Solid-State Circuits Conference, 2004. pp. 111-114

[11] T. Kulej and F. Khateb, "0.4-V bulk-driven differential-difference amplifier," Microelectronics Journal 46 (2015) 362–369, 1963, pp. 271-350.

[12] T. Kulej, "0.5-V bulk-driven CMOS operational amplifier," IET Circuits Devices Systems, Vol. 7, Iss. 6, pp. 352–360, 2013.

[13] M. Trakimas, S. Sonkusale, "A 0.5V Bulk-Input Operational Transconductance Amplifier with Improved Common-Mode Feedback ", ISCAS 27-30 May 2007 2007, pp. 2224 – 2227.

[14] A. N. Mohieldin , E. Sánchez-Sinencio "A Fully Balanced Pseudo-Differential OTA With Common-Mode Feedforward and Inherent Common-Mode Feedback Detector," IEEE Journal of Solid-State Circuits, Vol. 38, No. 4, Apr. 2003, pp. 663-668

[15] Y. Tsividis, "Operation and modelling of MOS Transistors" Oxford Publications, 2nd Ed. Ch. 8.

[16] P. R. Kinget, "Device Mismatch and Tradeoffs in the Design of Analog Circuits," IEEE Journal Of Solid-State Circuits, Vol. 40, No. 6, Jun. 2005.pp. 1212-1224

[17] B. J. Blalock, P. E. Allen, and G. A. Rincon Mora, "Designing 1-V Op Amps Using Standard Digital CMOS Technology" ," IEEE Transactions On Circuits And Systems -I, Vol. 45, No. 7, Jul.1998 pp.769-780

[18] G. Raikos, S. Vlassis, "0.8 V bulk-driven operational amplifier," Analog Integr Circ Sig Process (2010) 63:425–432.

Non-Intrusive FPGA-Based Profiler for Loop Execution Characterization

Pavan Kumar Nadimpalli
IIIT Bangalore
nadimpalli.pavankumar@iiitb.org

Subir K Roy
IIIT Bangalore
subir@iiitb.ac.in

Abstract—**Embedded system design involves meeting strict design goals such as performance, area and power consumption. In-order to meet these design goals embedded systems are implemented in programmable processors and application-specific hardware. Hardware/Software partitioning is thus, a critical step in the realization of embedded systems. The initial software description of the application is profiled to identify the critical sections of the software code which consume the largest percentage of execution time. These critical sections are then chosen as ideal candidates to be implemented as application specific hardware. It is reported that 90 percent of the execution time is spent in executing loops in typical embedded systems applications. In this paper we present a non-intrusive, low overhead FPGA based hardware profiler to identify at run-time the different loops and the time taken to execute these loops from the execution of different scenarios of the application software when compiled on the chosen programmable processor.**

Keywords: Embedded Systems, Hardware/Software Partitioning, Profiling, FPGA

I. INTRODUCTION

Embedded systems are pervasive. We interact with one or more embedded systems in our day to day life. With advancements in technology along with growing user demands the complexity of designing embedded systems is increasing. In order to deal with complexity and to meet the design goals, systematic approach is necessary along with the set of tools and methodologies. One approach in implementing embedded systems is a pure software approach in which the application runs on a one or more programmable processors. Even though this is easy to implement, it suffers from poor performance. The other approach is to go for a pure hardware implementation. Even though this approach offers good performance, it increases the cost and time to market. So, a hybrid approach is followed where a part of application is implemented to run on a processor while the remaining part is implemented for execution on a synthesized application specific hardware. We can achieve performance as well as flexibility using this hybrid approach provided we can automate the choice of apportioning the application to a processor and the hardware. Field-programmable gate arrays (FPGAs) are suitable platform to implement embedded systems because of their flexibility and lower time to market. With advancement in technology, the logic density and speed of operation of FPGAs is increasing while their cost is decreasing. The availability of soft-processor cores [1] in the form of RTL implementation in either Verilog or VHDL, enables their implementation

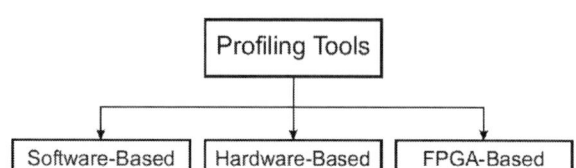

Figure 1. Classification of profiling tools [5].

along with application specific hardware as hybrid systems on FPGAs [2].

The critical step in the design of the embedded systems is to partition the application into hardware and software parts. This is called hardware/software partitioning[3]. A good partition leads to an implementation with good performance and lesser area. It is in this context that the role of profiling tools becomes very important [4]. Profiling provides run-time information about program execution. This profiling information is utilized in identifying the bottlenecks and can help in coming up with an optimal partition. Profiling tools can be classified into software based, hardware based and FPGA based profiling [5]. In software based profiling the application code is modified to add the instrumentation code and the Program Counter (PC) is sampled during periodic intervals. Because of the instrumentation of the application code software based profiling tools are inaccurate and suffers from overhead. *gprof* [6] is one of the most widely used software profiling tool. In hardware based profiling dedicated counters are specially implemented during the design of the processor to monitor and to count the occurrence of certain events. These counters can be accessed using special instructions. In FPGA based profiling, the soft-processor core on FPGA is augmented with additional hardware to monitor the program execution. There is no need to modify the original application code and neither is the original execution flow disturbed. Hence there is no overhead while being very accurate. Further this approach can also enable dynamic hardware/software partitioning [7] by utilizing the property of on the fly re-configurability of FPGAs.

In the studies that have been done earlier, it was observed that most of the execution time is spent in executing small sections of code as shown in Figure 2. In most of the embedded benchmarks it was observed that 90 percent of

2016 International Conference on VLSI Systems, Architectures, Technology and Applications (VLSI-SATA)

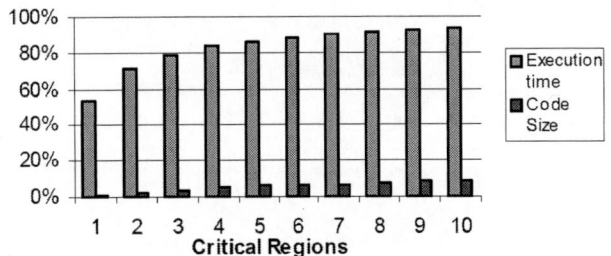

Figure 2. Average percentage of execution time spent in corresponding percentage of code size for the top N most critical code regions, for five MediaBench benchmarks.[14]

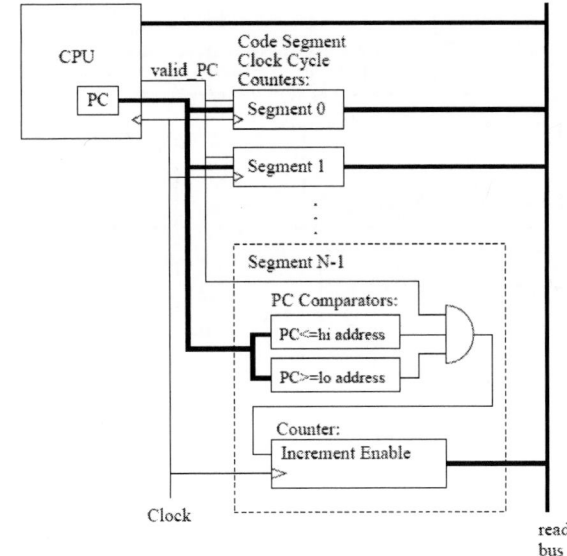

Figure 3. Architecture of SnoopP profiler [9]

execution time is spent in executing loops [8]. So the problem of identifying the critical regions becomes the p1roblem of identifying loops consuming the highest percentage execution time.

In this paper, we present a profiler for FPGA based embedded processors which provides detailed loop execution statistics. The profiling is done without altering the program's execution flow and changing or touching the original RTL implementation of the processor. It also has a low hardware overhead with respect to the FPGA resources.

The rest of the paper is organized as follows: In Section II we present prior work done in FPGA based profilers. In Section III we present the method of operation of profiler. In Section IV we present the architecture of our profiler. In Section V we present experimental results and in Section VI we present conclusion and directions for future work.

II. RELATED WORK

As shown in Figure 1, profiling tools can be classified into three categories. Software based profiling tools are widely used. GNU's *gprof* is a widely used software profiling tool. To use *gprof*, the application to be profiled is compiled with additional compiler flags to enable profiling options. The compiler automatically inserts instrumentation code to the original code to be able to enable generation of profiling information. It also generates interrupts at periodic intervals to sample the Program Counter (PC). The accuracy of the results depends upon the sampling frequency. Increasing the sampling frequency increases the accuracy, but, also increases the run-time overhead. Embedded applications can have strict timing constraints, where such overheads may not be acceptable. SnoopP [9] is a non-intrusive FPGA based profiler designed for use with Xilinx MicroBlaze processor [10]. Figure 3 shows the architecture of SnoopP profiler. It consists of a variable number of segment counters which can be configured by the user. The number of segments depend upon the number of code regions we wish to monitor. As shown in Figure 3 each segment consists of a comparator and a counter. The start and end address of the code regions we wish to monitor needs to be specified in a VHDL file prior to synthesis. SnoopP monitors the PC value and when the PC value lies in the address range specified in a segment, the corresponding counter for that

the segment is enabled to count the number of clock cycles spent in executing that particular code segment. The area required for the profiler increases with increase in the number of code regions we wish to monitor. For every application to be profiled the address ranges needs to be determined apriori and needs to to be entered in the VHDL file and further, needs to be re-synthesized.

Other profilers which are extensions of the SnoopP profiler are AirWolf [4] and Address Tracer [11]. AirWolf is an on-chip profiler developed for the Altera Nios II soft processor. There is no need to modify the source code. Software drivers are added at the beginning and end of a function to enable/disable the corresponding counters. Address Tracer combines SnoopP and AirWolf which enables it to count the cycles spent in executing a function and it's nested sub-functions.

Other recent work in FPGA based profiler is the Low-overhead and Extensible Architecture for Profiling (LEAP) [12]. It is developed as a part of the LegUp High-level Synthesis project [13]. It is based on the Tiger MIPS processor and is targeted for implementation on Altera Cyclone II FPGAs. Figure 4 shows the architecture of LEAP profiler. The OpDecoder block monitors the Instruction Bus and decodes instructions to identify a sub-routine call or a return from a sub-routine call. Unlike SnoopP where an individual counter is used for every sub-routine, LEAP uses hashing to generate a unique function number using the starting address of the function. The call stack is used to keep track of the function context. The data counter can be configured to perform instruction count profiling, or cycle count profiling, or stall cycle profiling. The counter storage is used to store the function number and the corresponding count.

Prior work in the area of FPGA based profiling discussed above focused primarily on profiling at a function call level,

978-1-5090-0037-1/16 $31.00 © 2016 IEEE   119

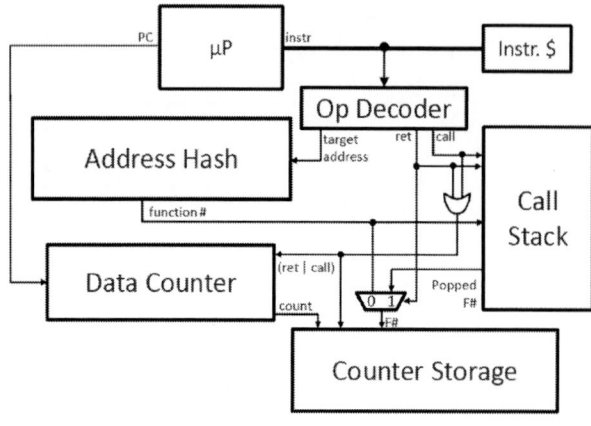

Figure 4. Architecture of LEAP [12]

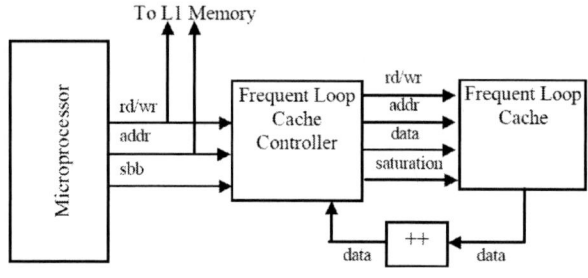

Figure 5. Frequent Loop detection Architecture [14]

or a sub-routine call level. The first on-chip profiler focused on profiling loop execution is the Frequent Loop Analysis Technique (FLAT) [14]. In FLAT a loop is identified by short backward branch (*sbb*) instruction that jumps to the first instruction of loop. Figure 5 shows the cache based architecture of FLAT. It uses Motorola's M*CORE microprocessor, which has the *sbb* signal, which is asserted when a short backward branch is taken. The frequent loop cache is used to store frequency counts. The cache is indexed using the *sbb* address. When a *sbb* is taken, the cache is read and if it's a hit the frequency is incremented and written back to cache. If it's a compulsory miss, then the new address is written to cache with a frequency count of one. If it's a conflict miss, then new address replaces the old address with a frequency value of one.

One drawback is that the FLAT technique is based on the availability of specific signals, such as the *sbb* signal, from the chosen processor. It is possible that the soft processor cores from different processor IP vendors may not have such signals available in the RTL implementation. We propose an approach which does not require such signals to be present, and thereby makes the detection of loops of various types and their profiling independent of soft processor cores.

Address	Instruction	Mnemonic	
//for(i=0;i<5;i++)			
1c4:	f8130008	swi r0, r19, 8	
1c8:	b800001c	bri 28	//1e4
//a++;			
1cc:	e8730004	lwi r3, r19, 4	
1d0:	30630001	addik r3, r3, 1	
1d4:	f8730004	swi r3, r19, 4	
1d8:	e8730008	lwi r3, r19, 8	
1dc:	30630001	addik r3, r3, 1	
1e0:	f8730008	swi r3, r19, 8	
1e4:	e8730008	lwi r3, r19, 8	
1e8:	32400004	addik r18, r0, 4	
1ec:	16439001	cmp r18, r3, r18	
1f0:	**bcb2ffdc**	**bgei r18, -36**	**// 1cc**

int i=0,a=0;
for(i=0;i<5;i++)
a++;

Figure 6. Code snippet consisting of simple for loop and corresponding assembly code

III. LOOP PROFILER ARCHITECTURE

In this section we give the functional description of the proposed profiler architecture. We also point to some important implementation details.

Loop profiling is done by monitoring the program counter value (pointing to the memory address from where the next instruction needs to be fetched) and instruction bus of the processor. A loop can be identified by locating the last instruction constituting the loop. This instruction has to be a branching instruction, for example, a jump instruction with a negative address offset, pointing to the first instruction constituting the loop body and thereby passing execution control to it. An example C code snippet consisting of a simple *for* loop and the corresponding assembly code generated by the GCC compiler for the MicroBlaze processor is shown in Figure 6. It can be seen that at address *1f0* there is a branch instruction pointing to an instruction located at the memory address *1cc*. Thus, a loop can be easily identified by a conditional branch instruction with a negative address offset. An instruction decoder is implemented to detect branching instructions in the instruction bus. It monitors the instruction bus and identifies backward branch instructions. When a loop is being executed, a counter corresponding to that loop is enabled to count the number of cycles spent in executing the loop body. A typical application program can consists of many loops. A loop can be uniquely identified by the address at which the backward branch instruction occurs. To associate a loop count value to a particular loop in the application code, a loop when it is identified for the first time, the corresponding branch instruction address in the form of a PC value is written into a Content Addressable Memory (CAM). CAM is a special type of memory used for fast search operations. Figure 7 shows the conceptual view of a CAM. Writing to a CAM is similar to writing to a RAM. However, it differs from a RAM during read operations, in that the input data is searched against stored

978-1-5090-0037-1/16 $31.00 © 2016 IEEE   120

2016 International Conference on VLSI Systems, Architectures, Technology and Applications (VLSI-SATA)

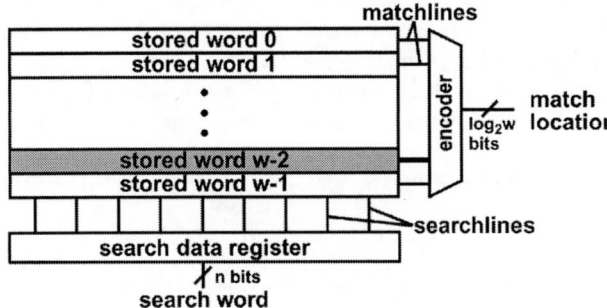

Figure 7. Conceptual view of a content-addressable memory containing w words. In this example, the search word matches location (w-2) as indicated by the shaded box.The matchlines provide the row match results. The encoder outputs an encoded version of the match location using $\log_2 w$ bits.[15]

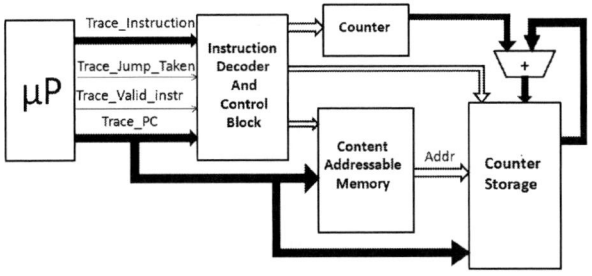

Figure 8. Loop profiler Architecture. [Control signals are indicated by unfilled arrows and data signals are indicated by block arrows]

data in every CAM location, with the address of the matching data being returned as an output. This is achieved in a single clock cycle by matching the input data concurrently with data in every address location. Whenever a loop is identified the address at which the loop occurs is searched in the CAM; if no match is found, the PC value is written into the CAM. The counter value along with PC value is written into counter storage RAM.

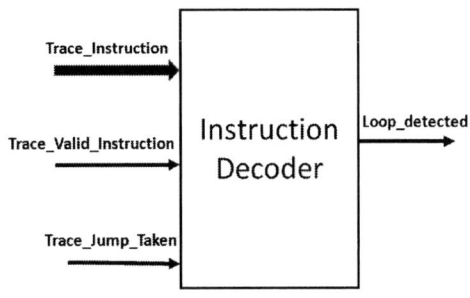

Figure 9. Block diagram of instruction decoder

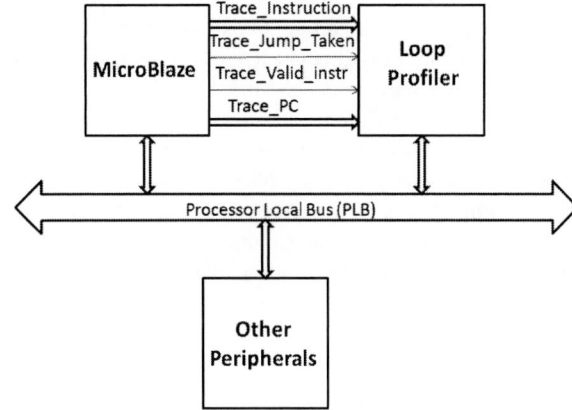

Figure 10. Interface of loop profiler with MicroBlaze

IV. PROFILING ARCHITECTURE AND EXPERIMENTAL PLATFORM

We have chosen the Xilinx SP605 Evaluation board having Spartan 6 FPGA as our experimental platform. We employ Xilinx MicroBlaze as our reference soft core processor. The hardware architecture of the proposed loop profiler is illustrated in Figure 8. The functionality of each module in the proposed profiler architecture is explained below.

The MicroBlaze processor has Trace interface through which the internal signals of the MicroBlaze can be accessed. some of these signals which are required by the profiling are directly connected to the processor. Instruction decoder monitors the instruction bus and detects a loop. The block diagram of instruction decoder is shown in Figure 9. The value on the instruction bus is valid only when *Trace_Valid_Instruction signal* is high. The instruction is decoded to check if it's a conditional branch instruction and with a negative offset. The *Trace_Jump_Taken* signal which is a part of MicroBlaze's trace interface is asserted when a branch is taken. It indicates that loop is detected. The execution flow is inside a loop when a loop is detected and branch is taken. The execution moving out of the loop is identified by loop being detected and branch not taken. The other modules are controlled by the control signals from this block.

The CAM acts as a look-up table for PC values corresponding to addresses with branch instructions. Whenever a loop is identified, the PC value at which the loop occurs is searched in the CAM. If the PC value is present in the CAM, it returns the CAM address which stores this value. The returned CAM address is used as an index to the Counter Storage RAM. If the PC value is not present in the CAM, it is written into a new address in the CAM. The depth of the CAM puts a limit on the maximum number of loops that can be profiled.

The loop counter captures the execution data corresponding to an active loop at runtime to generate the loop profiling measurements. It counts the number of cycles spent in executing a particular loop. Each time the particular loop is executed,

978-1-5090-0037-1/16 $31.00 © 2016 IEEE   121

the execution count is accumulated by adding it to the value corresponding to all the earlier executions of the loop which is stored in the Counter Storage RAM. Other modules in the proposed architecture associate this count to the particular loop. It is, thus, the Counter Storage RAM which acts as a repository for the profiling results of different loops. Whenever program's execution comes out of a loop any previous count corresponding to that loop is read and added to the current count and written back to the Counter Storage RAM.

The proposed architecture is portable as only the PC and Instruction bus of the processor are required to perform the profiling. This makes it easy to add this to any processor. The trace signals *Trace_Jump_Taken* and *Trace_Valid_Instruction* are required by our profiler along with Instruction bus and PC. If a processor doesn't have theses signals, they can be easily generated using simple hardware. The number of loops that can be profiled is limited by the number of positions in the CAM. The depth of the CAM can be configured according to the application that is to be profiled.

A. Interfacing profiler with MicroBlaze

The interfacing of the loop profiler to the MicroBlaze processor is shown in Figure 10. The profiler is added as a slave peripheral to the Processor Local Bus (PLB). The Trace signals from MicroBlaze which are required by the profiler are directly connected to the profiler.

V. EXPERIMENTAL STUDY AND RESULTS

For evaluating our proposed profiling architecture we have chosen *adpcm* benchmark which is a part of CHStone benchmark suite [16] and *crc, blit* and *fir* benchmarks from Powerstone benchmark suite[17].

To validate the results and to evaluate the accuracy, the results obtained by our profiler are compared with results obtained by our own implementation of SnoopP. The SnoopP counters are configured with addresses of the loop boundaries. The profiling results obtained by SnoopP and our profiler for *adpcm, crc, blit* and *fir* are tabulated in Table I, II , III and IV respectively . We can see that the cycle counts provided by SnoopP is slightly more than that of the count provided by our loop profiler in most of the cases. The reason for additional cycle count in SnoopP is that the counter in SnoopP starts counting as soon as the PC value falls in the address range. Whereas the loop profiler starts counting once the backward branch instruction is identified. As shown in Table I the cycle counts for Loop13 and Loop14 provided by both the profilers don't match. These are the loops in which a sub-routine is called inside the loop body. The results for Loop13 and Loop14 points out the limitation of SnoopP profiler in profiling loops with a sub-routine calls present inside the loop body. SnoopP doesn't count the cycles spent in executing the sub-routine, as the PC value will be outside the address range for the segment counter to be enabled. This is also the reason for large difference in cycle counts in case of other benchmarks presented in Table II, III and IV. Even if we were to assign a separate counter to count the cycles spent in sub-routine

we will not be able to attribute the count to any particular loop, as the sub-routine can be called by other sub-routines, or invoked within other loops. In our proposed profiler we counts such cycles attributable to called sub-routine along with cycles spent in executing other instructions within the loop. The device utilization summary of our Loop Profiler synthesized for Xilinx Spartan-6 FPGA is presented in Table V.

Table I
CYCLE COUNT PROFILE PROVIDED BY LOOP PROFILER AND SNOOPP FOR ADPCM BENCHMARK

Loop	Loop Profiler	SnoopP	Percentage Difference
Loop1	35,000	35,400	1.13
Loop2	37,400	37,800	1.06
Loop3	35,000	35,400	1.13
Loop4	27,000	27,400	1.47
Loop5	372	380	2.12
Loop6	372	380	2.12
Loop7	624	632	1.27
Loop8	418	426	1.89
Loop9	42,000	43,600	3.73
Loop10	464,758	17,354	185.60
Loop11	23,184	23,920	3.125
Loop12	47,872	48,736	1.78
Loop13	245,818	3,408	194.53
Loop14	213,632	3,608	193.35

Table II
CYCLE COUNT PROFILE PROVIDED BY LOOP PROFILER AND SNOOPP FOR CRC BENCHMARK

Loop	Loop Profiler	SnoopP	Percentage Difference
Loop1	61,440	63,488	3.27
Loop2	6,068	6,084	0.26
Loop3	95,232	21,000	127.73

Table III
CYCLE COUNT PROFILE PROVIDED BY LOOP PROFILER AND SNOOPP FOR BLIT BENCHMARK

Loop	Loop Profiler	SnoopP	Percentage Difference
Loop1	27,000	27,004	0.01
Loop2	27,000	27,004	0.01

Table IV
CYCLE COUNT PROFILE PROVIDED BY LOOP PROFILER AND SNOOPP FOR FIR BENCHMARK

Loop	Loop Profiler	SnoopP	Percentage Difference
Loop1	83,142	10,930	153
Loop2	1,917	114	177
Loop3	788,449	654	199.66
Loop4	697,498	594	199.65

Table V
SYNTHESIS RESULTS FOR OUR LOOP PROFILER ON XILINX SPARTAN-6 FPGA

Slice Logic Utilization	Used	Available	Utilization
Number of Slice Registers	135	54,576	1%
Number of Slice LUTs	359	27,288	1%
Number of Block RAMs	4	116	3%

VI. CONCLUSION AND FUTURE WORK

In this paper we propose a modified hardware profiler for profiling loops in the application layer model which can be realized as an embedded system. Comparing our approach with SnoopP, there is no need to manually enter the loop boundaries and also there is no need to re-synthesize for every application. Further, our approach also solves the limitation of SnoopP to profile loops with sub-routine call in the loop body. In comparison with the approach proposed in FLAT, there is no need to modify the RTL implementation of the processor; besides it is also more area efficient as compared to FLAT as there is no need to realize a cache controller. The proposed approach can be easily extended to profile sub-routines. The automation provided by our approach in identifying loop boundaries and the greater accuracy in the captured profile data for loops are the two key differentiating features of our proposed approach.

We believe that hardware profiling can play a crucial role in the design and implementation of current generation and future embedded systems. In future, we would like to extend this work to be able to profile sub-routines, as well as different execution threads for mapping onto multi-core processor architectures for multi-threaded embedded applications. We believe that this approach can also facilitate dynamic or on-the-fly hardware and software partitioning for many applications targeting re-configurable embedded systems.

REFERENCES

[1] J. G. Tong, I. D. L. Anderson, and M. A. S. Khalid, "Soft-Core Processors for Embedded Systems," in Proc. of the 18th International Conference on Microelectronics,December 2006, pp. 170 - 173.

[2] Ronald R. SaaS, Andrew Schmidt, "Embedded Systems Design with Platform FPGAs: Principles and Practices," Morgan Kaufmann Publishers, 2010.

[3] Frank Vahid, "What is hardware/software partitioning?" SIGDA Newsl.39(6):11, jun 2009.

[4] J. G. Tong, M.A.S.Khalid, "Profiling Tools for FPGA-Based Embedded Systems: Survey and Quantitative Comparison", Journal of Computers, Vol. 3, No.6, June 2008, pp. 1-14.

[5] J. G. Tong and M. A. S. Khalid, "Profiling CAD Tools: A Proposed Classification," in Proc. of the 19th International Conference on Microelectronics, December 2007, pp. 253-256.

[6] GNU gprof, https://sourceware.org/binutils/docs/gprof/

[7] G. Stitt, R. Lysecky and F. Vahid. "Dynamic Hardware/Software Partitioning: A First Approach." Design Automation Conference (DAC03), June 2003.

[8] Suresh, D.C., Najjar, W.A., Vahid, F., Villarreal, J.R., Stitt, G. "Profiling tools for hardware/software partitioning of embedded applications." Languages, Compilers and Tools for Embedded Systems (LCTES), 2003, pp. 189-198.

[9] L. Shannon and P. Chow, "Using reconfigurability to achieve real-time profiling for hardware/software codesign," in Proc. of the 12th International Symposium on FPGAs, pp.190-199, 2004.

[10] Xilinx Inc., "MicroBlaze Processor Reference Guide," UG081 (v12.0).

[11] El-Sayed M. Saad, Medhat H.A. Awadalla, Kareem Ezz El-Deen, "FPGA-Based Software Profiler for Hardware/Software Co-design", in 26th National Science Conference, Egypt, 2009, pp. D14.1-D14.8.

[12] Aldham, M.,Anderson, J.,Brown, S., AND Canis, A, "Low-cost harware profiling of run-time and energy in FPGA embedded processors." In Proceedings of the IEEE International Conference on Application-specific Systems, Architecture and Processors (ASAP), 6168, 2011.

[13] A. Canis, J. Choi, M. Aldham, V. Zhang, A. Kammoona, T. Czajkowski, S. D. Brown, and J. H. Anderson, "LegUp: An Open Source High-Level Synthesis Tool for FPGA-Based Processor/Accelerator Systems," ACM Transactions on Embedded Computing Systems (TECS), vol. 13, no. 2, p. 24, 2013.

[14] Ann Gordon-Ross, Frank Vahid, "Frequent Loop Detection Using Efficient Non-intrusive On-Chip Hardware" IEEE Transaction on Computers, Vol.54, No.10, Oct 2005, pp.1203-1215.

[15] K. Pagiamtzis and A. Sheikholeslami. "Content-addressable memory (CAM) circuits and architectures: A tutorial and survey," IEEE Journal of Solid-State Circuits, 41(3):712727, March 2006.

[16] Y. Hara, H. Tomiyama, S. Honda, and H. Takada. "Proposal and quantitative analysis of the CHStone benchmark program suite for practical C-based high-level synthesis." Journal of Information Processing, 17:242254, 2009.

[17] Malik, A., Moyer, W., Cermak, D. "A low power unified cache architecture providing power and performance flexibility." ISLPED, 2000.

A RISC-V Instruction Set Processor-Micro-architecture Design and Analysis

Aneesh Raveendran, Vinayak Baramu Patil, David Selvakumar, Vivian Desalphine
Centre for Development of Advanced Computing (C-DAC)
Bangalore, India
raneesh@cdac.n, vinayakp@cdac.in, david@cdac.in, viviand@cdac.in

Abstract— Micro-architecture design and analysis of a RISC-V instruction set processor has been articulated in this paper. Instruction Set Architectures (ISAs) for processors from Intel, AMD, Intel, MIPS etc. is protected through IP Rights and Infringements. Few ISAs do exist as open-source viz. Open RISC, SPARC, RISC-V etc. RISC-V ISA has been evolved from the efforts at University of California, Berkeley and has been open sourced as BSD license. This paper details the micro-architecture design and analysis of a 5-stage pipelined RISC-V ISA compatible processor and effects of instruction set on the pipeline / micro-architecture design. The design have been analyzed in terms of instructions encoding, functionality of instructions, instruction types, decoder logic complexity, data hazard detection, register file organization and access, functioning of pipeline, effect of branch instructions, control flow, data memory access, operating modes and execution unit hardware resources. The processor has been micro-architected, simulated using Blue-spec System Verilog, synthesized and analyzed on FPGA platform and 65nm and 130nm technology nodes for ASIC. The synthesis results are compared and analyzed with similar efforts on RISC-V ISA based processor core.

Keywords— *Processor Design, Processor Micro-architecture, Out-Of-Order Processor, RISC-V Instruction Set, RISC Processor, IEEE 754-2008 FPU Standard, Floating Point Co-processor.*

I. INTRODUCTION

Currently, the IC technology has advanced phenomenally to allow unprecedented implementations of computer systems on a single chip. Custom System on Chip (SoCs), where processors core(s) and cache are small part of the chip become ubiquitous; it is rare today to find an electronic product at any scale that does not include a processor core. Traditionally, instruction set architectures (ISAs) has been proprietary for commercial reasons and there is no known technical reason to release the ISAs as open source. Most of the ISAs from Intel, ARM, AMD, etc. are proprietary, guarded by intellectual property. Open source instruction set architecture like RISC-V [1] and Open RISC [2] based processors have got momentum in custom SoCs designs. The base user level RISC-V ISA from University of California, Berkeley had been released during '11 and the super-user level ISA have been released in '14.

Micro architecture implementation of Reduced Instruction Set Processor (RISC) is more complex and maintaining the Clock per Instruction (CPI) close to 1 is always challenging. The deviation of CPI from ideal value is due to several reasons such as branch miss-prediction, memory latency, pipeline hazards etc. Proposed paper analyses the processor micro architecture design and effect of ISA on pipeline and micro-architecture design with RISC-V ISA. The pipeline has 5-stages viz. Fetch, Decode, Register Select, Execute and Write Back. Pipeline is organized in such a way that integer and floating point execution units are independent to enable concurrent execution of integer and FP computations.

Section II briefs published literature on the topic, and Section III briefs RISC-V ISA. Section IV details pipeline organization and micro-architecture of our RISC-V Instruction Set Processor and Section V briefs internal micro-architecture of FP co-processor and Section VI analyzes the RISC-V ISA and its issues with pipeline stages design. Section VII narrates the testing, validation of processor core, logic synthesis and its analysis and section VIII concludes.

II. RELATED WORKS

Currently few open source ISAs (Open RISC, SPARC, RISC-V etc.) available for designing Low power RISC processor targeted on ASIC and FPGA. An implementation of LEON SPARC [3] processor at 28 nm has been evaluated with Ultra Low Voltage standard cell and memories. An instruction set extension on Open RISC ISAs [4] for digital signal processing is implemented and integrated to the base instruction set. In [5], an ISA design criteria has been proposed and synthesis of the processor instruction set from high level description of ISA has been reported.

Article [6] [7] [8] [9] analyses ISA features of RISC-V, brings out the technical specification for open licensed ISA and recommendations for an open standard for SoCs. Article [9], evaluates and compares RISC-V with EPIPHANY ISA [10] on the aspects of integer, load/store, branch and floating point instructions. Low-RISC [11] is developing fully open hardware systems based on RISC-V viz. Z-scale (a 32-bit, 3-stage, single issue, in-order pipeline), BOOM (64-bit out-of-order), FabScalar (superscalar architecture) and a vector processor (on-chip DC/DC converters on 28 nm technology node).

Shakthi project [12]proposes a RISC-V based processors as, C-Class (32-bit, 3-8 stage in-order variant aimed at 50-250 MHz),I-class(64-bit, 1-4 core, 5-8 stages, out of order, aimed at 200-1GHz), M-class(quad-threaded, up to 8 cores, up to 2.5 GHz), S-class(64-bit superscalar, multi-threaded at 1.2-3GHz with 2-16 cores), H-class(64-bit in-order, multi-threaded with 32-100 cores), T –class(security oriented 64-bit variants with tagged ISA) for various different applications. For the works [11] and [12] internal architecture details haven't been reported viz. FPU co-processor, pipeline design internals etc.

978-1-5090-0037-1/16 $31.00 © 2016 IEEE

Our architecture is a 32 bit, single core, 5-stage pipelined processor with single issue, out-of-order FPU/integer computation, and in-order commit / retire. Proposed core has been synthesized on FPGA and 65nm and 130nm technology nodes (Low Leakage &Standard Process) for ASIC. The synthesis results have been compared with similar efforts on RISC-V ISA compatible processors.

III. RISC-V INSTRUCTION SET ARCHITECTURE

A. ISA Characteristics

ISA provides user and supervisor level instructions to enable support for software stacks including Operating System. Base RISC-V ISA has fixed length 32-bit instructions that must be naturally aligned on 32-bit (word) boundaries. RISC-V ISA supports both integer and floating point operations. Floating point operations are compatible with IEEE 754-2008 standard [9].RISC-V ISA is defined as base integer ISA, which doesn't have any architecturally visible delay slots for branch operation and supports optional variable length instruction encoding.

B. Register File organization

RISC-V supports both integer (32, 32-bit) and floating point registers (32, 64-bit) for user and supervisor level processor modes. For single precision floating point data types, the lower 32-bit of floating point register is valid. The register 'R0' is hardwired to zero.

C. Instruction Classification

The RISC-V instructions are categorized based on the functionality and op-code (instruction [6:0]).Base instruction set (RV32) is classified into 8 categories.

TABLE I: INSTRUCTION GROUPING

Category	Op code	Instructions
Branch Inst.	11_000_11	BEQ, BNE, BLT, BGE, BLTU, BGEU
Load Instructions	00_000_11 00_001_11	LB, LH, LW, LBU, LHU FLW, FLD
Store Instructions	01_000_11 01_001_11	SB, SH, SW FSW, FSD
Arithmetic and Logic (ALU) Instructions	00_100_11	ADDI, SLTI, SLTIU, XORI, ORI, ANDI, SLLI, SRLI, SRAI,
	01_100_11	ADD, SUB, SLL, SLT, SLTU, XOR, SRL, SRA, OR, AND, MUL, MULH, MULHSU, MULHU, DIV, DIVU, REM, REMU
Atomic Operations	01_011_11	LR.W, SC.W, AMOSWAP.W, AMOADD.W, AMOXOR.W, AMOAND.W, AMOOR.W, AMOMIN.W, AMOMAX.W AMOMINU.W, AMOMAXU.W
System Instructions	11_100_11	SCALL, SBREAK, RDCYCLE, RDCYCLEH, RDTIME, RDTIMEH, RDINSTRET, RDINSTRETH, FRCSR, FRRM,FRFLAGS, FSCSR, FSRM,FSFLAGS, FSRMI, FSFLSGSI
Special, unconditional jump, Synchronization	01_101_11 00_101_11 11_011_11 00_011_11	LUI, AUIPC, JAL, JALR, FENCE, FENCE.I

Table I depicts the instruction grouping based on the op-code field and functionality. Both single and double precision floating point operations are classified into FPU related instructions. Each category of instructions is subdivided as subgroup based on the function field (bits [14:12]) in the instruction.

IV. PIPELINE ORGANIZATION AND MICRO ARCHITECTURE OF RISC-V INSTRUCTION SET PROCESSOR

RISC-V instruction set processor adopts typical RISC micro-architecture with in-order fetch, out-of order execution and in-order completion. Proposed architecture comprises of 5 pipeline stages viz. Fetch, Decode, Register Select (RS), Execute (EXE) and Write Back (WB). Fig 1 shows the architecture of 5-stage pipelined processor. In fetch stage, instruction is fetched from instruction memory from location indicated by Program Counter (PC). Instruction memory sends instruction to decode stage for instruction information extraction. The extracted information is forwarded to register select stage. From register select stage, the pipeline is split into three concurrent pipeline units, as integer instructions pass through integer execution pipeline, memory instructions pass through the memory pipeline and floating point instructions pass through the FP execution pipeline unit. Write back stage receives the responses from all of these three concurrent execution stages and based on the scheduling of the instruction, WB stage allows commit the instructions in-order.

A. Fetch Stage

Fig 2 shows the micro architecture of fetch unit. It computes the current PC and predicts the next PC value. Fetch unit has a 32 bit register to hold the current PC value, a 32-bit adder to calculate the next PC by adding 4 to the current PC. Next PC is predicted either as PC+4 or by using a sophisticated branch prediction scheme [12] [13].

Branch prediction scheme consists of Branch Target Buffer (BTB) and Branch Predication (BP) unit. BTB unit outputs a Boolean valid signal along with target PC address. BP unit provides Boolean information which contains whether the PC value is present or not. A multiplexer selects either PC+4 or predicted PC as next PC value based on the valid and prediction signals. The PC and next PC are given to the decoder stage for further processing. Branch predication unit is explained in detail in Section VI.

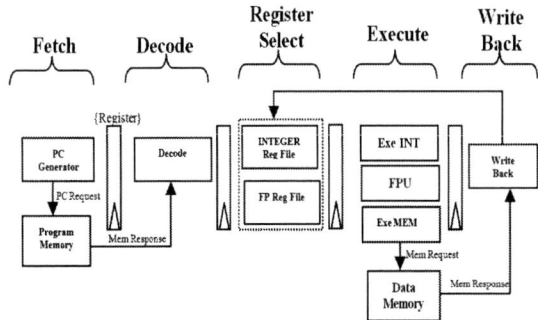

Fig 1: Top level architecture of RISC-V Processor

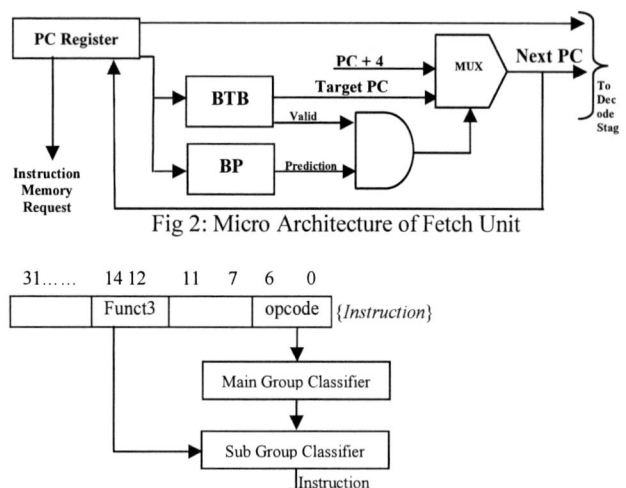

Fig 2: Micro Architecture of Fetch Unit

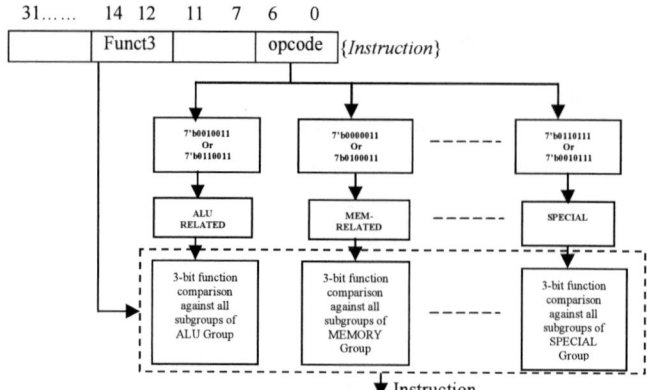

Fig 4: Micro-Architecture of instruction decoder

Fig3: Flow Diagram of Instruction Decoder

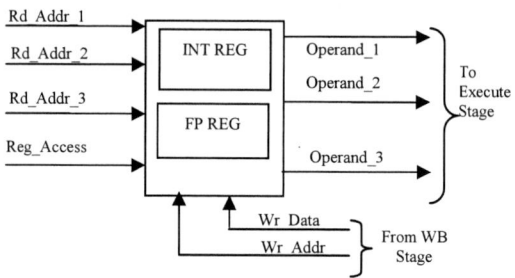

Fig 5: Register File Organization

B. Decode Stage

Instruction decoder receives the instructions from program memory with corresponding PC and predicted PC from fetch stage. Fig 3 depicts the flow diagram of instruction decoder. Based on the lower 7-bit (Op-code) of instruction, instructions are classified as per Table 1. Subgroup information is decoded from 3-bit field (function field). Address of sources and destination registers are of 5-bit field each and their position is fixed irrespective of type of instruction. If an immediate type instruction has occurred, the immediate data is sign-extended to 32-bit for further processing.

Fig 4 shows the micro-architecture of RISC-V instruction decoder. Op-code is given to main group classifier, which gives main group class as output as per Table 1. Main group classifier has seven parallel comparators (7-bit comparator) to decode main group class. Instruction op-code is compared against seven categories (Table 1) of op-codes concurrently. If any of the comparators' output goes high, corresponding main group signal is generated otherwise the instruction being decoded is considered as ILLEGAL. Further, the subgroup classifier determines the exact operation to be performed by the instruction. Subgroup is determined by using both 3-bit function field and main class. If 3-bit function field of instruction matches with 3-bit fields of the decoded main class, sub group classifier generates corresponding subgroup category under the decoded main class, otherwise the instruction falls under ILLEGAL.

Decoder generates control signals for further processing of the instruction. Control signals include, Register Access, Register update, and pipeline information (the pipeline on which the instruction is to be scheduled). Register Access can be integer register file access (INT-ACCESS) or floating point register access (FP-ACCESS) or NO-ACCESS.

Register update gives which register file needs to be written back after executing the instruction. Instruction processing is organised through two independent pipelines for integer ALU / Memory Address computation, and FPU. EXPIPE signal indicates the pipeline through which the instruction has to be processed.

Apart from decoding the instruction, decoder detects data dependency (Hazard Detection) between instructions which is explained in section VI. Decoded information is forwarded to register select stage in the form of packet along with PC and predicted PC.

C. Register Select Stage

RS stage accepts decoded information of instructions from decoder stage and selects the operand from either integer or floating point register file. RS stage contains an integer (32-No's of 32-bit wide) and floating point register file (32-No's of 64-bit wide).Register file is implemented as a Random Access Memory (RAM), which has a latency of one clock cycle with three read ports and one write port.

Fig5 shows the register file organization for RISC-V processor. Register file unit accepts three source addresses (Rd_Addr_1, Rd_Addr_2, Rd_Addr_3} and a control signal (Reg_Access) which specifies the access of Register files (Integer Access or Floating point Access). Register file output is 64-bit which holds the data from either integer or floating point register file and has one 64-bit write port for write back.

978-1-5090-0037-1/16 $31.00 © 2016 IEEE

Based on the pipeline information from decode stage, operands after register select are passed either to integer ALU / memory address computation pipeline or FP execution pipeline. An instruction scheduler FIFO which is part of integer / memory / FPU execute stage is used to schedule the instructions in the various pipelines to commit the instructions in-order.

D. Execute Stage

Execute stage consists of three concurrent units for Integer arithmetic and logic operations etc., operand memory address computation and Floating point computation. Integer execution performs arithmetic (Addition, subtraction, multiplication and division) and logical (AND, OR, XOR and shift) operations. Also, Integer arithmetic unit calculates the target address for unconditional or conditional jump and branch instructions. Integer execution unit executes the system related instruction such as SCALL, SBREAK instructions for supervisor level access of the instruction. An efficient data forwarding scheme is used to forward output of execution units to the input of the execution unit. Detailed micro-architecture for data forwarding scheme is explained in section VI.

Unit for Memory related instructions calculate the target data memory address for load and store operations. RISC-V ISA supports load or store operations on a byte, half-word and word data to and from data memory. Micro-architecture of an out-of-order FP execution unit is explained in section V.

E. Write Back Stage

Write Back (WB) stage commits the instruction from the pipeline and updates the register file with the results from execute units. WB reads the instruction from top of the scheduled instruction FIFO and based on the pipeline information, it reads either from integer or memory or floating point concurrent units.

V. MICRO ARCHITECTURE DESIGN OF OUT-OF ORDER FLOATING POINT UNIT

Fig6 depicts the top level architecture of floating point co-processor. Instruction decoder partially decodes the floating point instructions, after register select operations, the function fields with op-code are passed to the out-of-order co-processor. Floating point data operands to the co-processor are selected from the register select unit. Write back stage accepts the responses from co-processor and updates the destination register. To make the use of out-of-order, functional modules are developed individually, integrated as a single module with specific token for instruction scheduling.

A. FPU request

Input request packet contains operand (192-bit) for floating point operation, op-code and function fields. Fig7 shows the FPU request packet format. Each FPU request is stored in input FIFO of depth 2 which is further read by input decoder. Input operand width is fixed to 192-bit which holds 3 data operands, either a 32-bit (single precision or integer) or 64-bit (double precision) floating point.

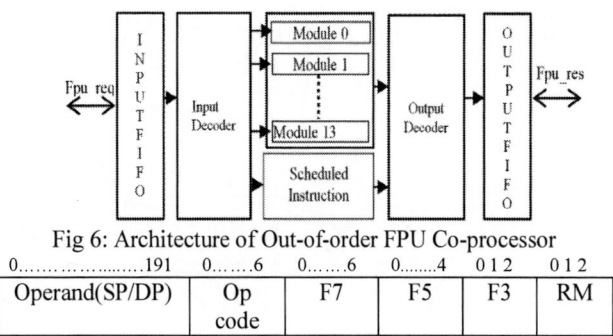

Fig 6: Architecture of Out-of-order FPU Co-processor

Operand(SP/DP) 0................191	Op code 0.......6	F7 0.......6	F5 0.......4	F3 0 1 2	RM 0 1 2

Fig 7: FPU request packet

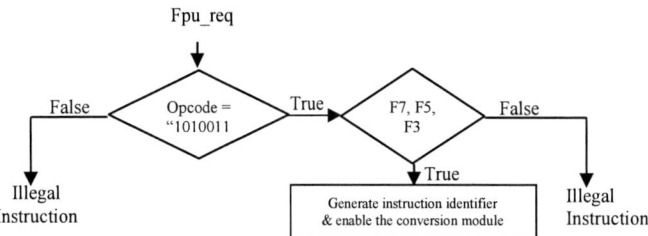

Fig 8: Flow diagram of Input decoder

Input request packet contains operand (192-bit) for floating point operation, op-code and function fields. Fig7 shows the FPU request packet format. Each FPU request is stored in input FIFO of depth 2 which is further read by input decoder. Input operand width is fixed to 192-bit which holds 3 data operands, either a 32-bit (single precision or integer) or 64-bit (double precision) floating point.

B. Input Decoder

Fig8 depicts the flow diagram of input decoder. Input Decoder in floating point pipeline reads the FPU request packet from the Input FIFO, decodes the floating point operation based on the op-code, F7, F5 and F3 fields. After decoding, corresponding individual hardware module is enabled and instruction token for the specific module is stored in the scheduled instruction FIFO. Illegal floating point operations are tagged with illegal token and stored in the scheduled instruction FIFO

C. Scheduled Instruction FIFO

Scheduled instruction FIFO stores the instruction tokens, which specifies present operations in the pipeline. Scheduled instruction FIFO has a depth of 6, which is equal to the maximum number of pipeline stages in the co-processor.

D. Functional Units

Functional units are for performing various floating point operations. Various FPU instructions are decoded to perform out-of-order execution in these functional units. Addition and subtraction operations share the same functional unit pipeline and have a clock latency of 5. Conversion operations have a latency of 5 clock cycles. Multiplication, division and square root operations have a clock cycle latency of 11, 31 and 34 clock cycles respectively. Operation results and exceptions are raised according to the IEEE 754-2008 standard.

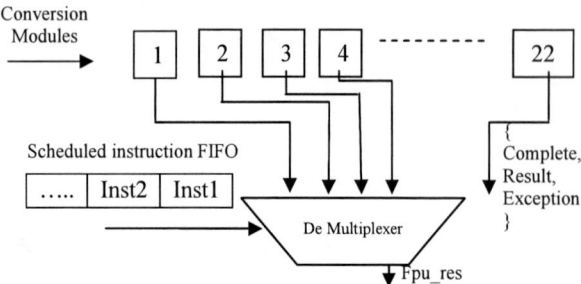

Fig 9: Flow diagram of Output decoder

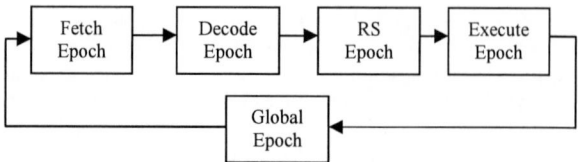

Fig 10: Epoch Register Mechanism

E. Output Decoder

Fig 9 shows the flow diagram of output decoder. Output decoder commits the oldest instruction from pipeline. To ensure in-order commit, output Decoder in the pipeline reads the scheduled instruction from the top of the scheduled instruction FIFO and polls the completion signal from the matched individual module. Once the floating point operation is completed, the result (32-bit or 64-bit), exceptions (5-bit) from the module is read and stored in output FIFO which has a depth of 2. On completion of instruction execution, first token from the top of the instruction scheduler FIFO is removed.

F. FPU response

Output FPU response packet holds maximum 64-bit result which can be of either 32-bit (single precision floating point or integer) or 64-bit (double precision floating point) result. Exceptions from each individual module are collected and stored in floating point Control and status registers.

VI. DESIGN ISSUES AND OPTIONS IN RISC-V PIPELINE

Proposed core is designed by considering various aspects like hazard, grouping of instructions, decoder logic complexity, register file access, and pipeline operations. RISC-V pipeline is associated with two hazards; data hazards and control hazards.

A. Data hazards

Data hazard (Read-After-Write (RAW)) in RISC-V pipeline occurred due to data dependency between instructions. RISC-V decoder stage has two 32-bit registers (named as scoreboard) keep the usage status of integer and floating point registers by the instructions in the pipeline by means of destination register address. In decode stage, an instruction will get decoded and destination register position will get updated in specified scoreboard. When an instruction completes its operation at the write back stage, corresponding bit position in scoreboard is reset. Performance impact of data hazard (RAW hazard) on pipeline stage is handled by data forwarding scheme. Data forwarding scheme is applicable only in case if both current and previous instructions are of integer type.

Data dependency in pipeline stages can be determined with 2-scoreboard registers and current instructions; pipeline access (integer, memory, FPU), register access, and register update information (instruction information). Output of data hazard detection unit specifies whether, integer execution output have to forward the data to input. Data forwarding unit is a combinational unit. Data hazard in RISC-V pipeline such as

RAW, WAW, WAR are handled in pipeline by either forwarding the data for RAW or by introducing bubbles in the pipeline for WAW hazard. Since, proposed core is in-order commit WAR hazard will not occur in our RISC-V pipeline.

B. Control hazards

Control hazard alias branch hazard causes a greater performance loss in RISC-V pipeline than data hazard. RISC-V ISA has unconditional jump and conditional branch instructions, causes clock cycle penalty during branch miss prediction. In our RISC-V pipeline, prediction is micro-architected by using sophisticated branch predication scheme and miss- prediction is micro-architected using EPOCH register mechanism.

1. Branch Miss Prediction

In RISC-V pipeline the branch target address is calculated in execute stage, meanwhile three successive instructions from branch instructions are entered into the pipeline, which have to be flushed or not based on the outcome of branch instruction. For jump instructions the pipeline needs to be flushed, which leads to a minimum penalty of 3 clock cycles. RISC-V branch instructions are conditional, if the branch condition is satisfied causes a penalty of 3 clock cycles and the pipeline has to be flushed. Epoch register mechanism is used to flush the instructions in processor pipeline. Basically the meaning of epoch is storing the history of something. Proposed processor contains a global epoch register common to all pipelines stages along with local epoch registers for each pipeline stage.

Epoch registers are used with single bit granularity. Fig 10 shows the orientation of Epoch register in each pipeline stage. The mechanism of epoch register concept is as follows: Initially all the epoch registers are initialized with logic zero. Since, the stages in the pipeline are sequential, instructions pass stage by stage. Along with the instructions, the global epoch value is circulated from one stage to the next stage till execute stage. When a branch instruction is encountered in pipeline, the target address of that particular branch is calculated in execute stage. In execute stage; the predicted PC is compared with the target address. If the target address does not match with the predicted PC, the global epoch register value is inverted and the PC is updated with the target address. Therefore, for the first miss prediction, global epoch becomes logic one. In the next clock cycle, the global epoch and local epoch registers of execute stage are compared. If the local epoch register value in execute stage mismatches with the global epoch register, the instruction will be discarded.

978-1-5090-0037-1/16 $31.00 © 2016 IEEE   128

2. Branch Prediction

Clock cycle penalty caused by flushing the pipeline is reduced by implementing a branch prediction scheme. Fig 2 shows the integration of Branch predictor to the fetch unit. A sophisticated branch prediction of [6] [7] [8] scheme is adopted in our processor to realize Branch prediction and Branch Target Buffer.

C. Grouping of instruction& Decode Level logic

RISC-V instructions are categorized based on the functionality and op-code (instruction [6:0]). Base instruction set (RV32) is classified into 8 categories which are elaborated in section III. Decoder uses fixed levels of different comparators to identify an instruction. Proposed decoder is organized in such a way that system instructions are having maximum level of logic with 7, 3, and 2-bit comparators

D. Register File Access and pipeline operations

RISC-V ISA supports, both integer and floating point registers. Operand accessed from either integer or floating point register file.It is determined at decoder stage. In effect, the register file access requires additional combinational module (multiplexer) for selecting the operand access.

VII. VERIFICATION AND ANALYSIS

A. Verification of core

Integer and floating pipelines of RISC-V Instruction Set Processor are developed by using Bluespec System Verilog (BSV), integrated with FPGAs Block RAM which serves as L1 cache. Floating point pipeline is verified at test bench level using pseudo random test vectors. For pseudo- random verification at test bench level, the input test vectors are generated using a high level language ("C") and applied to the Device Under Test (DUT) and also to the Matlab tool. The Matlab tool generated outputs are considered as the golden output and are compared against the outputs from the DUT. Core pipeline is verified with hex file generated using hand coded assembled instructions.

B. Implementation on FPGA

Processor core isprimarilysynthesized on Xilinx Virtex 6 "xc6vlx550t-2ff1759" FPGA. Table II depicts the devices utilization of proposed core on FPGA, open RISC and Rocket core (RISC-V implementation from UCB).

TABLE II: DEVICE UTILIZATION OF PROCESSOR CORE ON FPGA

Parameter	Proposed Core (with FPU)	Open RISC [17] (without FPU)	Rocket Core [18] (with Limited FPU ops)
Slice Registers	18340	2280	12388
Slice LUTs	46530	6744	46256
LUT-FF pair	11759	7063	5425
DSP48E1s	32	4	22

TABLE III: SYNTHESIS RESULTS OF CORE – 65nm AND 130nm

Parameter/Technology Node		65 nm		130 nm	
		LL @ 500MHz	SP @ 500MHz	LL @ 500MHz	SP@ 500MHz
Logic Cells	Proposed (with FPU)	173 K	171 K	1424 K	1287 K
	Rocket Core [18] (with Limited FPU ops)	121 K	121 K	1328 K	1178 K
Area	Proposed	561 mm^2	557 mm^2	2539 mm^2	2408 mm^2
	Rocket Core[18] (with Limited FPU ops)	451 mm^2	465 mm^2	2118 mm^2	1824 mm^2

C. Implementation on ASIC

Processor has been synthesized using Cadence Encounter RTL compiler using UMC 65nm and 130 nm with Low Leakage (LL), Standard Performance (SP), and High Performance (HP) libraries. Table III depicts the device of proposed core, open RISC and Rocket core.

Our instruction set processor consumes more hardware resource which is attributed to, that it is with independent hardware for FP add/sub, multiply, divide, square-root etc., out-of-order execute, in-order commit, deeply pipelined FP arithmetic units which improves throughput of FPU operations.

VIII. CONCLUSIONS

A 5-satge pipelined 32-bit instruction set processor compatible with RISC-V ISA has been micro-architected and analyzed the design issues / options for pipeline stages in terms ofgrouping of instruction, decoder logic complexity, register file access and control flow instructions. Processor includes a sophisticated branch prediction and advanced data hazard detection with data forwarding unit and IEEE 754-2008 compliant out-of-order execute and in-order-commit FPU.Both integer and floating point pipeline are verified at test bench level,core is implemented on Xilinx Virtex 6xc6vlx550t-2ff1759 FPGA and ASIC (UMC 65 and 130 nm library).

IX. FUTURE WORKS

In future, the core has been proposed to be enhanced with more rigorous data forwarding schemes between pipeline stages, concurrent load/store operations with load/store queue, and cache controller. With such features it is planned to test and verify the core with compiler generated code.

ACKNOWLEDGEMENTS

We express our sincere gratitude to Prof. AmruturBharadwaj, Department of Electrical Communication Engineering, Indian Institute of Science (IISc), Bangalore for helping us in synthesizing the core on various technology nodes.

REFERENCES

[1] A. Waterman, Y. Lee, David A. Patterson, KrsteAsanovi The RISC-V Instruction Set Manual,Vol. I: Base User-Level ISA. Tech. Rep. UCB/EECS-2011-62, EECS Department, UCB, May 2011.

[2] OpenRISC 1000 Arch. Manual, 1st Ed., OpenCores, 2012.

[3] Clerc S. Crolles, Abouzeid, F. Patel, D.A. Daveau, J.-M. Bottoni, C. Ciampolini L, Giner F, "Design and Performance Parameters of an Ultra-Low Voltage, Single Supply 32-bit Processor implemented in 28nm FDSOI Technology" Proc. of IEEE Intl. Symp.on Quality Electronic Design, pp. 366-370, 2015.

[4] Lopez-Parrado, A.Valderrama-Cuervo,"OpenRISC-based System-on-Chip for Digital Signal Processing" Proc. of IEEE symposium on Image, signal processing and Artificial Vision, pp. 1-5, Sept 2014.

[5] AndreyMokhov, Iliasov, Sokolov, Rykunov, Yakovlev, Romanovsky"Synthesis of Processor Instruction Sets from High-Level ISA Specifications", IEEE trans on computers, Vol:63, pp. 1551-1565, June 2014

[6] Liney Group article "The case of open instruction sets, open ISA would enable free competition in processor design". August 2014

[7] Krste Asanović & David Patterson, UC Berkeley, EE Times article " RISC-V Open standard for SoCs, The case of Open ISA" July 2014

[8] Krste Asanović David A. Patterson "Instruction Sets Should Be Free: The Case For RISC-V" 2014

[9] Andreas Olofsson "Analyzing the RISC-V Instruction Set Architecture" August 2014.

[10] User Manual "Epiphany Architecture Reference"

[11] Open source Processor Development "lowRISC" 2015

[12] Shakthi Processor series IIT Madras 2015

[13] T-Y Yeh and Y.N. Patt, "Two-Level Adaptive Branch Prediction", Technical Report CSE-TR-117-91, Computer Science & Engineering Division, Department of EECS, University of Michigan, (Nov. 1991).

[14] Chang, P.-Y., Maris Evers, Y.N. Patt "Improving branch prediction accuracy by reducing pattern history table interference", 1996 International Conference on Parallel architecture and compilation techniques, pp. 48-57, 1996

[15] IEEE standards Board and ANSI. IEEE Standards for Binary Floating-point Arithmetic, 2008, IEEE Std., 754-2008.

[16] J. Hennessy and D. D. Patterson, "Computer architecture, a quantitative approach". Morgan Kaufman, 2003.

[17] Damjan Lampret et al., "OpenRISC 1000 Architecture Manual", Architecture Version 1.0, Document Revision 0, December 5, 2012

[18] RISC-V processor core "Rocket Core" EECS Department, UCB, May 2011.

Reconfigurable Side Channel Attack resistant True Random Number Generator

Vijay Bahadur, David Selvakumar, Vijendran, Sobha.P.M
Centre for Development of Advanced Computing(C-DAC)
Bangalore, India
vijayb@cdac.in, david@cdac.in, vijendrann@cdac.in, sobhapm@cdac.in

Abstract—Random Number Generators (RNGs) play an important role in cryptography. The security of cryptographic algorithms and protocols relies on the ability of RNGs to generate unpredictable secret keys and random numbers. This paper presents an implementation of Side Channel Attack resistant Galois Ring Oscillator (GARO) based True Random Number Generator (TRNG) on FPGA. To study and prove the robustness of the random number generator against placement sensitivity, due to various physical properties of logic elements and thermal variations of FPGA, the design (single instance of GARO) was implemented at four different quadrants in the FPGA and the generated random bit streams were analyzed. Such designs enable resilience against side channel attacks by injection locking. Further, to prove that the implemented TRNG is resilient against side channel attack (Electromagnetic Injection (EM) Attack, Frequency Injection Attack) the frequency spectrum of GARO was captured and analyzed. It was observed that the output of GARO is not dominated by any single frequency unlike non-GARO based ring oscillator which makes it difficult to get locked due to EM / Frequency injection at the specific oscillator frequency. The output bit-stream has been sampled from multiple spatially distributed TRNG units by round-robin. National Institute of Standards and technology (NIST) statistical test suite has been used to benchmark the statistical properties of generated random bit streams and bit streams fulfills all the test suite requirements.

Keywords—Field Programmable Gate Array (FPGA), Ring Oscillator (RO), Electromagnetic Analysis, TRNG, Latch (L), Side Channel Attack (SCA), Electromagnetic Injection Attack on RNG, Frequency Injection Attack, Injection Locking, Xilinx ISE.

I. INTRODUCTION

Random Number Generators, an important security primitive in cryptography, are used to generate random numbers for different essential tasks like generation of secret keys, Initialization Vectors (IV), padding bits, seeds and nonce (numbers used once). An RNG uses a non-deterministic source (an entropy source), along with some processing function (an entropy distillation process) to produce randomness. Post-processing is needed to overcome any weakness in the entropy source that could result in the production of non-random bit streams.

RNGs can be classified as Pseudo Random Number Generator (PRNG), True Random Number Generator (TRNG) and Hybrid Random Number Generator (HRNG). A PRNG requires an input called the seed. The randomness of the output bit streams of PRNG completely depends on randomness of the input seed. This means that for a given seed the output random bit stream of PRNG would always be same at any instance of time. PRNG based system uses system clock, mouse movement, network traffic etc. to generate a seed. Hence, this system is vulnerable to security attacks [1].

In TRNG, no explicit input data such as seed is required since the entropy source is internal to the system itself. The internal entropy source is based on uncertain physical process. HRNG is a cascaded system with TRNG and PRNG, where the TRNG generates seed value for PRNG. HRNG would be useful when the rate of bit generation of TRNG is very less. In general, any TRNG system will have an entropy source for uncertainty and a sampling circuit which samples the random bit from entropy source without affecting the uncertainty. TRNGs are strongly recommended for applications that requires high level of security.

As Random numbers are very crucial in cryptography and other areas, security of RNGs become a very important factor. PRNGs are susceptible to attacks [1] which are generally software based. Hence, there is greater demand for hardware based TRNGs during recent times. Hardware random number generators are not completely secure, since attacks are possible on hardware based random numbers also. Side channel attack is one such possible attack [2] [3]. Side channels are covert channels which are present in the designed hardware and leaks information like power, timing, electromagnetic radiation, etc.

Various TRNG designs have been proposed and implemented on FPGAs. Each of these designs uses a different mechanism to extract randomness from an underlying physical phenomenon. Main source of randomness could be thermal or shot noise in the circuits, clock jitter, Metastability in circuits especially in flip-flops, Brownian motion, atmospheric noise and nuclear decay. FPGAs are popular platforms for implementing TRNGs because of their flexibility, small design cycle and re-configurability.

This paper is organized as follows: Section II explains the literature survey of FPGA implementations of Ring Oscillator (RO) based TRNGs. Section III explains our implementation of side channel attack resistant ring oscillator based TRNG on FPGA. A description of post-processing technique is presented in Section IV. Section V details the design options, implementation and design analysis on side channel attack resistance of the TRNG. The results of the NIST statistical

978-1-5090-0037-1/16 $31.00 © 2016 IEEE

tests on the random sequence generated by the proposed design are described in Section VI, Section VII briefs the performance metrics and RO based TRNG architecture is concluded in section VIII.

II. RELATED WORK ON RO BASED TRNGS

Most of the TRNG implementations on FPGAs are based on free running oscillators. Oscillators provide a simple and effective method to build TRNGs because no analog component like PLL is required and it provides a great source of entropy which is easily convertible into digital form [4] [5]. A simple digital oscillator may be built by chaining an odd number of inverter gates in a ring configuration. Due to the feedback path, the output of any of the inverters will oscillate from a logic one to a logic zero and vice versa. Hence, a square wave signal is obtained by tapping the oscillator at any point in the ring.

A theoretical basis of TRNGs based on sampling phase jitter in oscillator rings has been discussed by Sunar et.al. [4]. Source of entropy is accumulated phase jitter on oscillations of RO and the entropy distillation process is XOR gate which receives input from multiple rings and a sampling Flip-flop. This design has been implemented with large number of oscillator rings of fixed length and a resilient function as post - processing unit in FPGA by Schellekenset.al. [6].The proposed and implemented design in [6] uses more FPGA resources as it needs many oscillator rings to obtain the required randomness in the generated bits. Knut Wold [4] proposed a slightly modified design which introduces flip-flop pair at the end of each oscillator ring to remove metastability caused by sampling free running oscillator which has been illustrated in Fig. 1. This enhanced design eliminates the need for complicated post-processing required in Sunar's method detailed in [4].

Another TRNG which generates an order of magnitude higher entropy rate than the classical RO by Sunaret.al was proposed by M. Dichtl and Golic [7][8]. In this design, Galois and Fibonacci ring oscillators were introduced as shown in Fig. 2 and Fig. 3.

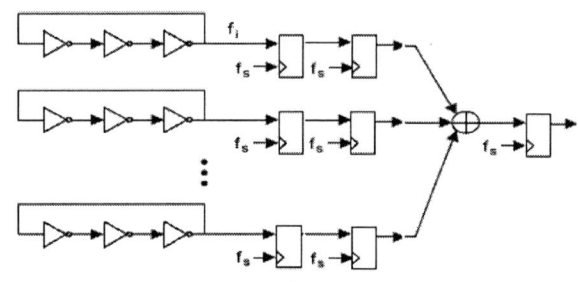

Fig. 1: Enhanced Classical RO based TRNG

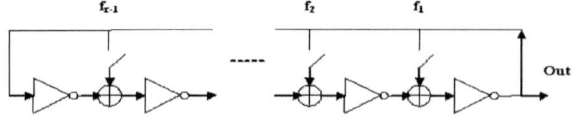

Fig. 2: Galois Ring Oscillator

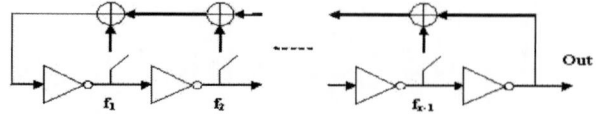

Fig. 3: Fibonacci Ring Oscillator

These ring oscillators consist of cascade of inverters with XOR logic gates corresponding to the feedback polynomial. The simple feedback in classical RO based implementation is replaced by complex feedback polynomial which introduces more entropy rate. The feedback connections are chosen so as to satisfy the basic condition that there are no fixed points in the corresponding state-transition function.Tarsaet.al[5], presented a Galois Ring Oscillator (GARO) based TRNG implementation by adding a flip-flop pair at the end of each Galois ring oscillator circuit as described in [9].

In 1998 John Kelsey [1] discussed various attack techniques like input based attacks, state compromise extension attack etc., which makes PRNG easily prone to attacks and the attacks are on software based implementations. There were no known attacks on TRNG before [2]. In 2009 Markettos, A.T. [2] proved that frequency injection attack on ring oscillator (relatively prime rings – 3 elements and 5 elements) based TRNG is possible and the bias of TRNG can be controlled by locking the frequency of ring oscillator. The frequencies are injected into the power supply lines. This attack has practical limitations like finding the working frequency of RO and locking frequency. Mutually-prime rings are more difficult to lock to the injection frequency [2] as compared to identical rings. Further, [2] indicates few defenses viz. combining TRNG feedback loop with logic or register elements, reduced asymmetry in the rings, and differential ring oscillators. In 2013, Pierre Bayon [3] presented the way to find the working frequency of ring oscillator by analyzing the electromagnetic radiation.

III. GALOIS RING OSCILLATOR BASED TRNG IMPLEMENTATION

The proposed implementation has random bit stream generator based on GARO which adopts a primitive polynomial of order 31.The design has been experimented with various primitive polynomials to form Galois Ring Oscillator and the results are discussed in section VI. Fig.4shows the architecture of the GARO based single random bit stream generator comprising 32 instances of 31inverter stages GARO.A GARO based TRNG, as shown in Fig. 4, and presented in [5] was implemented on a PCIe based Virtex5 FPGA card. The random bit stream is sampled at 100 MHz. The throughput of random bit generation is 100Mb/s.

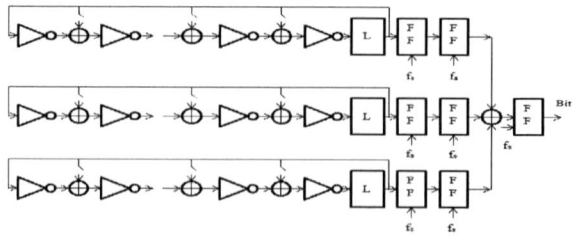

Fig. 4: Galois Ring Oscillator (1, 2...32 No.) based TRNG

978-1-5090-0037-1/16 $31.00 © 2016 IEEE

Table 1 shows the period of Galois ring oscillator implemented on Xilinx xc5vlx155 device. Since, the sampling frequency of TRNG is less than the ring oscillator frequency there is no possibility of sampling the same bit two times. The experimentation with GARO rings was done which includes the implementation of GAROs with different polynomials. Though internal routing delays in FPGA introduce different delays in each oscillator ring, the randomness resulted from implementing GAROs with different polynomial and randomness resulted from implementing GAROs with same polynomial was same. This is because while placing the GAROs in FPGA the Xilinx ISE tool chooses random LUTs to place logic and also routing of each GARO differs.

Ring oscillator circuit has been implemented with basic FPGA primitive. LUT (Look-Up Table) is used to implement an inverter and XOR operation. For the inverted logic, LUT1 is used with one input and one output configuration; for XOR logic, LUT2 is used with two input and one output configuration. Presence of LUT2 is based on the occurrence of logic '1' in the polynomial. In FPGAs, memory elements such as flip-flop, latch and RAM have an initial value. During the FPGA configuration, this initial value will be assigned to the respective memory elements. But other combination logic elements like LUT, MUX etc. will not have any initial value as they do not have any memory.

The ring oscillators are composed with only odd number of inverter chains which does not have any initial value. During simulation, this would result into unknown values. In order to initialize the ring oscillator with logic zero, a latch (L) is cascaded in the feedback path shown in Fig.4.This initial value will not bias the random bit stream because the random stream will be collected after the reset goes inactive and also the presence of latch in the GARO increases the jitter period. The presence of latch in the ring also forces the synthesis tool to infer ring oscillator structure shown in Fig.4without any modification.

The ring oscillator inverter logic was implemented by instantiating LUTs configured as an inverter, rather than the logic inference. Normally, Xilinx Synthesis Tool (XST) reduces the logic equation to avoid additional hardware utilization and the chains of odd inverters would finally become a single inverter. In order to keep the circuit as it is during implementation process, KEEP attribute was used in the HDL code.

Output of each oscillator ring is passed through two flip-flops (FF) in order to overcome the metastability due to asynchronous input to sampling circuit. The flip-flop pair at the end of each ring not only improves the randomness of the generated bits but also makes the output of the last flip flop stable. The output bits of last flip flop are the random bits and these random bits passed the NIST statistical tests for random bit stream as shown in next section.

TABLE I: RING OSCILLATOR DATA ON XILINX XC5VLX155

Parameters	Value
RO Length	31
No. of Rings	32
Period (ns)	5.8

The period of oscillator rings are sensitive to thermal variation which causes propagation delay differences in LUTs. This sensitivity of oscillator ring frequency to inherent physical properties of LUTs, variation in FPGA input voltage and thermal conditions cause the jitter in square wave signal generated by GARO. The entire design (32 No. of RO and each RO with 31 inverter stages) was implemented in four different quadrants of the FPGA at different instance of time to study the effect of placement/location of TRNG circuits inside FPGA on the randomness properties of the generated bit stream sequence. Xilinx location constraint (LOC) was used to place the design in four different quadrants. For each of the four implementations, the Xilinx tool adopts different internal routing process which causes different delays in oscillator rings. The resultant bit streams of the four different implementations were analyzed and it was observed that placing the design at different locations does not affect the randomness of the generated bit streams. The TRNG could be placed anywhere on the FPGA other than the above identified four quadrants as well. Based on these results it has been concluded that location of TRNG circuit won't influence the randomness of the bit stream.

In-order to increase the random behavior of TRNG module, a multi-core TRNG system is designed and the architecture is depicted in Fig. 5. Proposed multi-core TRNG system consists of 16 individual TRNG modules, a data or valid selector unit and a block RAM of size 4096x64 to store the random stream of data. Each TRNG module has a free running 32-GARO (31 stages) and the output of GARO is sampled at 100MHz. The estimated and measured GARO frequency of oscillation is 160MHz (approx.). Each GARO consists of a LFSR (primitive polynomial based) and a latch. Inverters in LFSR and latch are implemented using FPGA hardware macros. On a clock edge, random bit from 32-GARO modules get collected in a one bit register and bit-wise

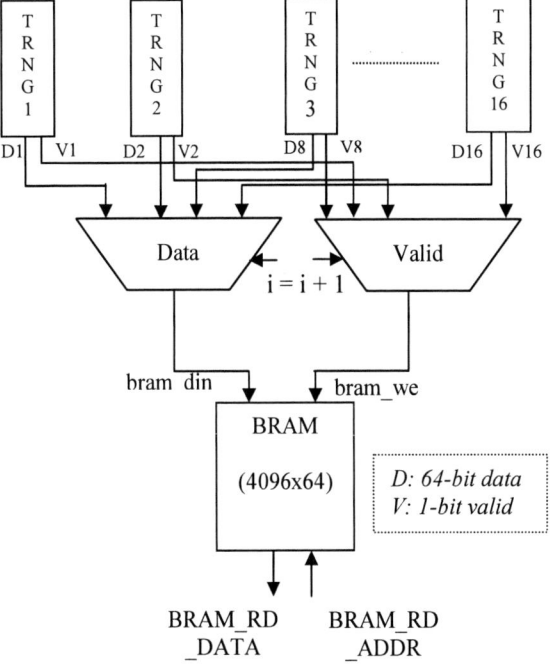

Fig. 5: Architecture of multi-core TRNG system

XOR of all 32 registers is done to obtain a single random bit. A 64-bit register is used to collect these random bits and form a 64-bit data. To mitigate the injection locking based attack and not compromising the randomness of the bit stream, such 64 bit random stream from each of 16 TRNG modules is collected in round-robin and stored in the BRAM (4096x64). The above description is without any post-processing for whitening the random sequence. Section IV discusses application of Von Neumann corrector at the 64-bit output.

IV. POST-PROCESSING TECHNIQUES

Fig.6 depicts the general architecture of an RNG implementation with post-processing. The quality of RNG depends on three components: a source of randomness, a sampler circuit and a post-processing block. In this paper a source of randomness is nothing but the clock jitter present in oscillator output, as discussed in previous sections. The sampler collects the entropy from the randomness source without disturbing the source of randomness. Post-processing is necessary to strengthen the design by reducing the bias in the generated random bits, but at an expense of reduced speed of output generation because post-processing techniques apply compression function on the digitized noise signal.

Simple and commonly used post-processing techniques are XOR and Von Neumann correctors. Complex post-processing techniques include linear compression functions based on linear codes etc. [10]. XOR correction is applied to reduce the bias of uncorrelated random bits at the output of sampler circuit. In this method, XOR of successive pairs of bits are done. XOR applied to pair of bits results in a reduction of output bit rate to half. It is also possible to XOR more than two bits which will further reduce the bias at the expense of low throughput.

Proposed design uses a single stage Von Neumann corrector to eliminate bias of uncorrelated random bits which are depicted in Table 2. In this method, successive random bits are compared, if both the bits are same, either zero or one, the pair is discarded. If the successive pairs of bit are not same, then the first bit in the sequence is considered and given as output and the second bit is discarded.

According to the calculation shown in [10], the output bit rate is approximately one fourth of input bit rate after applying Von Neumann correction technique but this method completely eliminates bias in the generated bit stream.

IV. RESISTANCE TO SIDE CHANNEL ATTACKS- DESIGN OPTIONS, IMPLEMENTATION AND DESIGN ANALYSIS

A detailed study on the security of the random number generator was also done. The paper [2] establishes that short length ring oscillators (mutually-prime), are prone to the frequency injection attack, but this attack has not been experimented on larger length RO. Our implementation which uses 31 stages GARO inverter chain is expected to be secure against such kind of attacks.

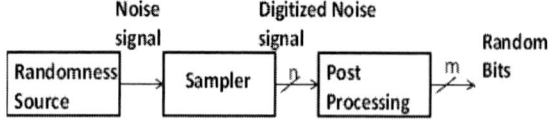

Fig. 6: General architecture of RNG

TABLE II: VON-NEUMANN CORRECTOR ON SAMPLE BIT SEQUENCE

Input	1	1	1	0	0	0	0	0	0	1	0	1	1	0
Output	Discard		1		Discard		Discard		0		0		1	

Bayon [3] claims that Electro Magneticinjection (EM) attack on ring oscillator based TRNG is possible on larger length RO also and claims that increase in RO length will not mitigate such attacks. The nature of the EM attack claimed by [10] largely depends on the knowledge of the active region in the FPGA, where TRNG is placed, and calculating the ring oscillator frequency using EM radiation map. If this information is known to the attacker, then he tries to lock the frequency of ring oscillator. If the ring oscillator frequency is locked then the bias of random bits are adversely affected.

In the proposed implementation, to make it infeasible for the attacker to find the active region in the FPGA, the TRNG design has been replicated across the FPGA. The output random bits are collected from each replicated design in fixed order. As described earlier, it was proved that the placement of the design on the FPGA does not affect the randomness of TRNG. With this enhanced design, multiple engines are part of the random bits streams. Therefore to attack the TRNG, the attacker has to find the location of each running engine and also frequency of each engine, which is very difficult. In this case frequency of each of 16 TRNG modules required to be gathered. To find the location and working frequency of any GARO the attacker needs the traces of EM radiation for that particular GARO.

Since, all TRNG engines are always active and placed close to each other, the EM radiation pattern of one engine can be affected by other engines. Also, consider the case that an attacker is able to lock the frequency of any one engine and is able to find the location of that engine. In this scenario, the attacker needs the samples of output bits with respect to the injected EM radiation. But in the proposed implementation, the bits are sampled from each running engine and at a particular instance the particular engine which is being sampled is unknown. Hence, we can conclude that the attacker will be not able to collect any useful traces of EM radiation.

But, an issue in replicating the TRNG design across the whole FPGA is that it will force the EDA tool to place TRNG engines much close to each other because of limited resources. In such cases the quality of random bits can be affected, which is undesirable. So, TRNGs are replicated across the FPGA, but not so close, so that, the nearby RO engines could not get locked to each other. In order to evaluate and ascertain that RO's frequencies are not locked to each other, the NIST tests were performed on the output random bits streams.

Von-Neumann post-processing technique has been implemented in the design. So, in case of the randomness of any one engine is being compromised (frequencies are locked), it will not affect the output randomness because output random bit bias is controlled by post-processing technique. We also performed NIST test on the random bits after these experiments and found that there was no degradation in the quality of the TRNG bits.

In order to experimentally evaluate and analyze the resilience against side channel attacks, spectrum of RO (31 stage inverter, single ring and no GF feedback), RO (31 stages inverter, 32 rings and no GF feedback) and GARO output (31 stages inverter, 32 rings and with GF feedback) was captured on a spectrum analyzer and it was observed that the first two cases the output spectrum had dominant fundamental and harmonics (odd). GARO had no dominant frequency and spectrum had mixed frequencies which make it inherently resilient against frequency injection on power supply lines / EM injection based side channel attacks. Snapshots shown as Figs. 7, 8 and 9 depict the results.

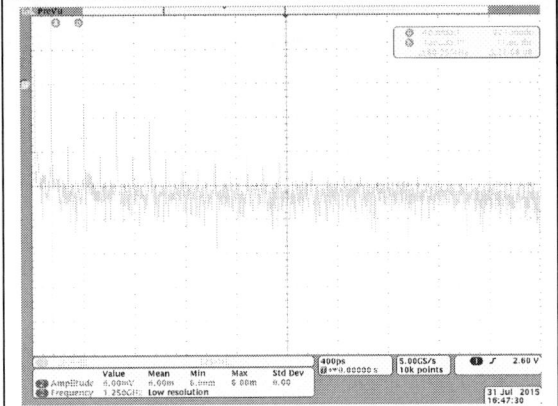

Fig. 7– Frequency Spectrum – One Ring, 31 stages, No GF Feedback – Fundamental Frequency – 40MHz

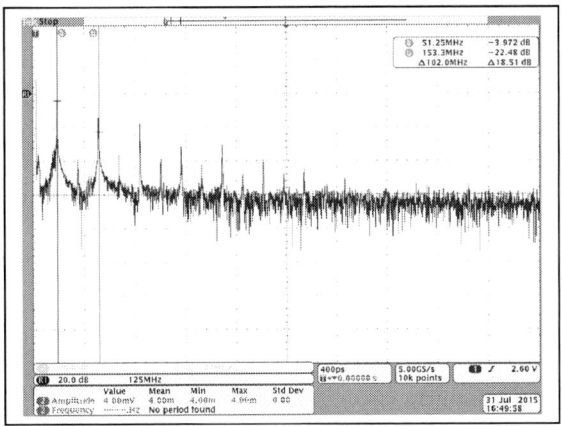

Fig.8– Frequency Spectrum– 32 Rings, 31 stages, No GFFeedback – Fundamental Frequency – 51MHz

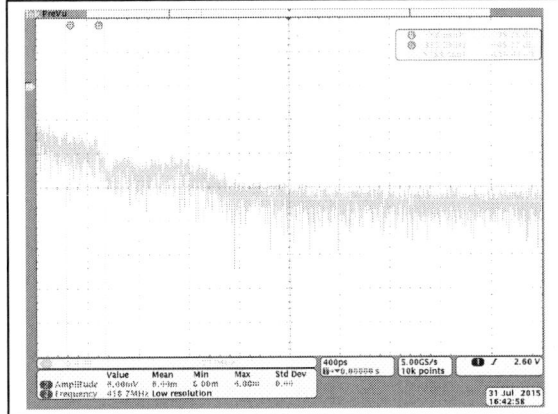

Fig. 9– Frequency Spectrum– 32 Rings, 31 stages, GFFeedback – Fundamental Frequency – Not Observable.

VI. TESTING AND ANALYSIS OF RESULTS

GARO based TRNG is implemented on PCI Express based Virtex-5 board. The performance of the proposed TRNG was evaluated with the help of NIST statistical tests [1]. NIST provides a set of 15 statistical tests which can be applied to a sequence of random bits to validate the randomness of the sequence. These randomness tests are based on statistical hypothesis testing. For each applied test, a P-value is calculated. P-value is the probability that the observed bit stream is less random than that of actual random bit stream. Based on P-value a conclusion is derived to accept or reject the null hypothesis that the generator is producing random values. Level of significance denoted by α, is the probability that the test will indicate a sequence as not random although it is actually random. The value of α is set to 0.01 for the statistical tests, which indicates that out of 100 sequences 1 sequence can be indicated as non-random though it is generated from a good random number generator.

To perform NIST tests, 100 sequences of random bits are collected from the random number generator. Requirement for minimum number of bits per sequence is different for different tests. Some of the tests need minimum 20,000 bits per sequence and others need minimum 1million bits per sequence. There is no theoretical upper limit for bit stream size for a sequence. To satisfy the requirement of all tests, 100 sequences with size of 1million bits were collected and tested. Table 3 depicts the characteristics of the polynomials used to form GARO and the testing results on FPGA (using NIST standard) with and without post-processing.

The NIST statistical results are shown in Table 4. The "PROPORTION" column indicates the number of sequences for which P-values are above the 0.01 confidence interval (CI)

TABLE III: NIST TEST RESULTS WITH VARIOUS POLYNOMIALS FOR TRNG

No.	Polynomial (32-bit)	Without post-processing	With Von Neumann corrector
1	4,7,8,10,11,12,13,18,19,20, 21,24,25,27,28,29,30,31,32	Passed	Passed
2	5,7,9,10,11,12,13,14,16,18, 19,21,26,29,30,32	*Failed*	*Failed*
3	1,2,4,6,7,8,11,13,14,16,19, 25,27,29,30,31,32	Passed	Passed
4	3,4,8,9,12,13,14,17,18,21, 24,27,28,31,32	Passed	Passed
5	2,3,4,6,10,11,15,16,17,20, 21,23,26,28,29,32	Passed	Passed
6	1,5,7,11,12,13,15,16,18,19, 20,21,25,29,30,31,32	*Failed*	*Failed*
7	2,3,5,6,8,10,11,12,16,18,22, 24,28,29,30,32	Passed	Passed
8	2,3,7,8,10,12,13,14,15,21, 23,25,29,30,32	Passed	Passed
9	1,2,6,8,11,12,14,16,22,23, 24,26,28,29,32	Passed	Passed
10	1,2,4,9,10,13,16,19,21, 22,23,27,28,29,30,32	Passed	Passed

978-1-5090-0037-1/16 $31.00 © 2016 IEEE

TABLE IV: NIST STATISTICAL TEST RESULTS FOR 100 SEQUENCES OF SAMPLES

```
RESULTS FOR THE UNIFORMITY OF P-VALUES AND THE PROPORTION OF PASSING SEQUENCES
------------------------------------------------------------------------------
  generator is <trng_03>
------------------------------------------------------------------------------
 C1  C2  C3  C4  C5  C6  C7  C8  C9 C10  P-VALUE  PROPORTION  STATISTICAL TEST
------------------------------------------------------------------------------
 11   5   8  17   9  15   9  10   7   9  0.236810    99/100   Frequency
 15  10   5  10  11   5   9  15  13   7  0.213309   100/100   BlockFrequency
 12   7  11  13   5   9  11  10  12   8  0.637119   100/100   CumulativeSums
 10   7  10  11  10  11  14  13   6   8  0.779188    99/100   CumulativeSums
  5   3  10  21   7   8  15  12  11   8  0.803996    99/100   Runs
  6   7   6   9   8  12  13  17  11  11  0.275709   100/100   LongestRun
 11  11   7  13   5  14   8  11   7  10  0.494092    97/100   Rank
 10  12   6   6  10   9   8   8  14   9  0.181557    99/100   FFT
 10  13   6   5  14   9  13   8  10  12  0.494092    97/100   NonOverlappingTemplate
  5   8  12   6   7  13   8  13  14  14  0.262249   100/100   NonOverlappingTemplate
  5   8   9   7   9  13  11  10  11  17  0.350405    99/100   NonOverlappingTemplate
  9  13  12   6  14  10   7  10   6  13  0.534146    99/100   NonOverlappingTemplate
  7  11  12  11   5  14  10   9  10  12  0.699313   100/100   NonOverlappingTemplate
 11  11   7   6   9   8   9  13  10   8  0.867692    99/100   NonOverlappingTemplate
  6  14  10  11  10   8   9   4  13  15  0.289667    99/100   NonOverlappingTemplate
  9  12  14  13   6   5   7  12   9  13  0.481199    98/100   NonOverlappingTemplate
  8  12  13  12  12  12   6   9   9   7  0.779188    99/100   NonOverlappingTemplate
 11   9   4  13   9  10  13  10   8  13  0.637119   100/100   OverlappingTemplate
  8  14   7  12  11   6  14   4  15   9  0.119867   100/100   Universal
 11   8  12   7  10  11  11   7  13  10  0.924076    98/100   ApproximateEntropy
 12   6   5   6   3   6  12   6   6   7  0.128379    68/69    RandomExcursions
  6   4   7   9   8   5   7   4  10   9  0.517442    69/69    RandomExcursions
  7   5   7   3   4   5  11   9   9   9  0.242986    67/69    RandomExcursions
  3   6   8   6  11   7  10   3   8   7  0.242986    69/69    RandomExcursionsVariant
  2   7   7   8   6  13   6   4  10   6  0.806450    69/69    RandomExcursionsVariant
  5   4   4  14   5   6  10  12   5   4  0.009422    69/69    RandomExcursionsVariant
  3   5   8   7   6   7   7  10   7   9  0.619772    69/69    RandomExcursionsVariant
  4   5   6  13   5   3   5  14   7   7  0.009422    69/69    RandomExcursionsVariant
 15  12   9  12   8   4  11   9  11   9  0.554428    99/100   Serial
 15  12  11   8   9  10  12   5  10   8  0.657093    99/100   Serial
 11   9  10   7  14  10  13   7   9  10  0.867692    99/100   LinearComplexity
------------------------------------------------------------------------------
The minimum pass rate for each statistical test with the exception of the
random excursion (variant) test is approximately = 96 for a
sample size = 100 binary sequences.

The minimum pass rate for the random excursion (variant) test
is approximately = 65 for a sample size = 69 binary sequences.
```

TABLE V: FPGA HARDWARE RESOURCES UTILIZATION

Logic Utilization	Utilized	Available	Utilization in %
Slice Registers	2360	97280	2%
Slice LUTs	17397	97280	17%
LUT-FF pairs	1584	18173	8%
Block RAM/FIFOs	4	192	2%

The range of acceptable proportions is determined using the Confidence Interval as given in (1).

$$\hat{p} \pm 3 * \sqrt{\frac{\hat{p}(1-\hat{p})}{m}} \qquad (1)$$

Where, $\hat{p}$ =1-α, and 'm' is the sample size.

For a sample size of 100 sequences, with α equal to 0.01, CI value ranges from 0.96 to 1.01. In Table 4, columns 1 to 10 indicate the range of P-value distribution for the tested samples. Uniform distribution of P-values indicates that the sequences that are generated are truly random. Polynomial 2 and polynomial 6 have failed in linear complexity test as indicated in Table 3.

VII. Performance Metrics

The TRNG module is synthesized on Virtex-5 xc5vLX155 FPGA and hardware utilization is depicted as Table 5.

The performance characteristics of the TRNG hardware engine that has been achieved without Von-Neumann corrector are as follows: Throughput – 100Mbps; Latency– 66 Clock Cycles (64 due to 64 bit registers and 2 clocks is due to the latches introduced to mitigate meta-stability).

VIII. Conclusions

This paper describes analysis of an implementation of a Side Channel Attack resistant GARO based TRNG system on Virtex 5 FPGAs. The GARO based TRNG has yielded good quality of random bit streams. The TRNG system is made of 31 stages GARO and 32 such are used and it has been

reported multiple such rings are difficult to attack using frequency injection attack. Further, each ring has been implemented with different/same primitive polynomials in Galois Field and the TRNG output passed all the NIST tests for randomness. An experimental analysis revealed that the frequency spectrum of GARO output is not dominated by single frequency and it's harmonics unlike the RO without Galois Field primitive polynomial based feedback in the RO. Due to that phenomena it is not feasible to lock GARO frequency with single frequency based frequency injection attack. By experimentation it was found that implementing the design at different locations in the FPGA does not influence the statistical properties of the generated random bit stream and proved the implementation is location insensitive. Replicating the design across the whole FPGA makes attacks on RO based TRNG very difficult as it is extremely difficult to lock many engines by frequency injection. This is due to the fact that the oscillator's frequency will differ due to variation in delay on account of differing routing/logic delays. Further, we adopted the Von-Neumann correction for whitening the sequence.

Future work includes implement and benchmark the side channel attack resistant TRNG with various post-processing techniques viz. BCH coding, Hash functions etc.Further, testing shall be carried out to ascertain the resilience against frequency injection (on power supply/EM radiation) attacks on the scheme proposed in this paper. Besides, TRNG designs using cascading Fibonacci FIRO and GARO oscillator in series can also be experimented and evaluated for randomness properties.

References

[1] John Kelsey, B. Schneier, David W, Chris Hall, "Cryptanalytic Attacks on Pseudorandom Number Generators", Fifth Intl. Workshop Proc., Fast Software Encryption, Vol. 1372, LNCS, pp. 168-188, 1998.

[2] Markettos, Simon Moore,'The frequency injection attack on ring-oscillator-based true random number generators'. Proc. of 11th Int. Workshop on Cryptographic Hardware and Embedded Systems, pp. 317-331, September 2009.

[3] Bayon, Bossuet, Aubert, Fisher, "Electromagnetic Analysis on Ring Oscillator-Based True Random Number Generators", IEEE Intl.Symp. On Circuits and Systems, pp. 1954-1957, 2013.

[4] B.Sunar, W.J. Martin, Stinson D.R, "A Provably Secure True Random Number Generator with Built-in Tolerance to Active Attacks", IEEE Trans. on Computers, pp. 109-119, 2007.

[5] Tarsalonut Gabriel, Gigi-Daniel Budariu, ConstantinGrozea: "Study on a True Random Number Generator design for FPGA", 8th Intl. Conf. on Communication, pp. 461-464, 2010.

[6] Dries Schellekens, Bart Preneel, Verbauwhede, I, "FPGA Vendor Agnostic True Random Number Generator", Intl. Conf. on Field Programmable Logic and Applications, pp. 1-6, 2006.

[7] MarkusDichtl, Jovan DjGolic, "High-Speed True Random Number Generation with Logic Gates Only", 9th Intl. Conf. on Cryptographic Hardware and Embedded Systems, pp. 45-62, 2007.

[8] Jovan Dj.Golic, "New Methods for Digital Generation and Postprocessing of Random Data", IEEE Trans.on Computers, pp. 1217-1229, 2006.

[9] Knut Wold, Chik How Tan, "Analysis and Enhancement of Random Number Generator in FPGA Based on Oscillator Rings", Intl. Conf. on Field Programmable Logic and Applications, Article ID 501672, 2008.

[10] Siew-Hwee Kwok, Yen-Ling Ee, Guanhan Chew, KanghongZheng, KhoongmingKhoo, Chik-How Tan, "A comparison of Post-Processing Techniques for Biased Random Number Generators", Proc. of the 5th Intl. Conf. on Information security theory and practice: security and privacy of mobile devices in wireless communication, pp. 175-190, 2011

[11] NIST "A Statistical Test Suit for Random and Pseudo Random Number Generators for Cryptographic Applications", SP-800-22, Rev. 1a, 2010

978-1-5090-0037-1/16 $31.00 © 2016 IEEE

RF Tracking Test System Design for Closed Loop Testing of Ku-Band Antenna

*Rahul Mishra, Sowbhagya, S. Sudhakar, Sudeesh B., A.M. Nagalakshmi, S. Udupa

ISRO Satellite Centre, Vimanapura Post, Airport Road, Bangalore, India
Email: rahulm@isac.gov.in, Tel: +9108025082360

Abstract— **Ku band antennae are one of the unique features of GSAT11 spacecraft. As the pointing accuracy of the satellite platform is around 0.1°, improved antenna pointing is achieved using the steerable Deployment Pointing Mechanism (DPM), Radio Frequency (RF) Tracking based measurement system and Attitude and Orbit Control Electronics (AOCE) resident controller. Testing of these subsystems in open and closed loop has been carried out for the first time in zero-g environment. This test has been instrumental in evaluating/validating the closed loop performance of all the subsystems along with the characterization of the RF tracking system and its system response. This paper discusses the design approach and development of RF Tracking Test System and improvements carried out therein to enable smooth testing and lessons learnt. Here, AOCE simulator design has been effected for accelerating the product design cycle and in-parallel carrying out the interface testing along with the algorithmic validation and system characterization.**

Keywords— **RF Tracking, closed loop, embedded systems, DPM, Data Acquisition System, controller.**

I. INTRODUCTION

As the frequency band of operation increases the pointing requirement becomes more challenging which in turn calls of for greater modelling, evaluation, simulation and testing challenges on ground eg.[1]. Ku-band antenna on satellite has a narrow pointing requirement, for near null pointing, which necessitates finer angular steering (achieved by stepper motors in DPM). This DPM steering is carried out using the RF-Beacon (Beacon is situated on the ground) tracking sensor which operates in a closed loop control along with the DPM.

The closed loop control is realized with the AOCE-Antenna Driver system (*figure-1*) in conjunction to achieve the pointing till the error is limited within the Roll/Pitch threshold band. This antenna pointing is required to achieve a gain margin of 1dB and to attain beam to beam isolation between the spot beams. Additionally, such a closed loop antenna pointing, limits the beam hovering thus improving the overall QoS.

Tracking Receiver subsystem interacts with the AOCE using MIL-1553 Bus based communication protocol with AOCE acting as the Bus-Controller and Tracking Receiver as an RT (Remote Terminal). Based on the Pitch/Roll errors received from Tracking Receiver Remote Terminal (TRRT), AOCE conditions the error inputs and consequently controls using the commanded data to the Antenna Drive Electronics (ADE) for steering the Antenna. The control is proportional control with respect to the reference and the tracking receiver data.

The data received from the TRRT and commanded to ADE are displayed by the data acquisition system which interacts with the AOCE simulator software and displays on the GUI and other software utilities required by the user.

The rest of the paper is organised as follows: Section-II mentions the main contribution of the work. Section-III illustrates the main components of closed loop and its architecture. Section-IV briefly describes the hardware design and Section-V describes in detail the main control software logic incorporated for the closed loop tracking. Subsequently, Software design, test setup design and conclusion sections are discussed.

II. CONTRIBUTION OF THE WORK

AOCE simulator cum test system is designed for RF tracking system evaluation in a zero-g environment which is carried out for the first time. The control software logic has been realised for the system characterization (phase linearity of the receiver, tracking in null shifted conditions etc.), algorithm and RF interface validation.

III. CLOSED LOOP TEST ARCHITECTURE

RF Tracking Test Set-up (*figure-1*) consists of various subsystems which are explained as follows:

A. AOCE-Simulator

AOCE simulator software interfaces with two subsystems namely ADE and Tracking Receiver. For communicating with ADE, a miniaturized [2] USB based hardware (called ADETS) is designed for error data, Relay/pulse commanding, and drive pulse duration selection along with the DPM actuation. Additionally, it performs the Monitoring data acquisition pertaining to ADE. The Test System simulates the Bus Controller to interact with the TRRT and provides for commanding Antenna Selection and Phase Control Word (PCW for shifting the phase) to it and acquires the Analog words and TRRT parameters for Telemetry monitoring.

B. Tracking Receiver

TR [3] provides the Azimuth and Elevation errors using coherent detection and sends to AOCE-BC along with the other Monitoring parameters through TR Remote Terminal on MIL1553.

RF Tracking Test System

Figure 1: RF Tracking Test System

The error scales upto the $\pm 0.4°$ range from the bore-sight. The command-able PCW is used for phase optimization between error and sum signals.

C. ADE

ADE is the DPM drive electronics which caters to the azimuth and elevation motor rotation in a single data command based on the AOCE registered error depending on the open/closed loop mode of operation. It uses H-bridge based circuit and error data bit pattern to generate the specific pulse sequence streams to actuate the motors in desired direction.

D. DPM

One DPM consists of two stepper motors for azimuth and elevation rotation of a single reflector. For this RF tracking test one DPM-Reflector assembly out of the four used systems is used. The ADE drive pulses to DPM are provided through long harness so as to provide for the height of CATF Antenna Positioner.

E. Reflector Antenna/RF Beacon

Reflector along with the feed system and antenna switch matrix provides the sum and error signals to the tracking receiver which in turn provides corresponding errors in terms of DC voltages to AOCE-BC.

F. Data Acquisition software and Display System

Driver, applications (ADE and MIL1553-BC) and GUI software [2,4] have been developed in C, CPP and Qt languages. DPM controller software logics have been incorporated after the TRRT data acquisition and processing. The test objectives targeted are DPM rotation, polarity check, error signal slope, S-curve validation, errors' threshold, null offset, crosstalk etc of the RF tracking system.

IV. HARDWARE DESIGN

Hardware design of AOCE simulator consists of two subsystems namely ADE Interface system and Bus controller simulator system. A hardware software co-design [4,5] approach has been followed in design of these systems and

their associated driver and application software designs. A brief hardware description is as follows:

A. ADE interface system (AIS)

AOCE control software commands the AIS for Pitch/Roll motor rotation related operations with and without the TRRT data input based on the closed-loop/open-loop/auto/hold modes of operation.

AIS is a microcontroller (*figure-2*) based system that acquires the commands from AOCE simulator and telemeters the motor coil actuation response data back to the data acquisition-DAQ/GUI software.

B. BC simulator system(BCSS) for TRRT communication

BCSS for TRRT is realized with the help of ISA based 1553 simulator board. This card can simulate four Mil-std-1553B terminals. DDC's BU 61580 has been used as protocol chip. All the four terminals have separate dual mil-std-1553 channels. The card has 12 discrete outputs and five discrete inputs. All the discrete I/Os can be read/written through the memory mapped I/O ports. All the MIL 1553 devices are placed in the Memory bus of the CPU. The card (*figure-3*) is packaged on a full size ISA PCB with ISA bus interface.

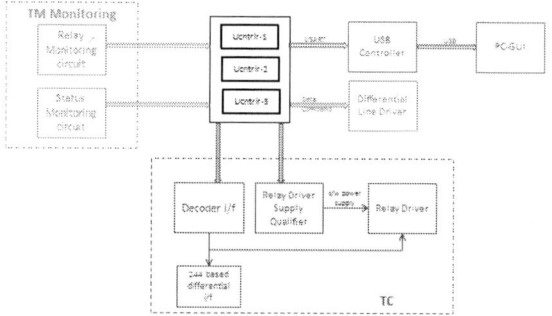

Figure 2: AIS Hardware block diagram

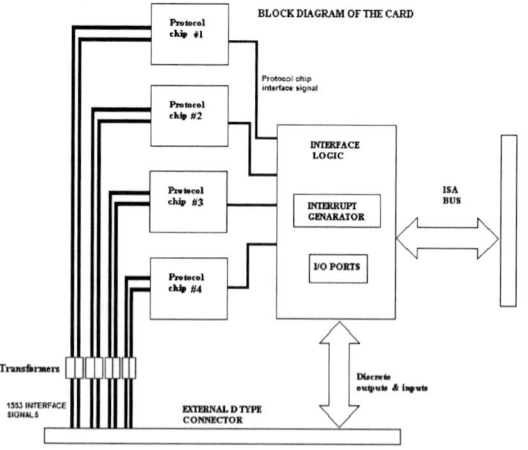

Figure 3: BCSS architecture block diagram

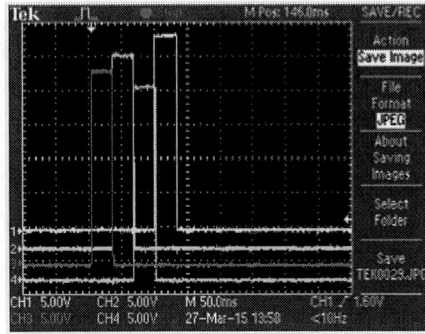

Figure 4: sequence of pulsed actuation of DPM (open loop)

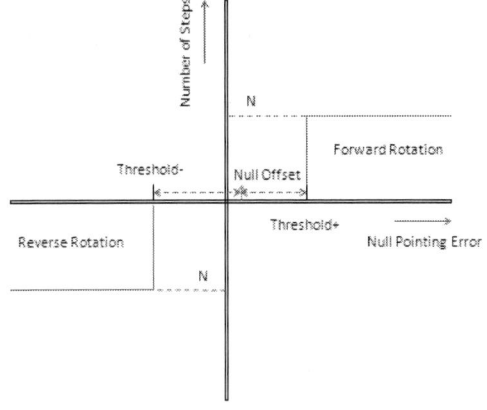

Figure 5: Controller Threshold and Null Offset incorporation

V. CONTROL LOGICS DESIGN

RF Tracking antenna rotation is provided with various modes of operation as explained below:

A. Open Mode

In open loop motors are operated independent of the tracking receiver outputs and enables motors to operate any number of steps within the azimuth/Elevation range of operation. After Pitch/Roll motors are commanded *power-on* and o*pen mode* driving is selected, motor driving parameters such as *pulse on/off duration, drive polarity, number of steps* etc. are commanded and the *drive rotate (figure-4)* command is initiated.

In open loop the evaluation is done using the photogrammetry/theodolite based measurements in offline. This operation is required during the initial reflector deployment or after loss of lock with the primary lobes of the antenna due to disturbance.

B. Closed Loop Mode

In closed loop, tracking receiver is brought in loop with the motor and controller commands the programmable (1 to N) number of pulses in clockwise or anticlockwise directions based on the azimuth/elevation orientation error *(figure-5)*.

In closed loop positive and negative thresholds are kept so as to prevent motor oscillations after achieving the desired antenna pointing. Additionally, provision is added to account for the null offset of the Tracking Receiver and to add the effect of it to the controller software. The errors are acquired from the Tracking Receiver every 32ms and based on that motor actuation pulses to the AIS are programmed for on-time, off-time and number of steps in cycles that are multiples of 32ms.

The control software incorporates the provision of conditioning the input error information using the moving average filter in loop for smoothing the error data and hence reducing the possibility of noise triggered hunting at both the thresholds. The window size of the filter is configurable for the testing.

Phase control word setting of tracking receiver provides for taking into account the thermal effects of the reflectors and the waveguide differences. Hence it is provided for setting in terms of temperature or directly in hex counts. Also, the antenna select command helps in multiplexing the same TR for all the reflectors on-board, but it is not used during the single reflector characterisation.

The closed loop mode of operation is the normal mode of operation in the orbit in which the payload will be functional for communication application.

C. Hold Mode

In hold mode, the selected motor coil is continuously provided with the actuation voltage for the required duration of position holding and motor is evaluated for its détente at that position.

D. Auto Mode

In this mode, the motor is commanded for forward/reverse rotation until the open/close interface of micro-switch is encountered in that direction. The micro-switch thus limits the range of operation of both the motors' operation range. Micro-switches, additionally, provide a reference point to initiate open loop motor actuation.

VI. GUI SOFTWARE AND DRIVER DESIGN

A. BCCS driver and application software

Init1553 *(figure-6)* application reads/writes from/to TRRT data through device driver based on the scheduler table. It communicates with the GUI software with message queues for commanding and monitoring the data.

Additionally, Monitor terminal have been incorporated for MIL1553 data and time analysis along with the debugging and monitoring the bus activity. Bus Monitor software interacts with the driver software to acquire the 1553 bus data and provides for the data logging and reading functions.

B. ADE Test System driver software

The Application software [2] is required to carry out read and write data functions with the USB controller using D2XX driver with the pre-configurable parameters in a scheduler.

The software reads the frames received from each microcontroller and fetches the relevant information, processes the data and posts them to the GUI. ADE exchanges data with the init1553 and GUI softwares using message queues and shared memory.

C. GUI Software Design

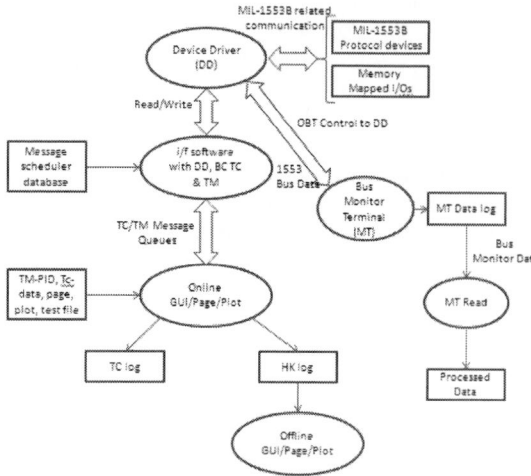

Figure 6: BCSS software organisation

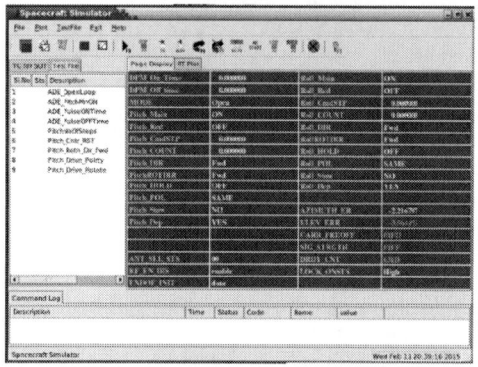

Figure 7: GUI snapshot

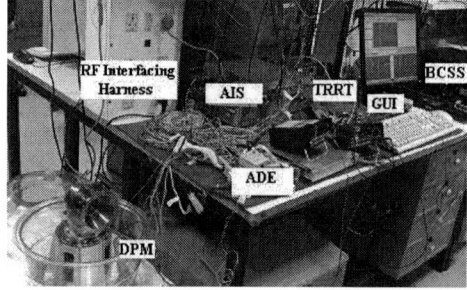

Figure 8: Integrated AOCE simulator with DPM

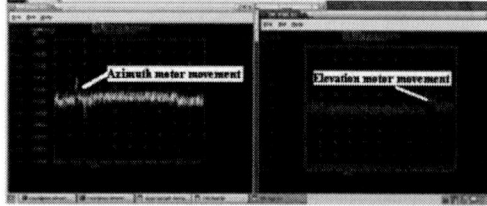

Figure 9: Azimuth and Elevation errors' plot (real time)

GUI (*figure-7*) is based on Qt. It caters to the command, TM display provisions, plotting, data logging and automated testing provisions required for testing and report generation. The Telemetry data is displayed on the page with their updated values. The test file is displayed at left margin with various provisions, required for flexibility in testing, displayed on the top of the GUI page. The plotting (*figure-9*) window facilitates the multi-plot display, logging and print.

VII.   TEST SET-UP AND RESULT DISCUSSION

The RF tracking test system/AOCE simulator is interfaced with the DPM (*figure-8*) and RF tracking system in CATF and both the open loop and closed loop tests and evaluation were carried out successfully. The TRRT card provides the 2-axis error voltage to the AOCE-Simulator-BC through MIL-1553B interface and ADE is driven by AOCE-Simulator-AIS. Since the DPM is located at the CATF platform, ADE to DPM harness (RF interfacing harness) is provided with a long harness with minimum effect on the ADE Drive pulses. The software and subsystems were operational and stable for long hours. Based on the test experience some lessons were learnt to further improve upon the testing time such as commandable threshold and offset settings and recording the last motor coil position in the database.

The Azimuth and Elevation errors in closed loop mode can be observed in the plot (*Figure-9*). In closed loop mode the error decrement with the DPM can be clearly seen. The DPM is commanded in Open loop for a sufficiently large azimuth and Elevation errors and in closed loop mode selection these large errors can be seen to be diminishing till the error threshold levels set in the software. The error correction is simultaneous in both the axes. Additionally, the error linearity of the receiver also has been verified to be 9.5V/degree. The threshold values are commandable and the results are presented for 150mV threshold values.

VIII.   CONCLUSION AND FUTURE WORK

This work envisages automation, miniaturization and reusability of the test systems required for various on-board packages in future.

REFERENCES

[1]   V.Y.Lo, "Ka-Band Monopulse Antenna-Pointing Systems Analysis and Simulation," TDA Progress Report42-124 February 15, 1996.

[2]   Rahul M. etal, "Microcontroller Based Miniaturized Test System for Antenna Drive Electronics", ISCIT-VSSC symposium, March-2015.

[3]   Pankaj K. Gupta et.al., "Two Channel Monopulse Tracking Receiver for Onboard Antenna Tracking System", ICCICT, October 19-20, 2012.

[4]   RahulM.etal.,"Automated Test System Using Mixed Signal Microcontroller for Wheel Interface Module", ISCIT-VSSC symposium, March-2015.

[5]   Wayne H. Wolfe, "Hardware-Software Co-Design of Embedded Systems", Proceedings of the IEEE, Vol.82, No.7, July1994.

[6]   Giovanni De Micheli etal.,"Hardware/Software Co-Design" Proceedings of the IEEE, Vol.85, No.3, March1997.

2016 International Conference on VLSI Systems, Architectures, Technology and Applications (VLSI-SATA)

Tunable Distributed Harmonic Voltage Controlled Oscillator for Generating Second and Third Harmonic Microwave Signals in 180nm CMOS

Kalyan Bhattacharyya

Department of Electronics & Communication Engineering
Amrita School of Engineering, Coimbatore
Amrita Vishwa Vidyapeetham University
Coimbatore 641112, India
b_kalyan@cb.amrita.edu

Abstract— **Novel tunable distributed harmonic Voltage Controlled Oscillators (VCOs) are reported here for generation of 2nd or 3rd or both 2nd and 3rd harmonic frequency microwave/millimeter-wave signals, designed using Distributed Oscillator (DO) and achieved frequency tuning by body bias. The above generation of signals is become possible by the use of combination of innovative gain cell (i.e. amplifier), CPW lines and selection of the value of peaking inductor in the gain cells. The highest power output of the oscillator is at the third harmonic frequency of 41.84 GHz using low cost industry-standard 180nm CMOS which has a cut-off frequency of 50 GHz.**

Keywords— *Distributed Oscillator; Voltage Controlled Oscillator; Body Bias; Coplanar Waveguide*

I. INTRODUCTION

Distributed Oscillator based Distributed Voltage Controlled Oscillator was reported in reference [1] and a harmonic frequency oscillator was reported in reference [2]. Reference [2] has reported a harmonic oscillator and that "for an Nth-harmonic oscillator, N oscillators with proper phase relation are arranged in parallel," which is a cumbersome and complex large design [2]. The two US patents can generate only even harmonics [3], [4] of the fundamental frequency, [3] is for second and fourth harmonics and [4] is for fourth harmonic only. Tunable Distributed Harmonic Voltage Controlled Oscillator reported here is for novel and simple generations of second [5] or third harmonic [5], [6] or both second and third harmonics [5] of fundamental frequency. The two US patents have completely different designs using a pair of simple FET (hence no inventing steps) as amplifiers and used microstrip and slot lines [3], [4].

Distributed Oscillator (DO) [1] reported here operate in the forward gain mode of a Distributed amplifier (DA). Design uses in each stage an innovative gain cell of cascade of n-MOSFETs Common Source (CS) with peaking inductor (Fig. 1) [6], coplanar waveguides (CPWs) which forming gate and drain transmission lines and a feedback path between output and input. As the forward wave travels down the gate line, each n-MOSFET transfers signal to drain line through its transconductance. Signals add in drain line in forward propagating direction. The drain line output is then fed back to the gate line input. The forward wave in the gate line is collected across a matched termination and backward wave is

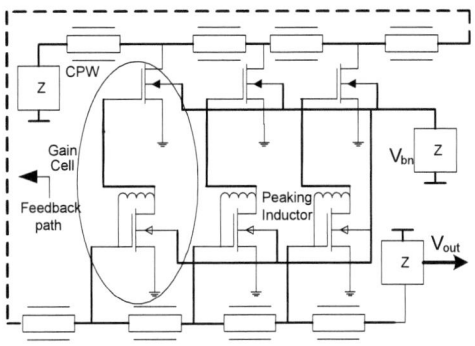

Fig. 1. Simplified schematic of tunable distributed harmonic voltage controlled oscillator with body biasing (V_{bn}), shown 3-stages out of 4-stages.

absorbed by a matched termination in the end of drain line [1], as shown in Fig. 1. Traditional non-DA oscillators use relatively complicated circuits with large on-chip inductors and capacitors. The use of innovative gain cell between gate and drain lines and 90Ω coplanar waveguides matched to 50Ω impedance after FETs loading results in fast rise and fall times [1], and thus a high oscillation frequency in microwave/millimeter wave band is achieved in low cost industry standard 180nm CMOS.

A 4-stage oscillator design with an innovative gain cell [6] (i.e. amplifier) and CPWs as inductive transmission lines is shown in Fig. 1, will be mentioned as DVCO-H henceforth [5]. The highest power output of the oscillator is at the third harmonic frequency of 41.84 GHz. It is also reported in [2] that "for harmonic oscillators", "the high Q-factor, high device gain and low phase noise are more reachable". Fig. 1 showing the tunable DVCO with cascade of two CS [5], [6], [7] n-MOSFETs with peaking inductor in gain cells for harmonics generation and body bias [8] of n-MOSFETs for frequency tuning.

The contents of the paper is section II includes the design of the harmonic VCO including design of gain cell, Section III

978-1-5090-0037-1/16 $31.00 © 2016 IEEE 141

presents results of harmonic generations with discussion and section IV having the conclusion of the paper.

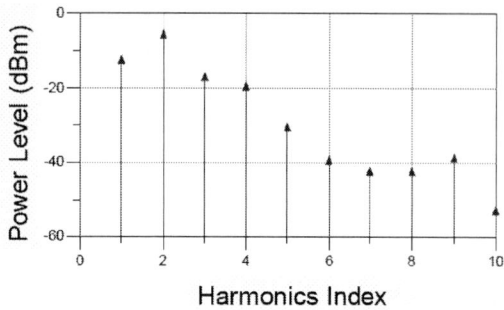

Fig. 2. For DVCO-H, result of 2nd harmonic is dominating.

II. DESIGN OF DISTRIBUTED HARMONIC VCO

A. Design of Gain Cell

A novel distributed harmonic VCO based second and third harmonic frequency generation design DVCO-H uses an innovative gain Cell in four stages [6]. Shown in Fig. 1, for the DVCO-H, the gain cell is a cascade of two CS n-MOSFETs with gate and drain of bottom n-MOSFET is connected through a peaking inductor (drain of bottom n-MOSFET is also connected to gate of top n-MOSFET) [5]. MOSFETs of gain cell would have contribution in C_{in} and C_{out} and hence for the frequency of oscillation [6].

B. Design of DVCO-H

For DO, uniform inductive values of CPWs are used between four gain cells. Characteristic impedance Z_{line} (= $(L_{line}/C_{line})^{1/2}$) of the coplanar transmission lines in our oscillator circuits has been set to 90 Ω for impedance matching to a 50 Ω external load after device loading [1].

Due to parasitic elements, the performance of the MOSFETs in common source configuration degrades at microwave frequencies, creating challenges for the RFIC designers. The high gate resistance R_G has many undesirable effects like improper impedance matching thereby reducing the power transferred, increasing the Noise Figure of the FET as thermal noise is introduced by it and also reducing the maximum frequency of oscillation f_{max} [9]. However the gate resistance R_G is reduced by using multi-fingered, double-side connected gate [9]. Each n-MOSFET in the gain cells of DO, was designed with a double-sided connected gate with multiple 5-micron fingers to reduce gate resistance R_G [6]. The distributed amplifier achieves a higher gain–bandwidth product by absorbing the parasitic capacitances C_{in} and C_{out} of transistors into transmission lines. In a DA, the FET gain cells are connected with series inductive elements L_G and L_D produced by the transmission line effect of CPWs at the gate and drain lines. These series inductive elements, along with shunt capacitances C_{in} and C_{out} (total input and output capacitances of transistor) at the gate and drain nodes of the FETs respectively, are coupled by the transconductance of the gain cells. The C_{in} and C_{out} along with the series inductive coplanar transmission line elements form a T-type low pass

filter structures, L_G-C_{in}-L_G and L_D-C_{out}-L_D respectively for gate and drain lines for each stage of the oscillator [6]. These

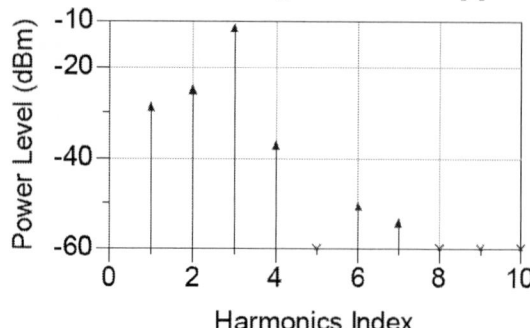

Fig. 3. For DVCO-H, result of 3rd harmonic is dominating.

T type structures are repeated for each gain cell used in the oscillator as in Fig. 1.

Because of the peaking inductor with specific value of inductance in the gain cells, as well as the gain cells so designed, the oscillator of the invention oscillates to generate a high frequency second or third harmonic signal with highest power level among the harmonic frequencies. For the circuit of Fig. 1, for inductance value of around 2nH, second harmonic is dominating, where as for inductance value of around 1nH, third harmonic is selected with highest power level among the harmonics. For peaking inductor value of 1.25nH, both second and third harmonics are generated with equal power [5].

During body biasing of n-MOSFETs, the change in body bias directly changes the drain- bulk capacitance C_{DB} and the gate-drain capacitance C_{GD} and indirectly changes the gate-source capacitance C_{GS} [8] through the change in the threshold voltage V_T [8]. Body bias variation of MOSFETs employs the bias dependent change of C_{DB} and C_{GD} resulting in the change of C_{out} and of C_{GS} and C_{GD} resulting in change in C_{in} [6], where C_{in} and C_{out} are respectively the total input and output capacitances of MOSFET [6]. The change in parasitic capacitances with body bias results in the change of frequency of oscillation [5], [6].

III. RESULTS OF HARMONIC GENERATIONS

A. Second Harmonic Dominating

TABLE I

Body Bias (V)	Second Harmonic Dominating with highest Power					
	Fundamental		*Second Harmonic*		*Third Harmonic*	
	Frequency (GHz)	Power (dBm)	Frequency (GHz)	Power (dBm)	Frequency (GHz)	Power (dBm)
0	9.01	-11.57	18.02	-5.12	27.03	-14.87
-1.0	9.11	-11.73	18.22	-5.75	27.33	-13.89
-1.5	9.17	-11.66	18.34	-6.01	27.51	-13.55
+0.6	8.99	-11.59	17.98	-4.69	26.97	-16.09

The highest power output of the oscillator of Fig. 1 is 0.307 milliWatts (-5.12 dBm in Table I) at the second

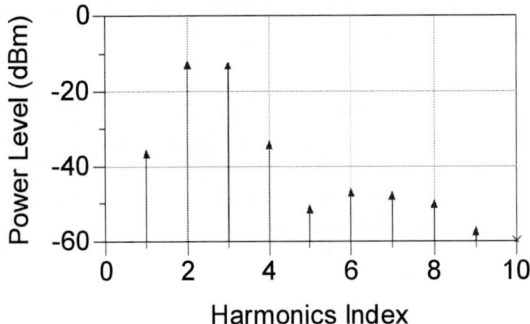

Fig. 4. For DVCO-H, results of both 2nd and 3rd harmonics are dominating.

harmonic frequency of 18.02GHz in K band as shown in Fig. 2, for selecting the peaking inductors value as 2nH. Second harmonic frequency is tunable by ~ 360 MHz by forward (0.6V) and reverse (-1.5V) body bias n-MOSFETs as presented in Table-I [5].

B. Third Harmonic Dominating

By changing the value of peaking inductors from 2nH to 1nH for the DVCO-H, the oscillator of the invention oscillates to generate third harmonic signal of 32.27 GHz with power output of 0.054mW (highest among the harmonic frequencies), as shown in Fig. 3. The first or fundamental frequency was 10.76 GHz with a power output of only 0.0029 mW and the second harmonic frequency was 21.52 GHz with a power output of 0.027 mW. Fig. 3 showing the spectrum of 4-stage DVCO-H showing power levels in harmonics, clearly showing the maximum power level is in 3rd harmonic frequency [5].

C. Design Issues for Second and Third Harmonic Generations

For the above second and third harmonic generation (reported in Figs. 2 and 3), the bottom n-MOSFETs each had a gate width of 35μm (having 7 fingers, each finger of 5μm) and the top n-MOSFETs each had a gate width of 55μm (having 11 fingers, each finger of 5μm) [5]. However, later it is reported that the width of n-MOSFETs are reduced further for generation of higher frequency.

For peaking inductor value in the range of 1 to 1.1nH, third harmonic is getting selected with highest power among harmonics. For peaking inductor value 2 to 2.25nH, second harmonic is getting selected with highest power among harmonics [3]. There is no change of value of peaking inductor from the above mentioned values for second and third harmonic generation respectively in the rest of the paper.

D. Both Harmonic Generations

For peaking inductor value of 1.25nH both second and third harmonics are getting selected with equal power, as shown in Fig. 4. When the peaking inductor value was 1.25 nH, the oscillator of the invention produce both second and third harmonic signals with a high power output of about -12

dBm (0.063 milliwatts). The first harmonic signal of -35.81 dBm (0.00026 milliwatts) was about 23 dB suppressed below

TABLE II

Bo-dy Bias	Third Harmonic Dominating with highest Power					
	Fundamental		*Second Harmonic*		*Third Harmonic*	
(V)	Frequ-ency (GHz)	Power (dBm)	Frequ-ency (GHz)	Power (dBm)	Frequ-ency (GHz)	Power (dBm)
0	13.74	-28.12	27.47	-24.47	41.21	-10.67
-1.0	13.86	-28.19	27.73	-24.59	41.49	-9.10
-1.5	13.95	-28.58	27.89	-24.68	41.84	-7.52

the second and third harmonic signals. Either of the second and third harmonic signal can be advantageously used for communication applications [5].

E. Change in Width of MOSFETs and Results

Next the bottom n-MOSFETs each had a gate width of 25μm (having 5 fingers, each of 5μm) and the top n-MOSFETs each had a gate width of 35μm (having 7 fingers each of 5μm) are considered (may be referred as second typical oscillator) for next set of results.

During second harmonic generation, a 22.96 GHz second harmonic frequency with -6.93 dBm (0.2027 milliwatts) power level is obtained.

The third harmonic frequency achieved is 41.84 GHz of 0.177 mW (-7.52 dBm) by reverse body bias of -1.5V of n-MOSFETs, and details presented in Table-II [5]. It is seen from Table-II that the third harmonic frequency of 41.21 GHz at zero body bias was tunable up to 41.84 GHz by reverse body bias of -1.5V of n-MOSFETs. It is also seen from Table-II that at the zero body bias, the third harmonic signal was around -10 dBm (0.1 milliwatts) and the first harmonics was -28.12 dBm (0.00154 milliwatts) i.e. about 18 dB suppressed below the third harmonic. The second harmonic was -24.47 dBm (0.00357 milliwatts) i.e. about 14 dB suppressed below the third harmonic. Total tuning of third harmonic by body biasing was from 41.21 GHz to 41.84 GHz ie 630 MHz [5].

F. Thermal Stability

The thermal stability of the second typical oscillator (i.e. with reduced value of the width of MOSFETs as in Table-II)

TABLE III

Te-mp	Third Harmonic Frequency Oscillator					
	Fundamental		*Second Harmonic*		*Third Harmonic*	
(°C)	Frequ-ency (GHz)	Power (dBm)	Frequ-ency (GHz)	Power (dBm)	Frequ-ency (GHz)	Power (dBm)
25	13.74	-28.12	27.47	-24.46	41.21	-10.67
-25	13.62	-27.21	27.24	-21.03	40.87	-9.31
50	13.77	-28.69	27.55	-27.31	41.32	-13.65

978-1-5090-0037-1/16 $31.00 © 2016 IEEE

during 3rd harmonic frequency generation was studied over a range of temperatures i.e. room temperature of 25°C is

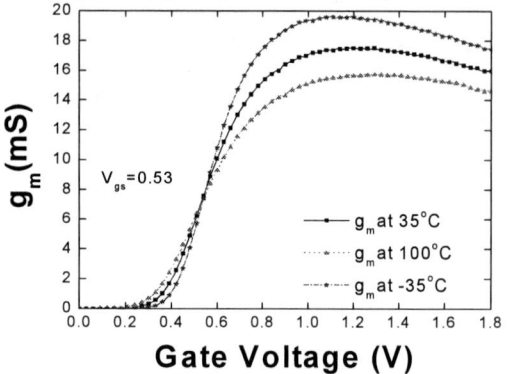

Fig. 5. Measured transconductance of n-MOSFET from -35°C to 100°C.

changed from -25°C to 50°C, and the results (with zero body bias) are reported in Table-III [5].

As can be seen from Table-III that output power level has started increasing at -25°C due to increase of mobility at lower temperature and subsequently the increase of transconductance [10]. The measured results of transconductance for n-MOSFET after the Zero-Temperature-Coefficient Point of transconductance (ZTCgm) [11] at Vgs = 0.53V, is the increase of peak at lower temperature of -35°C i.e. the top curve in Fig. 5. Mobility decreases at higher temperature [10] and hence the decrease in transconductance is seen at 100°C in Fig. 5 and following this trend, the output power of the oscillator has decreased at 50°C from that of room temperature in Table-III. It is seen from Table-III that the oscillator has stable performance for frequency and output power during the temperature variations.

G. Discussion on Results

The oscillator is able to generate third harmonic frequency when the peaking inductor value is 1nH and the second harmonic when the peaking inductor value is 2nH and both second and third harmonics of fundamental frequency when the peaking inductor value is 1.25nH in the gain cells. The highest power output of the oscillator is happening at the third harmonic frequency of 41.84 GHz and tunable by body bias up to 630 MHz (i.e. up to 41.21 GHz). The third harmonic generation of this oscillator has been checked with temperature variations from room temperature of 25°C and results showing that this harmonic oscillator has achieved the complete thermal stability.

IV. CONCLUSION

Novel generation of 2nd and 3rd harmonics is presented, designed using ADS 2008 Harmonic Balance (HB) simulation with UMC (foundry) RF MOSFETs models. Body bias type

tuning attracts no extra devices and extra chip area as compared to MOS Varactors. The change in input and output parasitic capacitances, caused due to change in body bias voltage, used for frequency tuning of DVCO-H.

Acknowledgment

This work is a single author patent application of this author from the Department of Electrical Engineering of Indian Institute of Technology (IIT) Bombay, the details of the patent application has been mentioned in reference [5] and the all kinds of supports of technical resources and finance of IIT Bombay are thankfully acknowledged.

References

[1] H. Wu and A. Hajimiri, "Silicon-Based Voltage-Controlled Oscillators," IEEE Journal of Solid-State Circuits, vol. 36, March 2001, pp. 493-502.

[2] Shih-Chieh Yen and Tah-Hsiung, "ChuAn Nth-harmonic Oscillator Using An N-push Coupled Oscillator Array with Voltage-clamping Circuits, Digest IEEE IMS, USA, 2003, pp. 2169-2172.

[3] Planar High Frequency Oscillator, by Masayoshi Aikawa, Takayuki Tanaka, Fumio Asamura and Takeo Oita, (2006, Dec 5). Patent US 7145405 B2

[Online]. Available: uspto

[4] High Frequency Oscillator by Transmission Line Resonator, by Masayoshi Aikawa, Takayuki Tanaka, Fumio Asamura and Takeo Oita, (2006, Oct 10). Patent US 7119625 B2

[Online]. Available: uspto

[5] Kalyan Bhattacharyya, "Tunable distributed harmonic voltage controlled oscillator for generating second and third harmonic microwave signals", Indian Patent Application 2392/MUM/2010, date of publication of abstract in Indian Patent Office Journal is 06/09/2013 (pending).

[6] Kalyan Bhattacharyya, "CMOS Ku/K-band Distributed Oscillators Using Cascade of CPW Coupled n-FETs Gain Cells with Record Performance of Phase Noise and Ka-band Third Harmonic Generation Technique," Digest 2010 IEEE Wireless and Microwave Technology Conference (WAMICON), Florida, USA, April, 2010, pages 4.

[7] Xin Guan and Cam Nguyen, "Low-Power-Consumption and High-Gain CMOS Distributed Amplifiers Using Cascade of Inductively Coupled Common-Source Gain Cells for UWB Systems," IEEE Trans. on Microwave Theory and Techniques, vol. 54, August 2006, pp 3278-3283.

[8] M. Jamal Deen, R. Murji, A. Fakhr, N. Jafferali and W.L. Ngan, "Low-power CMOS integrated circuits for radio frequency applications," IEE Proceedings - Circuits, Devices and Systems, Oct 2005, pp. 509-522.

[9] Yuhua Cheng, M. Jamal Deen and C.H. Chen, "MOSFET Modeling for RFIC Design," IEEE Trans. on Electron Devices, vol. 52, July 2005, pp 1286-1303.

[10] S.M. Sze, Physics of Semiconductor Devices, 2nd ed., John Wiley & Sons: Singapore, 1981, pp. 451-453.

[11] Wen-Lin Chen, Sheng-Fuh Chang, Kun-Ming Chen, Guo-Wei Huang and Jen-Chung Chang, "Temperature Effect on Ku-Band Current-Reused Common-Gate LNA in 0.13μm CMOS Technology," IEEE Transaction on Microwave Theory and Technique, Volume: 57, No. 9, September 2009, pages: 2131 - 2138.

Performance Enhancement of Slot Synchronization in W-CDMA

Mridula Korde, Assistant Professor,
Shri Ramdeobaba College of Engineering and Management, Nagpur
kordems@rknec.edu

Abstract— The W-CDMA system plays an important role in the 3G cellular systems because of its compatible networking architecture to the present popular Global System for Mobile Communications (GSM) systems as well as the salient features of a Code Division Multiple Access (CDMA) system, including multi-path fading tolerance, high system capacity, low power consumption, good performance and coverage. In CDMA system, a procedure used by a Mobile Station (MS) to search for the best cell site and achieve code, time, and frequency synchronization with it is referred as cell search. Fast cell search is essential to reduce switch on delay (initial cell search), increase stand by time (idle mode search) and maintain good link quality (active-mode search). This is particularly true for the Wideband Code Division Multiple Access-Frequency Division Duplex (W-CDMA-FDD) system since it employs nonsynchronous base stations instead of synchronous ones of other CDMA systems to extend coverage from outdoor to indoor. Cell search is critical for achieving code and time synchronization. The main process of achieving cell search is divided into three stages followed by code verification, tracking and carrier frequency adjustment: 1) slot synchronization, 2) frame synchronization and code group identification, 3) primary scrambling code identification. This paper addresses slot synchronization in downlink as cell search algorithm in W-CDMA. In the three step search, the receiver searches for the slot timing by correlating the received signal with the Primary Synchronization (PSCH) code using Matched Filter (MF). This paper propose a design for a hierarchical matched filter for Primary Synchronization Code (PSC).The aim is to reduce the hardware complexity using Field Programmable Gate Array (FPGA) logic resources required to realize the filter.

Index Terms—WCDMA, CDMA, 3G, Matched Filter, Synchronization

I. INTRODUCTION

The third generation (3G) cellular systems have been standardized to provide higher data rates, better quality, and larger capacity than the second generation (2G) systems. The wide-band code-division multiple-access (WCDMA) system plays an important role in the 3G cellular systems because of its compatible networking architecture to the present popular GSM systems as well as the salient features of a CDMA system, including multi-path fading tolerance, high system capacity, low power consumption, good performance and coverage [1].

In WCDMA systems, each cell site is assigned a unique long scrambling code in the downlink for identification. The process that a mobile station tries to identify and synchronize to a cell is called cell search [2]. The main process of achieving cell search is divided into three stages followed by code verification, tracking and carrier frequency adjustment: 1) slot synchronization, 2) frame synchronization and code group identification, 3) primary scrambling code identification [3] [10] [11]. The overall synchronization processes are implemented by pipelining. When mobile stations are switched on, going through handover or during idle and active modes, they have to synchronize to the scrambling code used in the best serving base station [12].

A great deal of researches has been contributed to design and analysis of the cell search in the WCDMA system. Pipelining three stages was proposed to achieve faster search than serial execution at the cost of higher complexity and power consumption [5], [9]. A fast method using adaptive filters was proposed to help code phase acquisition in additive white Gaussian noise (AWGN) channels [10]. Another scheme was proposed to enhance initial cell search against frequency errors by using correlation properties of synchronization codes [11]. A method of coherent slot detection was also presented to improve search performance under frequency offset [12]. A differential detection method was also proposed to make cell search robust to wide range of initial frequency offset [13]. Partial symbol de-spreading (PSD) with noncoherent combining was used to overcome large-frequency error [5].

II. CELL SEARCH IN W-CDMA

A. Synchronization Channels in WCDMA

Cell search is performed in two scenarios: when a MS is switched on (initial cell search) and during active or idle mode (target cell search). Target cell search is used to find handover candidates during a call. Cell search design is important and needs to be completed in minimum delay as it impacts the system performance.

Each cell in a CDMA system is identified by its downlink scrambling code which is of length 38,400 chips. The 38,400 chips form a radio frame which is divided into 15 slots. Each slot in the radio frame is of 2,560 chips [4].

Figure 1 show the slot and frame structure of the three synchronization channels used in cell search: the Primary-Synchronization Channel (P-SCH), Secondary-Synchronization Channel (S-SCH) and the Common Pilot

978-1-5090-0037-1/16 $31.00 © 2016 IEEE

Channel (CPICH). The P-SCH together with the S-SCH is also called Synchronization Channel (SCH).

In the P-SCH, a 256 chip sequence is transmitted at the start of each slot. The same P-SCH sequence is used by all the BSs and is transmitted once every slot. As the same sequence is used by all the transmitting stations, only one matched filter is sufficient to detect the slot boundary value. The S-SCH is used for carrying 15 different sequences, one in each slot, for the different code groups and is repeated after every frame. These sequences are used in identifying the code group. The CPICH is used to carry the downlink common pilot symbols scrambled by the scrambling code of the BS. Each slot of this channel is divided into 10 symbols, each of 256 chips in length [6].

B. Stage 1: Slot Synchronization

During stage 1 of the cell search procedure the MS uses the SCHs Primary Synchronization Code (PSC) to acquire slot synchronization to a cell. This is typically done by a single matched filter matched to the PSC which is common to all cells. The slot timing of the cell can be obtained by detecting peak values in the matched filter output. The starting position of the synchronization code may be determined from observations over one slot duration. However, decisions based on observations over a single slot may be unreliable, when the signal-to-noise ratio (SNR) is low or if fading is severe. Reliable slot synchronization is required to minimize cell search time.

In order to increase reliability, observations are made over multiple slots and the results are then combined. This ensures that the correct slot boundary is identified.

C. Stage 2: Frame Synchronization and Code Group Identification

During stage 2 of the cell search procedure, the MS uses the SCHs Secondary Synchronization Code (SSC) to achieve frame synchronization and identify the code group of the cell found in stage 1. This is done by correlating the received signal with all possible SSC sequences and identifying the maximum correlation value. Since the cyclic shifts of the sequences are unique, the code group as well as the frame synchronization is determined.

D. Stage 3: Scrambling Code Identification

During stage 3 of the cell search procedure, the MS determines the exact primary scrambling code used by the cell. The primary scrambling code is typically identified through symbol-by-symbol correlation over the CPICH with all codes within the code group identified in stage 2.

This three stage cell search algorithm helps in simplifying the synchronization process of the MS with the BS

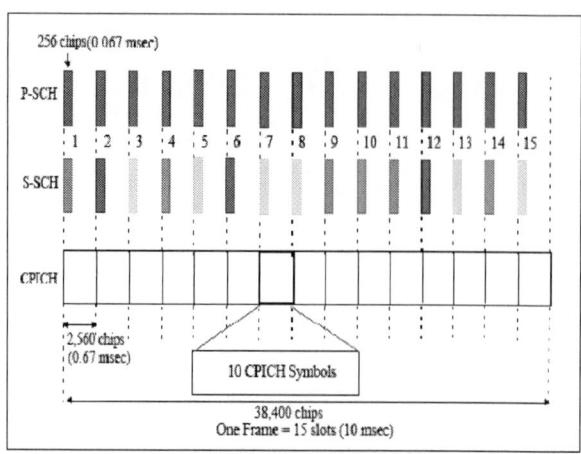

Figure 1. Synchronization Channels in WCDMA

III. SLOT SYNCHRONIZATION DESIGN

W-CDMA and CDMA2000 systems both use DS-CDMA technology for 3G. In these systems, spreading codes are used to differentiate physical channels from the same transmitter, and scrambling codes are used to differentiate transmitters. In synchronous systems such as CDMA2000, the mobile station searches the code timing for initial code acquisition. On the other hand, in an asynchronous W-CDMA system each base station uses its own scrambling code, the mobile station searches for both timing and code uncertainty for initial code synchronization. The use of 512 complex Gold codes in the W-CDMA makes it impractical to exhaustively search all the possible codes for the code timing. This problem can be simplified busing the three-step cell search scheme in the W-CDMA system, adopted by Third Generation Partnership Project (3GPP) specifications. Three-step cell search scheme operation in the W-CDMA can be processed by using three channels. Primary Synchronization Channel (P-SCH), Secondary Synchronization Channel (S-SCH), and Common Pilot Channel (CPICH). P-SCH provides information on slot timing which consists of 2560 chip timing candidates. Among the many synchronization signals, P-SCH is a common sequence of 256 chips that is periodically transmitted by each cell. The terminal looks for this common sequence to establish the slot boundary of at least one of the cells. In the first step search, the receiver searches for the slot timing by correlating the received signal with the P-SCH code using matched filter. This section proposes a design for matched filter for Primary Synchronization Code detection based on FPGA.

A. Matched Filter

A matched filter is a filter whose frequency response is designed to exactly match the frequency spectrum of the input signal. The operation of matched filters is the same as correlating a signal with a copy of itself. Unlike a transmission filter where the continuous data stream entering the filter is modified to form a new continuous data stream, a matched filter output must be considered to be a flow of individual results. These results must then be analyzed to identify the individual point where the match occurred. Figure 2 shows the basic operation of a matched filter.

A matched filter can be realized with the same structure as Finite Impulse Response (FIR) filter with +1 and -1 coefficient. The structure is similar to the manual method; the code sequence of interest is tested at each bit position along the data pattern.

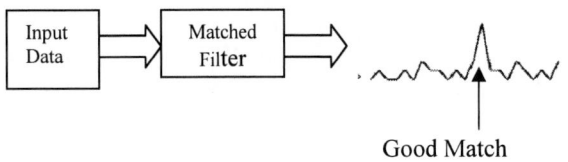

Good Match

Figure 2. CDMA Matched Filter Basic Operation

IV. MATCHED FILTER DESIGN FOR IMPROVEMENT IN SLOT SYNCHRONIZATION (STAGE 1)

In stage 1, the P-SCH matched filter is used to detect the slot boundary. Due to low operating signal-to-noise ratio, the matched filter outputs have to be noncoherently accumulated over more than one slot to get reliable decision statistics. After accumulation, a number of candidates are identified as possible slot boundary candidates.

The MS first needs to acquire the PSC which is common to all the BSs. These codes are of length 256 chips. The matched filter output is given by [13]

$$Y = \sum_{j=0}^{255} R_j \, Cp_j \qquad (1)$$

IV. Generation of Primary Synchronization Codes

A. Implementation in Verilog

The primary synchronization sequence thus constructed is passed through hierarchical matched filter as well as conventional matched filter.

Verilog code is used to implement the design structure and simulation is done with Xilinx ISE. The RTL coding of the design was verified for functionality by functional simulation using Xilinx simulation tool. The functional simulation approach was hierarchical. Each unit of the design was coded into a separate module. Each module was tested for functionality independently. The top level module was tested using test benches. Gate level architecture is used for design of all the functional blocks. Also, static timing analysis is used to verify timing closure in the final design layout. Each unit of the design was coded into a separate module. The top level module was tested using FM test benches. Main logic blocks are made up of AND gate, full adder, and D Flip-Flops.

Synthesis is the process of translating HDL based design into a gate level description which is ready for

optimization. Synthesis of simulated design gives the implementation of design using the primitive resource and gates available in target device, decreases design time by eliminating the need to define every gate and reduces number of errors that can occur during a manual translation of hardware description to design schematic.

The exact implementation depends on the synthesis tool used and the target Device Architecture. The FPGA used for synthesis is Xilinx Spartan 3 and the device selected is XC3S400-PQ208. The Spartan™-3 family of Field-Programmable Gate Arrays is specifically designed to meet the needs of high volume, cost-sensitive consumer electronic applications. The eight-member family offers densities ranging from 50,000 to five million system gates.

Synthesis result is obtained where logic utilization for both conventional matched filter and hierarchical matched filter are compared. The various parameters like No. of slice Flip-Flops, No. of LUTs, No. of bounded IOBs, Equivalent Gate Count are studied.

The synthesis result is shown in Table 1.

TABLE I.　SYNTHESIS RESULT OF MATCHED FILTER TABLE STYLES

Spartan 3family XA3S400 Device Speed Gate -4	Slices	Slice flip flops	4 Input LUTs	Bonded IOBs	Equivalent Gate Count	Overall Hardware Utilization
Conventional Matched Filter	476 Out Of3584	256 Out of 7168	80 Out of 7168	201 Out of 141	11,702	35.6%
Hierarchal Matched Filter	849 Out of 3584	1340 Out of 7168	6350 ut of 7168	22 Out of 141	7,934	13.8%

B. RTL Schematic of Filter Design

RTL Schematic is obtained using Xilinx 9.1i tool. RTL schematic view of conventional matched filter and hierarchical matched filter are obtained and description of internal blocks is shown below.

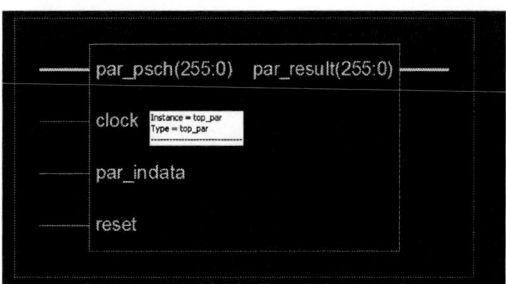

Figure 6. RTL Schematic of Conventional Matched Filter

2016 International Conference on VLSI Systems, Architectures, Technology and Applications (VLSI-SATA)

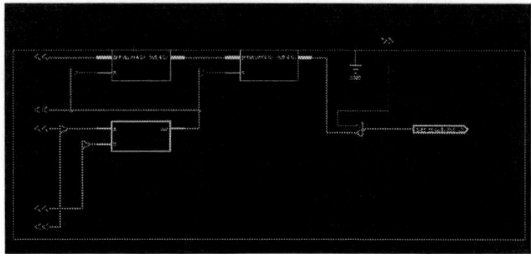

Figure 7. RTL Schematic of Internal Blocks of Matched Filter

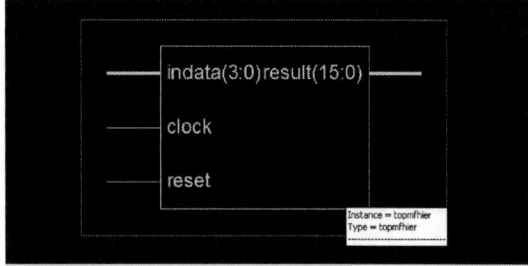

Figure 8. RTL Schematic of Hierarchical Matched Filter

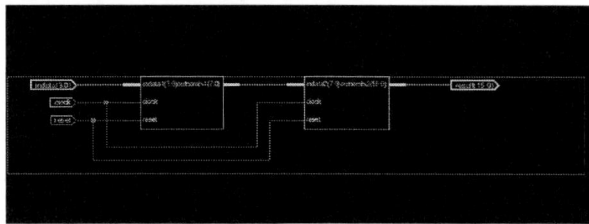

Figure 9. RTL Schematic of Two Internal Concatenated Blocks of
Hierarchical Matched Filter

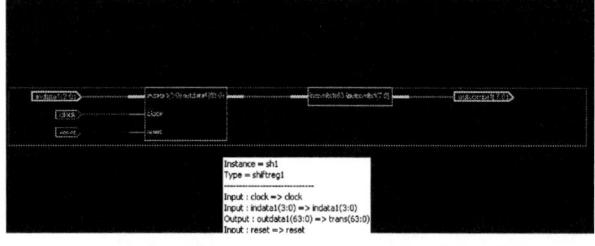

Figure 10. RTL Schematic of Level Two of Hierarchical Matched
Filter

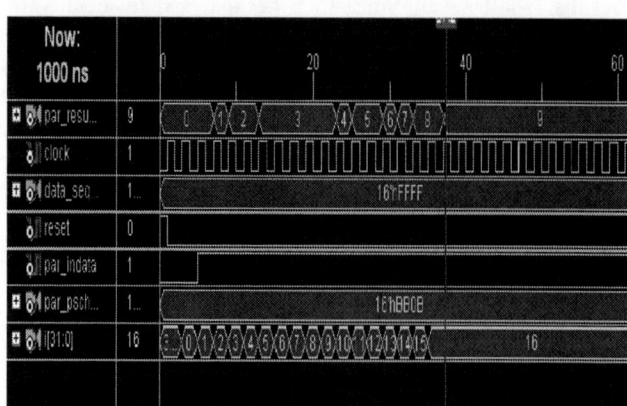

Figure1: Simulation Result of Conventional Matched Filter

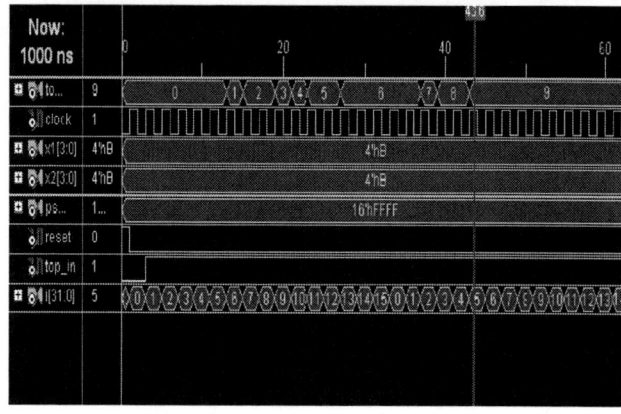

Figure 2. Simulation Result of Hierarchical Matched Filter

A. Simulation Result of Filter Design

Simulation Results are obtained using Xilinx ISE 9.1i tool for both filters as shown below in Figure 11 and Figure 12. When hierarchical matched filter was compiled and simulated, five clock cycles are required initially to process data. So the result which is expected at 16th clock cycle is observed at 21st clock cycle

B. Design Utilization Summary

Device utilization summary gives total gate count, required for design, which helps to check area required for the particular design. Verilog codes of both the filters are synthesizable and corresponding design summaries are shown in Figure 13 and Figure 14.

978-1-5090-0037-1/16 $31.00 © 2016 IEEE 148

Device Utilization Summary [estimated values]			
Logic Utilization	Used	Available	Utilization
Number of Slices	476	3584	1
Number of Slice Flip Flops	256	7168	
Number of 4 input LUTs	635	7168	
Number of bonded IOBs	201	141	14
Number of GCLKs	1	8	1

Figure 3. Design Utilization Summary of Conventional Matched Filter

Device Utilization Summary			
Logic Utilization	Used	Available	Utilization
Number of Slice Flip Flops	1,340	7,168	18%
Number of 4 input LUTs	80	7,168	1%
Logic Distribution			
Number of occupied Slices	849	3,584	23%
Number of Slices containing only related logic	849	849	100%
Number of Slices containing unrelated logic	0	849	0%
Total Number of 4 input LUTs	80	7,168	1%
Number of bonded IOBs	22	141	15%
IOB Flip Flops	4		
Number of GCLKs	1	8	12%

Figure 4. Design Utilization Summary of Hierarchical Matched Filter

V. CONCLUSION

Thus, a design of hierarchical matched filter for primary code synchronization channel is presented in the work. This is an important step in receiving a synchronized data for W-CDMA mobile systems. A conventional matched filter implementation would require 256 taps and a large adder circuit. This would increase the delay as well as power consumption at the receiver which is not desirable. A hierarchical structure is used for performing the matched filter operations, which will need lesser number of taps, reduced circuitry, and lower power consumption. In any hardware implementation, there is strong economic imperative to minimize the number of complexity of the arithmetic operations employed in the data path. The aim is to reduce the hardware complexity using FPGA logic resources required to realize the filter. As seen from Table 1, hierarchical matched filter can be designed with less hardware as compared to traditional matched filter. A lower equivalent gate count (7934) is achieved on a Xilinx Spartan 3 family XC3S400 Device as compared to the traditional matched filter when the same constraints were used in the synthesis of both the designs. Design of the conventional matched filter implementation requires 256 additions and multipliers whereas design of the hierarchical matched filter implementation requires 30 additions and 32 multipliers. Thus, the hierarchical matched filter achieves a saving of a factor of 8.533 in terms of additions and 8 in terms multiplications.

REFERENCES

[1] M. Kiessling and S. A. Mujtaba, "Performance enhancements to the UMTS (W-CDMA) initial cell search algorithm," in Proc. IEEE Int. Conf. Commun., 2002, vol. 1, pp. 590–594.

[2] C.-F Li, Y.-S. Chu, W.-H. Sheen, F.-C. Tian and J.-S. Ho, "A low power ASIC design for cell search in W-CDMA system," IEEE J. Solid-State Circuits, vol. 39, no. 5, pp. 852–857, May 2004.

[3] Chi-Fang Li, Member, IEEE, Yuan-Sun Chu, Jan-Shin Ho, and Wern-Ho Sheen, Member, IEEE, "Cell Search in WCDMA under Large-Frequency and Clock Errors: Algorithms to Hardware Implementation", IEEE Transactions on Circuits and Systems I: Regular Papers, Vol. 55, No. 2, March 2008.

[4] Johan Nystrom, Karim Jamal, Yi-Pin Eric Wang', Riaz Esmailzadeh , "Comparison Of Cell Search Methods for Asynchronous Wideband CDMA Cellular System" 2005 International Conference on Wireless Networks, Communications and Mobile Computing

[5] Y.-P. E. Wang and T. Ottosson, "Cell search in W-CDMA," IEEE J. Select. Areas Commun, vol. 18, no. 8, pp. 1470–1482, Aug. 2002.

[6] C.-F Li, Y.-S. Chu, W.-H. Sheen, F.-C. Tian and J.-S. Ho, "A low power ASIC design for cell search in W-CDMA system," IEEE J. Solid-State Circuits, vol. 39, no. 5, pp. 852–857, May 2004.

[7] Chi-Fang Li, Member, IEEE, Yuan-Sun Chu, Jan-Shin Ho, and Wern-Ho Sheen, Member, IEEE, "Cell Search in WCDMA under Large-Frequency and Clock Errors: Algorithms to Hardware Implementation", IEEE Transactions on Circuits and Systems—I: Regular Papers, Vol. 55, No. 2, March 2008.

[8] Shailendra Mishra, "Performance Analysis of 3-stage Cell search Process in a WCDMA System", International Journal of Business Data Communications and Networking, Volume 3, Issue 3,2007

[9] Shailendra Mishra, "Performance Enhancements to the WCDMA Cell Search Algorithms" Journal of Information Technology and Applications Vol. 1 No. 2 September, 2006, pp. 59-78

[10] E. Dahlman, P. Beming, J. Knutsson, F. Ovesjö, M. Persson, and C. Roobol, "WCDMA The radio interface for future mobile multimedia communications", IEEE Trans. Veh. Technol., vol. 47, pp. 1105- 1118, Nov. 1998.

[11] J. Moon and Y.-H. Lee, "Cell search robust to initial frequency offset in WCDMA systems," in Proc. IEEE Int. Symp. Personal, Indoor and Mobile Radio Commn. Sep. 2002, vol. 5, pp. 2039–2043.

[12] Paul Hard y, Andy Miller, and Maria George Ken Chapman, "CDMA Matched Filter Implementation in Virtex Devices," Application Note: Virtex™ Series XAPP212 (v1.0), March 31, 2000.

[13] Sanat Kamal Bahl, "A Comparative Study of W-CDMA CellL Search Design," Journal of Circuits, Systems and Computers, vol. 14, no. 1, pp. DOI: 10.1142/S021812660500212X, February 2005.

Design of 3C-SiC Symmetric and Asymmetric Double Gate MOSFET

Sudarshana Jilowa[1], Sandeep Singh Gill[2], Gurjot Kaur Walia[3]

Department of Electronics and Communication Engineering

[1]Student Master of Technology

[2]Professor

[3]Assistant Professor

Guru Nanak Dev Engineering College

Ludhiana, India

[1]sudjilowa@gmail.com

Abstract—The superior material properties such as higher breakdown field, thermal conductivity and wider bandgap make Silicon Carbide (SiC) appropriate for transistor applications. However, there are certain parameters where SiC have yet to achieve required performance levels. SiC have number of polytypes having mature technology. Being one of the SiC polytypes, 3C-SiC has higher mobility and saturation velocity. This paper presents symmetric 3C-SiC Double Gate MOSFET (DG MOSFET) and asymmetric 3C-SiC DG MOSFET. The two designs were simulated for different applied voltage at various lengths. From simulated results it can be seen that asymmetric 3C-SiC DG MOSFET shows increment in on-current values and decrement in off-current values for 22 nm, 40 nm and 60 nm channel length.

Keywords— 3C-SiC, Double Gate MOSFET, On-current, Off-current

I. INTRODUCTION

With enormous improvements in Silicon (Si) devices for better processing techniques, scaled device designs and material quality have led to great advancements in semiconductor industry in the last five decades [1]. However, many Si devices have arrived at their physical limits in performance. Silicon Carbide (SiC) has emerged as an attractive semiconductor to replace Si devices [2]. Although earlier commercialization of SiC devices have been suffering because of the poor quality of SiC wafers, resulting in low device yields, but with the strong efforts and programs conducted by government for the advancement of SiC in the last two decades, SiC single-crystal wafers were developed [3].

SiC has high breakdown field and higher thermal conductivity than Si [4]. SiC also have superior material properties and electrical characteristics such as lower on-resistance [5]. The key characteristics of SiC includes higher switching speed (which allows the use of smaller capacitors, inductors and transformers, bringing down the overall system weight and size), high thermal conductivity (which allows operation at high power density and enables efficient heat removal from the device) and wide energy gap which offers higher operating temperature and lower leakage current. SiC occurs in many crystalline structures or polytypes. All of them have different crystal lattice, carrier mobilities and bandgaps [6]. 3C-SiC is one of the polytypes having diamond crystal lattice.

Symmetric and asymmetric 3C-SiC Double Gate MOSFET has been designed on various lengths which represent significant improvements in the performance over conventional double gate devices. The simulated results have been analyzed using Cogenda Visual TCAD software.

The paper is organized in this way; 3C-SiC details and its properties discussed in section II, in section III proposed device design, and then in section IV simulated results, section V summarizes and concludes this paper.

II. 3C-SIC

Being one of the polymorph of SiC, 3C-SiC is an appropriate material for transistor applications because 3C-SiC has the advantage of having high saturation velocity and mobility, in addition to the other advantages of SiC [7]. 3C-SiC devices can work in hostile condition where Si devices cannot. The large bandgap of SiC results in a much higher operating temperature and radiation hardness. The growing of 3C-SiC on silicon substrate is the key challenge because it causes high density of dislocations in the lattice. And also leads to thermal expansion in two materials. The Vapour Liquid Solid (VLS) process is used to reduce the thermal expansion and dislocations as well.

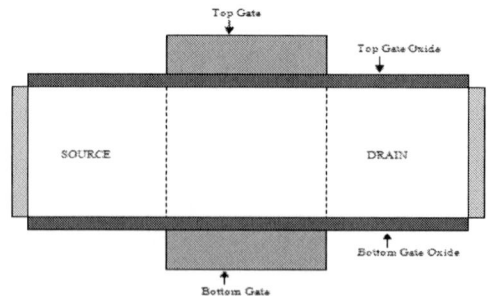

Fig. 1: 3C-SiC DG MOSFET

2016 International Conference on VLSI Systems, Architectures, Technology and Applications (VLSI-SATA)

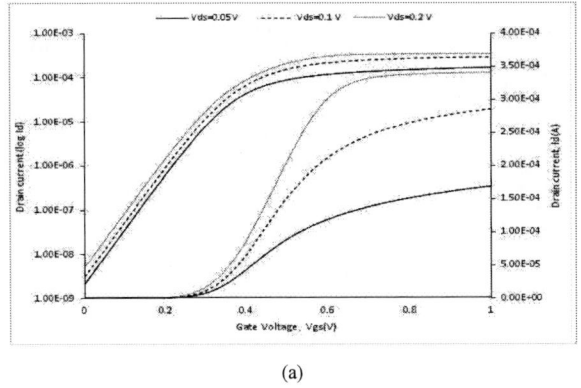

(a)

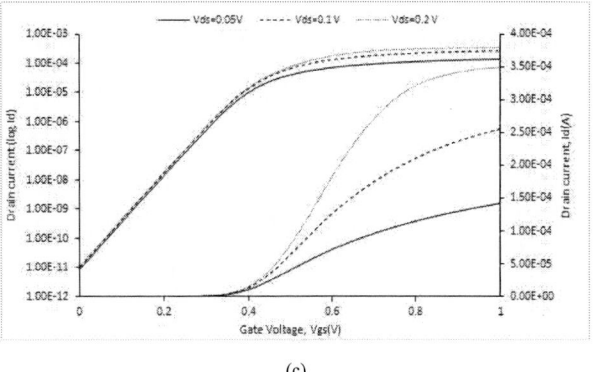

(c)

(b)

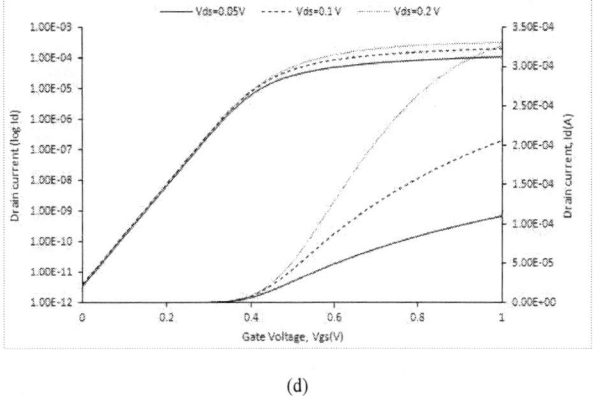

(d)

Fig. 2: Plot showing the transfer characteristics of symmetric 3C-SiC DG MOSFET at different lengths (a) 22nm (b) 40nm (c) 60nm (d) 100nm

Double Gate MOSFET (DG MOSFET) has been generally recognized as one of the most promising device. It has a unique property of electrical coupling of the two gates, which leads to a better gate control over the channel thus reducing the leakage current, Drain Induced Barrier Lowering (DIBL) effect and short-channel effect. The main idea of a DG MOSFET is to have a channel of very small width and to control the channel by applying gate contacts to both sides of the channel [8]. To continue better results in device performance new materials with better transport properties are intensively examined [9]. Si devices usually become unreliable at 150° C, and rarely function above 200° C, while SiC devices have shown to operate above 300° C. Major advances have been made in the fabrication of SiC electron devices in recent years.

III. DEVICE DESIGN

The proposed device design is shown in Fig.1. The substrate is composed of 3C-SiC, the top and bottom gate of n-polysilicon or p-polysilicon. Aluminium (Al) is used for drain and source ohmic contact. The gate width of the device is 5 nm, the oxide thickness is 2 nm, the width of the 3C-SiC substrate is 10 nm, and the thickness of the Al metal contacts is 2 nm. Doping concentration is $1e^{18}$ cm^{-3}. This configuration is further dissevered as symmetric and asymmetric 3C-SiC DG MOSFET. Symmetric 3C-SiC DG MOSFET have both gates of same material and work function. And switching of both gates has been done simultaneously by interconnecting both

gates and giving them one gate voltage. While asymmetric 3C-SiC DG MOSFET has different gate material. Other parameters were kept same as depicted earlier.

IV. DEVICE SIMULATION AND ANALYSIS

The simulation was performed using the following doping values: the n and p doping for 3C-SiC substrate doping is $1e^{18}$ cm^{-3}. Next, the device structure was built and simulated.

The simulation was done using Cogenda Visual TCAD software. Transfer characteristics and output characteristics have been analyzed on different gate lengths with different drain and gate voltages respectively.

Fig. 2 shows the relationship between the drain current, I_d and the gate voltage, V_{gs}, at different drain voltages for different channel lengths with logarithmic and linear configurations. Fig. 2(a) shows that the drain current gradually increases with the drain voltage increment in 3C-SiC symmetric DG MOSFET with 22 nm channel length. Here leakage effect is negligible.

Using the same configurations transfer characteristics of the device was also simulated for different lengths and the results are shown above. Fig. 3 shows that I_d increases with increasing gate and drain voltages (here values of drain current are shown for $V_{gs} = 0.5$ V to 1 V taken from bottom to top).

978-1-5090-0037-1/16 $31.00 © 2016 IEEE   151

2016 International Conference on VLSI Systems, Architectures, Technology and Applications (VLSI-SATA)

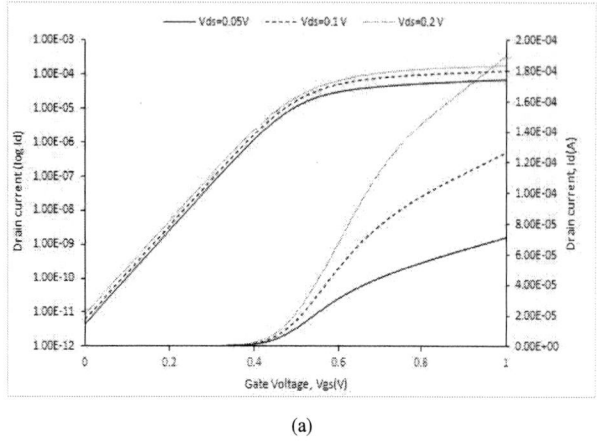

(a)

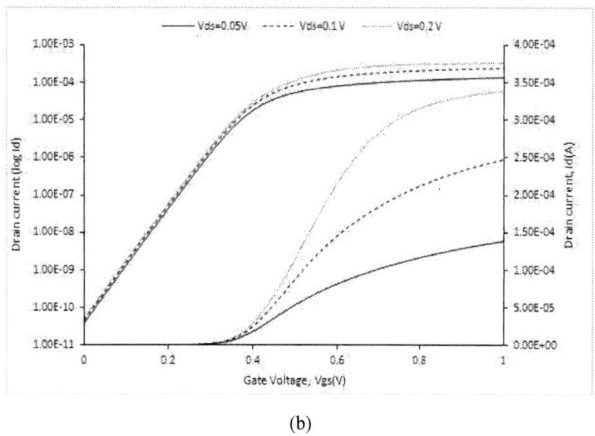

(b)

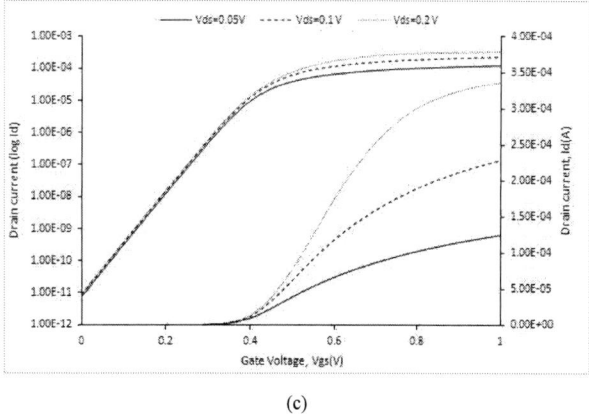

(c)

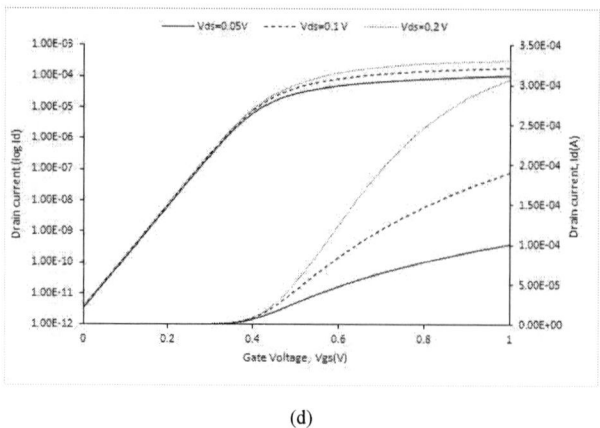

(d)

Fig. 3: Plot showing transfer characteristics asymmetric 3C-SiC DG MOSFET at different lengths (a) 22 nm (b) 40 nm (c) 60nm (d) 100 nm

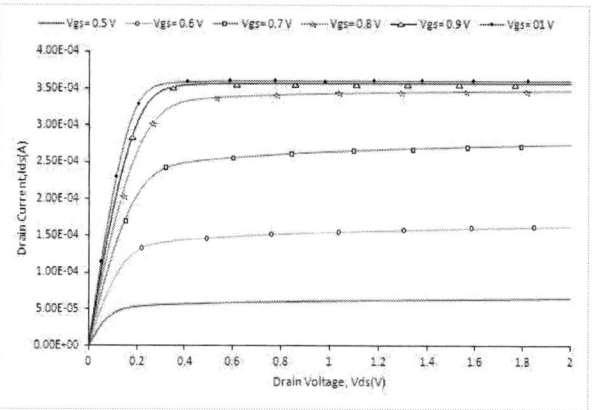

(a)

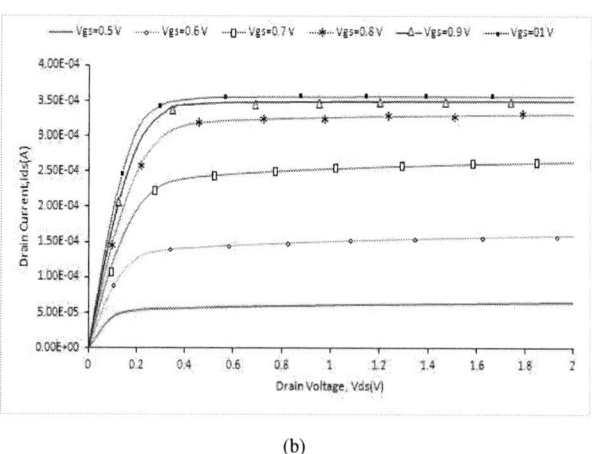

(b)

Fig. 4: I_d-V_{ds} plot for various gate voltages (a) symmetric 3C-SiC DG MOSFET (b) asymmetric 3C-SiC DG MOSFET

Fig. 3 indicates I_d -V_{gs} plots of 3C-SiC asymmetric DG MOSFET on logarithmic scale and linear scale with same parameters as given for symmetric DG MOSFET.

The simulation was carried out for different lengths on 0.05 V, 0.1 V and 0.2 V drain voltages V_{ds}. The impact of this on device performance is studied by extracting several device parameters. For low power consumption I_{off} should be less and for device speed application I_{on} should be more. Hence this establishes a trade-off between I_{on} and I_{off} leading to better I_{on}-

I_{off} ratio. Fig.4 shows drain current behavior for symmetric and asymmetric 3C-SiC DG MOSFET on various gate to source voltages. It shows linear behavior of drain current with increase

978-1-5090-0037-1/16 $31.00 © 2016 IEEE   152

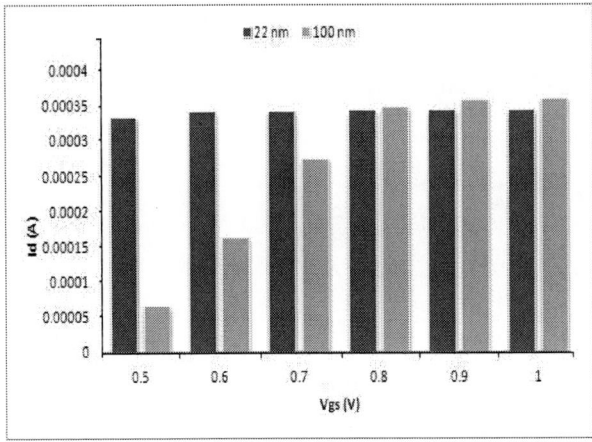

Fig. 5: Drain current variation on 22nm and 100nm at different gate voltage in symmetric DG MOSFET

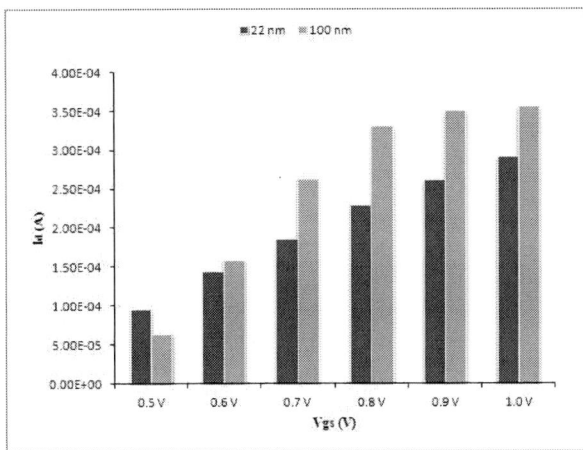

Fig. 6: Drain current variation on 22nm and 100nm at different gate voltage in asymmetric DG MOSFET

in the gate potential. Although after some time it attains saturation. Fig. 5 and Fig. 6 depicts the effect of gate potential on 22nm and 100nm channel length to show the behavior of the device on short channel length and long channel length device. In Fig. 5, the variation of drain current in symmetric DG MOSFET is very less for short channel device and more for long channel length device. Following tables present the effect of channel length on performance parameters of symmetric and asymmetric 3C-SiC DG MOSFET. Table I shows comparison between symmetric and asymmetric 3C-SiC DG MOSFET at 22 nm channel length. There is an increment of 0.28% in on-current of the device in asymmetric 3C-SiC DG MOSFET. Subthreshold swing is constant for both designs. Also there is a decrement in the off-current in asymmetric design. For better performance of the device, on-current should be more because it is an important aspect related to speed and off-current should be low to reduce the amount of power consumption.

At 40 nm 23.4% increment was seen for asymmetric 3C-SiC DG MOSFET shown in Table II. Also there was an increment of 4.8% in subthreshold swing and decrement in off-current.

TABLE I.   COMPARISON BETWEEN SYMMETRIC AND ASYMMETRIC 3C-SIC DG MOSFET AT 22 NM CHANNEL LENGTH

Parameters	3C-SiC DG MOSFET		Variation
	Symmetric	Asymmetric	
$I_{on}(A)$	$3.48e^{-04}$	$3.49e^{-04}$	0.28% increment
$I_{off}(A)$	$3.16e^{-09}$	$9.27e^{-12}$	10^3 times decrement
SS (mV/ decade)	70	70	constant
I_{on}/I_{off}	1×10^5	2×10^8	10^3 times increment

TABLE II.   COMPARISON BETWEEN SYMMETRIC AND ASYMMETRIC 3C-SIC DG MOSFET AT 40 NM CHANNEL LENGTH

Parameters	3C-SiC DG MOSFET		Variation
	Symmetric	Asymmetric	
$I_{on}(A)$	$1.45e^{-04}$	$1.79e^{-04}$	23.4% increment
$I_{off}(A)$	$8.14e^{-11}$	$6.24e^{-12}$	10^1 times decrement
SS (mV/ decade)	62	65	4.8% increment
I_{on}/I_{off}	5×10^7	1×10^8	10^1 times increment

TABLE III.   COMPARISON BETWEEN SYMMETRIC AND ASYMMETRIC 3C-SIC DG MOSFET AT 60 NM CHANNEL LENGTH

Parameters	3C-SiC DG MOSFET		Variation
	Symmetric	Asymmetric	
$I_{on}(A)$	$3.46e^{-04}$	$3.49e^{-04}$	0.86% increment
$I_{off}(A)$	$1.13e^{-11}$	$9.27e^{-12}$	10^1 times decrement
SS (mV/ decade)	62	62	constant
I_{on}/I_{off}	3×10^6	2×10^8	10^2 times increment

TABLE IV.   COMPARISON BETWEEN SYMMETRIC AND ASYMMETRIC 3C-SIC DG MOSFET AT 100 NM CHANNEL LENGTH

Parameters	3C-SiC DG MOSFET		Variation
	Symmetric	Asymmetric	
$I_{on}(A)$	$3.27e^{-04}$	$3.26e^{-04}$	0.3% decrement
$I_{off}(A)$	$4.0e^{-12}$	$3.35e^{-12}$	Approximately same
SS (mV/ decade)	60	60	constant
I_{on}/I_{off}	1×10^8	1×10^8	constant

From Table III, 0.86% increment and from Table IV, 0.3% decrement in on-current can be depicted for asymmetric DG MOSFET. In asymmetric design for 22 nm the ratio I_{on}/I_{off} is 10^3 times more than symmetric design, at 40 nm it is 10^1 times more, at 60 nm it is 10^2 times and at 100 nm it is constant. From Table IV it can be seen that the variation in symmetric and asymmetric 3C-SiC DG MOSFET is very less. From all tables it can be seen that higher channel length has less effect on the off-current. On the other hand, the increase in channel length enhances the subthreshold swing of the device. For asymmetric DG MOSFET device the on-current to off-current ratio I_{on}/I_{off} is stable which shows that there is very less difference between on-current and off-current.

V. CONCLUSION

The two device designs presented in this paper, symmetric 3C-SiC DG MOSFET and asymmetric 3C-SiC DG MOSFET, represent significant improvements in the performance over conventional devices. It has been observed that symmetric 3C-SiC DG MOSFET's drain current has stable behavior at short channel length device in spite of different gate potentials. Although for long channel length device it shows increment at early stage while it attains stability subsequently. On the other hand asymmetric 3C-SiC DG MOSFET's drain current has almost linear characteristics for short channel length device and long channel length device. The asymmetric 3C-SiC DG MOSFET shows increment in on-current values and decrement in off-current values for 22 nm, 40 nm and 60 nm channel length. For 100 nm both designs have approximately same response.

Additionally, the minimum amount of leakage current can be estimated from the off-current values. This is accomplished by the high thermal conductivity of SiC, showing low power consumption by the device. The increased values of on-current of both the devices represent better switching characteristics which makes 3C-SiC DG MOSFET more promising design for integrated device technology.

REFERENCES

[1] R.J. Callanan, A. Agarwal, A. Burk, M. Das, B. Hull, F. Husna, A. Powell, J. Richmond, S. Ryu, and Q. Zhang, "Recent Progress in SiC DMOSFETs and JBS Diodes at Cree," IEEE 15th Annual Conference on Industrial Electronics (IECON), Orlando, 2008, pp. 2885-2890.

[2] J.C. Zolper, "Emerging Silicon Carbide Power Electronics Components," IEEE Twentieth Annual Conference on Applied Power Electronics (APEC), Austin, 2005, pp. 11-17.

[3] J. Wang, T. Zhao, J. Li, A.Q. Huang, R. Callanan, F. Husna, and A. Agarwal, "Characterization, Modeling, and Application of 10-kV SiC MOSFET," IEEE Transactions on Electron Devices, vol. 55, pp. 1798 – 1806, August 2008.

[4] Y. Mikamura, K. Hiratsuka, T. Tsuno, H. Michikoshi, S. Tanaka, T. Masuda, K. Wada, T. Horii, and T. Sekiguchi, "Novel Designed SiC Devices for High Power and High Efficiency Systems," IEEE Transactions On Electron Devices, vol. 62, pp. 382-389, October 2014.

[5] L.D. Stevanovic, K.S. Matocha, P.A. Losee, J.S. Glaser, J. Nasadoski, and S.D. Arthur, "Recent Advances in Silicon Carbide MOSFET Power Devices," IEEE Twenty-Fifth Annual Conference on Applied Power Electronics (APEC), Palm Springs, 2008, pp. 401-407.

[6] R. Singh, and M. Pecht, "Commercial Impact on Silicon Carbide," IEEE Industrial Electronics Magazine, vol. 2, pp. 19 – 31, September 2008.

[7] S. Tiwari, T. Undeland, S. Basu, and W. Robbins, "Silicon Carbide Power Transistors, Characterization for Smart Grid Applications," IEEE 15th International Conference on Power Electronics and Motion Control, Novi Sad, 2012, pp. LS6d.2-1 - LS6d.2-8.

[8] T. Nakamura, M. Aketa, Y. Nakano, M. Sasagawa, and T. Otsuka, "Novel Developments Towards Increased SiC Power Device and Module Efficiency," IEEE Conference on EnergyTech, Cleveland, 2012, pp. 1-6.

[9] K. Wada, S. Nishizawa, and H. Ohashi, "Design and Implementation of a Non-Destructive Test Circuit for SiC-MOSFETs," IEEE 7th International Power Electronics and Motion Control Conference, Harbin, 2012, pp. 10 – 15.

A Hierarchical Cluster-Based Model with Run-Time Reconfigurable Resource Allocation on FPGAs

Amin Yoosefi

Departments of Electrical and Computer Engineering
Graduate University of Advanced Technology
Kerman, Iran
a.yoosefi@student.kgut.ac.ir

Hamid Reza Naji

Departments of Electrical and Computer Engineering
Graduate University of Advanced Technology
Kerman, Iran
hamidnaji@ieee.org

Abstract— Programmability, flexibility and parallel computational capabilities are some of the features making field-programmable-gate-arrays (FPGAs) advantageous over application-specific-integrated-circuits (ASICs). Thanks to the dynamic partial reconfiguration, FPGA provides a virtual hardware resource wherein hardware tasks can swap in and out of the hardware dynamically at runtime. In this paper, we extend the FPGA infrastructure by providing it with a hierarchical cluster-based model similar to multi-core systems. In the proposed model, FPGA is hierarchically clustered into one master node at the top of the system model and several cluster nodes, connected through a dedicated network. To support parallel reconfiguration, each node is provided with a dedicated configuration controller. In addition, a runtime reconfigurable resource allocation approach is proposed. In the proposed approach, reconfigurable resources join and leave clusters at runtime dynamically based on runtime conditions, providing reconfigurable resource sharing.

Keywords—field-programmable-gate-array (FPGA); dynamic partial reconfiguration; task scheduling; clustering; multi-Core

I. INTRODUCTION

Field-programmable-gate-arrays (FPGAs) are well known for their programmability, flexibility, parallel computational capabilities, and low Time-to-Market cost. Thanks to these features, many devices are implemented in FPGAs.

Newer FPGA devices provide even more capabilities by supporting dynamic partial reconfiguration (DPR). In FPGAs supporting DPR, the reconfigurable area is partitioned into tiles, called reconfigurable logic units (RLUs). Each RLU can be reconfigured dynamically at runtime without affecting the logics implemented in the remaining RLUs. This feature provides a virtual hardware in which hardware tasks can swap in and out of FPGA resources dynamically at runtime.

Reconfiguration in modern high-end FPGAs with millions of reconfigurable resources can be very time consuming on the order of hundreds of milliseconds [1]-[2]. Although FPGA technology provides features that may improve the performance and decrease the hardware costs, it suffers from the reconfiguration overhead which may have an inverse impact on the performance and energy consumption.

To address the reconfiguration overhead, an extensive research has been performed. Some researches propose to employ task scheduling algorithms applying techniques such as configuration caching, prefetching and replacement policy in order to reduce the number of times required to reconfigure the hardware or to hide the reconfiguration latency [3]-[6].

In configuration caching technique, the reconfigurations likely to be reused in the near future are kept inside the hardware. However, in cases where a new configuration is needed and there are not enough resources available on the hardware, an existing configuration should be selected as a victim to be replaced with the requested configuration. The replacement policy decides about selecting the potential victim. With prefetching, configurations are loaded into the reconfigurable hardware in advance of when they are needed.

In [7], a communication-aware scheduling algorithm is proposed, which tries to enhance the system performance by reducing communication distances during scheduling.

Some researches try to reduce the real size of configuration data to be transferred. In DPR, as discussed previously, a single portion of the reconfigurable hardware is reconfigured, while the rest are untouched. As a result, only the configuration related to that portion has to be transferred, so the size of transferred data is reduced. Configuration compression is another example. The first proposed configuration compression technique [8] targets the Xilinx 6200 series [9] and is partially reconfigurable at a very fined-grain level.

On the other hand, since FPGAs at the market have limited number of configuration controllers, they do not support parallel reconfiguration. In these FPGAs, configurations are loaded in the reconfigurable hardware one at a time. This is a limitation to the FPGA technology when multiple configurations are ready to be loaded at the same time and this may significantly reduce the parallel time. One solution to this limitation, proposed by [10], is adding more reconfiguration controllers to carry out several reconfigurations in parallel. In this paper, we extend [10] in order to support dynamic reconfigurable resource allocation.

To the best of our knowledge, however, no one has considered providing FPGAs with a parallel architecture similar to multi-core systems. In this paper, a hierarchical

978-1-5090-0037-1/16 $31.00 © 2016 IEEE   155

cluster-based model of FPGA is proposed, composed of one master node at the top of system model and several cluster nodes, connected through a dedicated network. Provided with a configuration controller, each node can reconfigure its tasks independently. The model also applies a hierarchical task scheduling scheme for partitioning tasks between clusters with the purpose of load balancing.

In the multi-core systems, each core can execute only one task at a time. However, in the proposed model subtasks inside a cluster node can execute in parallel, thanks to FPGA technology, but the reconfigurations inside a cluster are in a sequence, similar to multi-core systems.

In addition, a runtime RLU allocation approach is proposed. Based on the runtime conditions, RLU resources are allocated to clusters at runtime dynamically.

The rest of the paper is organized as follows. Section II describes the task model used in this research. The details related to the proposed parallel mode is presented in section III. In section IV, the concluding remarks are made.

II. TASK MODEL

An input task to the system is considered as a set of subtasks communicating with each other. We use a task graph to represent the structure of a task. Here, a task graph is a *directed acyclic graph* or a *DAG* defined as a tuple $G = (V, E, W, C)$, where V and E denotes the nodes and the edges set in G, respectively.

Nodes in a DAG represent subtasks while the edges show the communications or just dependency relations between subtasks. An edge $e_{ij} \in E$ directed from n_i to n_j indicates that the node n_j is data dependent on the node n_i. This means that the node n_j must wait for the required data to be produced by n_i before starting its execution. The positive weight $w(n)$ assigned to a node $n \in V$ indicates the computation cost of that node, and the nonnegative weight $C(e_{ij})$ assigned to an edge $e_{ij} \in E$ represents the communication cost from n_i to n_j .

Although subtasks within a single task can communicate with each other, in this paper it is assumed that there is no communication among tasks. In other words, tasks can execute in parallel independently regardless of their arrival order to the system. For example, in Fig. 1 tasks are shown in squares and subtasks within a single task are represented as circles. As shown in Fig. 1-a, tasks T_2, T_4, and T_1 arrive at times 2, 8 and 12 ms respectively. In Fig. 1-b the DAG of task T_4 is shown, which is composed of subtasks ST_1, ST_2, ST_3, ST_4 and ST_5.

Let $T = \{T_1, T_2, ..., T_N\}$ be the set of all input tasks to the system. Each task T_i is specified with a triple $T = (Pt_i, Nres_i, At_i)$, where Pt_i is the parallel time for task T_i; $Nres_i$ is the minimum number of RLUs giving the minimum configuration delay; and At_i is the arrival time for task T_i.

For each task, Pt_i and $Nres_i$ are calculated at compile time and are used at runtime. As a result, there is no overhead in calculating these parameters at runtime. Since the total execution time for a task T_i can vary depending on the number of RLUs available, we use the worst-case Pt_i for T_i. In this paper the worst-case Pt_i is calculated when there are only two

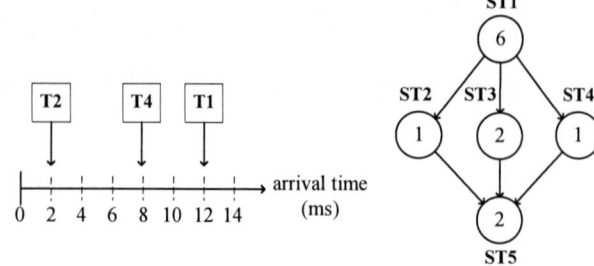

Fig. 1. An example of the task model, a) incoming tasks to the system along with their arrival times, b) DAG of task T_4 composed of subtasks ST_1, ST_2, ST_3, ST_4 and ST_5.

RLUs available. This is because at least two RLUs is needed for buffering task configurations, similar to the work in [11].

As an example, consider Fig. 2 in which the DAG in Fig. 1 has been scheduled for different number of RLUs. In this example, the reconfiguration delay is assumed to be 2 ms .It is also assumed there is one configuration controller available. As you can see, the obtained parallel time in the worst-case i.e., when there are only two RLUs available, is 16ms (Fig. 2-a). According to this, the value of Pt for task T_4 is 16ms. For three and four RLUs, the parallel times 14 and 12 ms are obtained respectively (Fig. 2-b and Fig. 2-c). For more than four RLUs available, the parallel time is still 12ms and adding more RLUs has no effect on the parallel time (Fig. 2-d). As a result, the value of Nres parameter for task T_4 is 4. According to this, task T_4 is represented as follows:

$$T4 = (Pt:16, Nres:4, At:8) \qquad (1)$$

For a task T_i, the only parameter that is not known until runtime is the arrival time to the system (At_i).This means the proposed system model is capable of managing tasks as they arrive at runtime, dynamically.

In this paper a reconfigurable area refers to an area organized as several reconfigurable logic units (RLUs). The reconfigurable area is characterized by the number of RLUs contained (Cap). It is assumed RLUs are the same size, so reconfiguration delay for all RLUs is constant R. It is also assumed that each subtask occupies exactly one RLU, so the reconfiguration delay for all subtasks is constant R, as well.

III. THE PROPOSED MODEL

In this section the proposed model is discussed in detailed. Also, the architecture to support the runtime RLU allocation is proposed.

The proposed hierarchical model consists of one master node at the top of system model and several cluster nodes communicating through a dedicated network. Each cluster node is allocated a reconfigurable area for implementing the hardware tasks assigned to it. This allocation can be static or dynamic, which are discussed in the following sections.

A. Static Model

In static model each cluster node is allocated a fixed dedicated number of RLUs and this allocation does not alter at runtime. The proposed static architecture model is shown in Fig. 3.

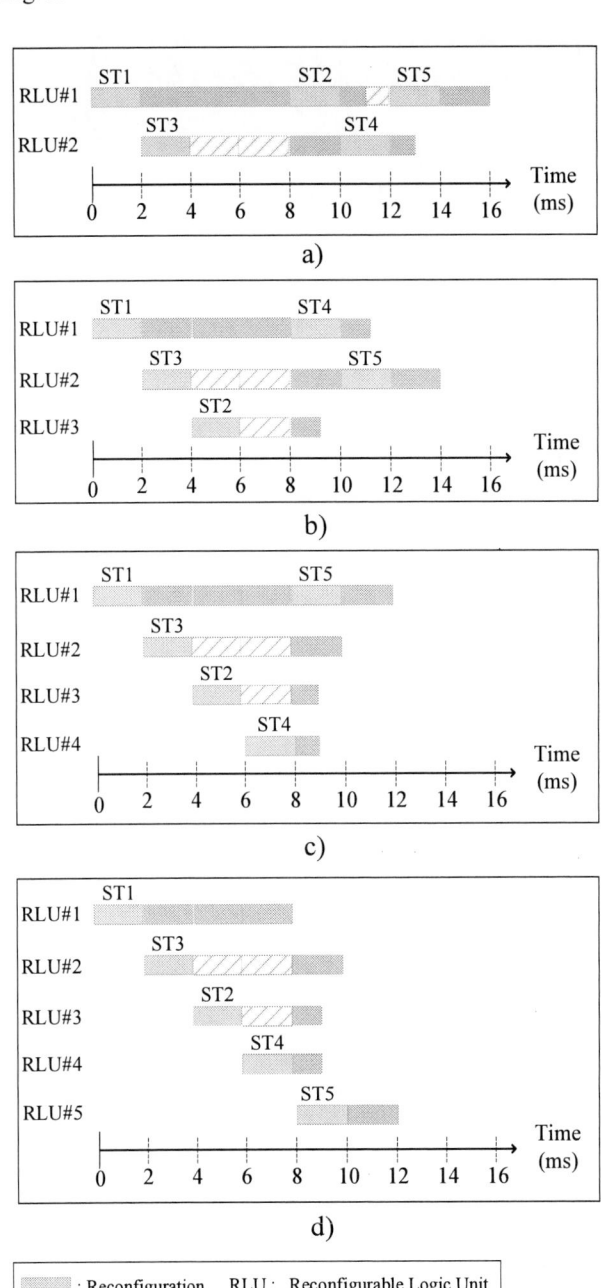

a)

b)

c)

d)

: Reconfiguration   RLU : Reconfigurable Logic Unit

: Computation   ST : Sub-Task

: Idle

Fig. 2. Scheduling the DAG in Fig. 1-b on a) two RLUs, b) three RLUs, c) four RLUs, and d) five RLUs.

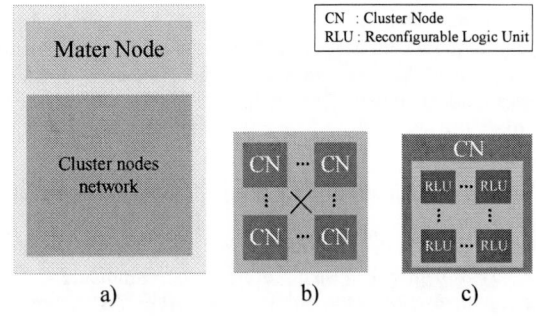

Fig. 3. a) The proposed static model, b) Cluster nodes network, c) The dedicated reconfigurable area allocated to a cluster node.

The number of RLUs available in a cluster node represents the capability of that node. The more RLUs in a cluster node, the more capability provided with the node. In a homogenous system, RLUs are divided between cluster nodes evenly.

Since in static model RLUs are allocated at compile time and the number of allocated RLUs is fixed during runtime, managing RLUs at runtime does not require any complicated control mechanism. However static model suffers from the fragmentation problem, which may impact on the overall performance.

To demonstrate the fragmentation problem consider the following example. Suppose the following tasks T_1, T_2, T_3 and T_4 arriving at times 0, 2, 3 and 8 ms. The state of system at the time 8ms is shown in Fig. 4. When task T_4 enters the system at 8ms, cluster nodes CN_1, CN_2 and CN_3 are busy running tasks T_1, T_2 and T_3 respectively. Task T_4 needs at least 5 RLUs to be scheduled efficiently, while there is 3 RLUs available in the free cluster CN_4. The node CN_4 picks up T_4 and schedules it for execution on three RLUs, whereas there are five RLUs available in the system.

In addition, a policy should be taken for determining each node's capability. When determined, the capability of the nodes are fixed and cannot be changed at runtime, which can be a limitation to the static model.

B. Dynamic Model

To address the limitations in static model, the dynamic model is proposed. In the dynamic model, RLU allocation takes place dynamically based on a cluster node's requirement. In other words, RLUs are shared among the cluster nodes at runtime.

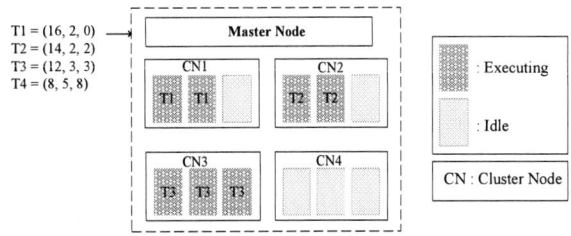

Fig. 4. The fragmentation problem in the proposed static model

The proposed dynamic architecture model is shown in Fig. 5. In dynamic model, all the reconfigurable areas in clusters are merged into one unique reconfigurable area locating outside the cluster nodes network (Fig. 5-c). RLU allocation takes place from this area based on the requirements of the requesting nodes.

To support dynamic RLU allocation, the architecture shown in Fig. 6 is proposed. Inside each cluster, RLUs are connected through a BUS network (Fig. 6-a). If two subtasks residing on separate RLUs inside a cluster require to communicate data, this communication occurs through the data BUS. As shown in Fig. 6-b, each RLU is connected to all clusters and the master node through a multiplexer. At runtime, the master node determines for each RLU to be allocated to which cluster node by setting its multiplexer's select lines appropriately. An RLU leaves a cluster area by being connected to the master node.

There is a similar structure for reconfiguration process within a cluster node. Whenever a subtask configuration has to be loaded into a cluster, the configuration controller retrieves configuration from an external memory and loads it into the memory associated with the specified RLU through the configuration BUS.

The structure of an individual cluster node is shown in Fig. 7. A cluster node is composed of a local scheduler, a configuration controller and a reconfiguration area.

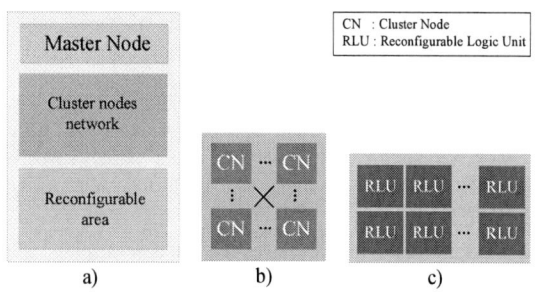

a)

b)

c)

Fig. 5. a) the proposed dynamic model, b) cluster nodes network, c) the reconfigurable area

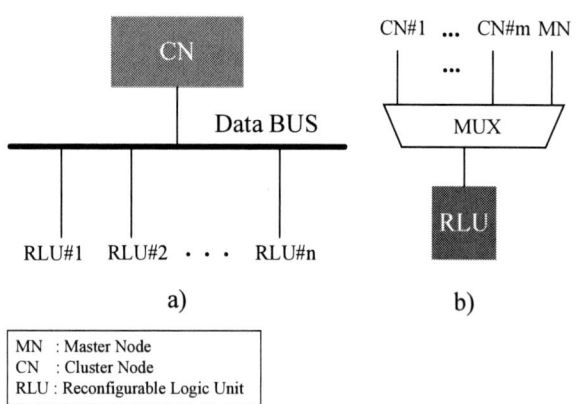

a)

b)

Fig. 6. The proposed architecture for supporting runtime RLU allocation, a) the architecture within a cluster, b) the architecture within an RLU

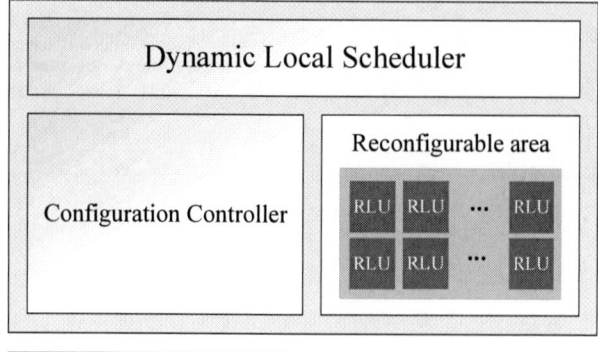

Fig. 7. The proposed structure for a cluster node

The local scheduler is any runtime task scheduler in the literature that selects a task from the task queue and schedules it on the cluster's reconfigurable area. The local scheduler can apply prefetching and replacement techniques in order to hide the most of the reconfiguration delays, as stated before.

To support parallel reconfiguration, each cluster is provided with a configuration controller. The configuration controller component loads the configuration of a subtask based on the local scheduler request. Although reconfiguration process among clusters occurs independently, within an individual cluster subtasks are loaded one at a time. That is because, each cluster node is provided with one single configuration controller.

The master node could be any cluster node or a specific node. The architecture for the master node is shown in Fig. 8. In addition to a local scheduler, the master node also includes a global scheduler which has two primary roles. First, it is responsible for partitioning and redistributions of tasks between cluster nodes. Second, it manages the RLU allocation to the clusters dynamically at runtime. In other words, RLUs allocation is centralized controlled by the master node. The MN can also schedule and execute tasks like a cluster node

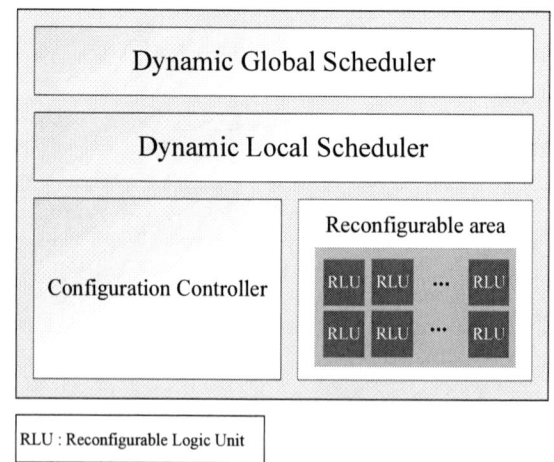

Fig. 8. The proposed structure for the master node

through the LS.

As stated before, cluster nodes and the master node are connected through a dedicated network. In the proposed system model, the network is used just for message passing between nodes. In other words, each cluster retrieves and loads subtasks' configurations independently through its dedicated configuration controller without involving the network. As a result the load on the network is not a considerable amount. Since in this paper tasks do not communicate to each other, we do not have any data transfer in the network. However, this network can be used for data transfers, as well, without adding considerable load compared to exchanging subtasks.

Scheduling a task is defined as determining the start times of execution and reconfiguration for subtasks included in the task. The location (here, RLU) where a subtask is reconfigured and executed is also specified in the scheduling. In this paper, scheduling is performed hieratically at two levels.

At the upper level, incoming tasks are partitioned and distributed between the clusters such that each node is loaded evenly. This process is controlled by the master node and it is repeated at runtime over and over as needed.

In this paper, the primary goal in partitioning is to distribute N ready tasks $T_1, T_2, ..., T_N$ among m clusters $C_1, C_2, ..., C_m$ such that each node is equally loaded. For each task T_i, the worst-case total execution time, i.e. Pt_i, is considered in partitioning process.

At the lower level, tasks assigned to a cluster node are split into subtasks and the subtasks are scheduled for execution on the cluster's reconfigurable area, performed by the local scheduler.

For each assigned task, the cluster node compares the number of RLUs required by the task, i.e. the Nres parameter, with the number of RLUs currently allocated to the cluster (Cap). If more RLUs are needed, the cluster node sends its request to the master node. The master node handles the request and allocate RLUs to the requesting cluster if needed. However, when there are enough RLUs available in the cluster's reconfigurable area, the cluster node releases the remaining free RLUs by sending a message to the master node. At any time, each cluster node maintains as many RLUs as it is needed by releasing the remaining free RLUs to be used by other cluster nodes.

IV. CONCLUSIONS

To make the most of the parallel computational capabilities provided by FPGA technology, the destructive impact of reconfiguration overhead should be eliminated as far as possible. To address the limitation, an extensive research has been performed. However, no one has provided FPGA with a parallel structure similar to multi-core systems. In this paper, a hierarchical clustered model is proposed for FPGA.

The proposed model consists of one master node and several cluster nodes, connected through a dedicated network. Each cluster node contains a dedicated configuration controller. Within a cluster, Tasks are reconfigured one at a time through the dedicated configuration controller, while reconfigurations among the clusters are in parallel.

In this paper, two architecture models are proposed: static and dynamic models. In static model, RLU allocation takes place at compile time and does not alter during execution time. Although RLU management in static model is simple, there is a limitation with the static model, called fragmentation. To address this limitation, the dynamic model is proposed, wherein clusters are allocated and unallocated RLUs at runtime based on runtime conditions. The architectures for supporting dynamic RLU allocation is also presented.

Also, a two-level hierarchical task scheduling scheme is used for the proposed model. At one level, the Global Scheduler, located in the master node, is responsible for partitioning tasks between nodes equally. At the other level, the tasks are split into subtasks and scheduled by local schedulers located on the cluster nodes.

REFERENCES

[1] Altera, Inc. Stratix-IITM Device Handbook, Volumes 1 and 2, Altera, Inc., San Jose, 2005.

[2] Xilinx, Inc. Virtex-II Platform FPGAs: Complete Data Sheet, Xilinx, Inc., San Jose, 2004.

[3] Z. Pan and B. E. Wells, "Hardware supported task scheduling on dynamically reconfigurable SoC architectures," IEEE Trans. Very Large Scale Integr. (VLSI) Syst., vol. 16, no. 11, pp. 1465-1474, Nov. 2008.

[4] J. Clemente, J. Resano, C. González, and D. Mozos, "A hardware implementation of a run-time scheduler for reconfigurable systems," IEEE Trans. Very Large Scale Integr. (VLSI) Syst., vol. 19 ,no. 7, pp. 1263-1276, Jul. 2011.

[5] R. Kalra and R. Lysecky, "Configuration locking and schedulability estimation for reduced reconfiguration overheads of reconfigurable systems," IEEE Trans. Very Large Scale Integr. (VLSI) Syst., vol. 18, no. 4, pp. 671-674, Apr. 2010.

[6] Z. Li, K. Compton, and S. Hauck, "Configuration caching for FPGAs," IEEE Symposium on FPGAs for Custom Computing Machines, 2000.

[7] Y. Sheng, Y. Liu, R. Li, and X. Xiao, "A communication-aware scheduling algorithm for hardware task scheduling model on FPGA-based reconfigurable systems," Journal OF Computers, vol. 9, no. 11, pp. 2552-2558, Nov. 2014.

[8] S. Hauck, Z. Li, and E. J. Schwabe, "Configuration compression for the Xilinx XC6200 FPGA," IEEE Symposium on FPGAs for Custom Computing Machines, 1998.

[9] Xilinx, Inc. XC6200 Field Programmable Gate Arrays Product Description, Xilinx, Inc., San Jose, 1997.

[10] Y. Qu, J. P. Soininen, and J. Nurmi, "A parallel configuration model for reducing the run-time reconfiguration overhead," in Proc. IEEE DATEC, Mar. 2006, pp. 965-970.

[11] H. R. Naji, B. E. Wells, and L. Etzkron, "Creating an adaptive embedded system by applying multi-agent techniques to reconfigurable hardware," Future Generation Computer Systems, vol. 20, pp. 1055–1081, Feb. 2004.

An Efficient VLSI Architecture for Data Encryption Standard and its FPGA Implementation

J. G. Pandey, Aanchal Gurawa, Heena Nehra, A. Karmakar

CSIR - Central Electronics Engineering Research Institute

Pilani-333031, India

jai@ceeri.ernet.in

Abstract—To achieve the goal of secure communication, cryptography is an essential operation. Many applications, including health-monitoring and biometric data based recognition system, need short-term data security. To design short-term security based applications, there is an essential need of high-performance, low cost and area-efficient VLSI implementation of lightweight ciphers. Data encryption standard (DES) is well-suited for the implementation of low-cost lightweight cryptography applications. In this paper, we propose an efficient VLSI architecture for DES algorithm based encryption/decryption engine. Depending upon the encryption/decryption needs, the same set of architecture performs both encryption and decryption operations. In the implementation of DES algorithm, a chain of multiplexer-based architecture is used to implement the substitution operations (S-Boxes). The proposed architecture is modeled in the VHDL design language and synthesized in the Xilinx Virtex-5 xc5vfx70t field-programmable gate array (FPGA) device. Hardware synthesis result shows that the proposed design utilizes only 1.07 % slice LUTs, 0.31 % slice registers and 29.22 % of bonded IOBs of the FPGA device fabric.

Keywords—DES; encryption/decryption; block cipher; lightweight cryptography; VLSI architectures; FPGAs.

I. INTRODUCTION

In modern authentication-based applications, such as: bank transactions, electronic mail, audio/video conferencing etc., secure communication is very essential. The secure communication requires a mechanism, which insures that no unauthorized person can access the communicated information over the unsecure medium. Thus, to accomplish the task of information security; cryptography is must [1,2].

Cryptography is a science that enables the confidentiality of communication through an insecure channel [1]. The basic cryptographic processes consist of conversion of plaintext into a ciphertext by the process of encryption and retrieval of the plaintext from the ciphertext by decryption process. The cryptographic process is used for authentication in many applications such as: in bank cards, wireless telephones, e-commerce, pay-TV, etc. [2]. Encryption/decryption is also required for the access control in many systems, such as car-lock systems, lifts, metro trains, etc. Nowadays it is widely used for electronic payment in prepaid telephone cards, e-cash cards etc. [1,2].

Some of the applications including health-monitoring and biometric data based recognition applications need short-term security. In addition, with advent of embedded computing and omnipresent smart embedded devices, there is a need of high-performance, area-efficient and low cost very-large-scale integration (VLSI) implementation of lightweight ciphers such as data encryption standard (DES). For the implementation of low-cost lightweight cryptography, the DES algorithm is very well suited [3,4]. DES algorithm is a symmetric block cipher and it provides adequate level of security with low hardware cost [5]. Though, the 56-bit key limits the security level, yet, brute-forcing this key space using software requires a few months alongwith several computing engines [4,6]. Although, DES has evolved into the advanced encryption standard (AES) [7], nonetheless, many applications continue to rely on DES for cryptography and information security. Therefore, the designers and implementers continue to support for efficient architecture for the DES in many short-term security applications [6]. Thus, there is always a need of optimized, high-performance hardware implementation of DES and other lightweight ciphers [3].

Based on the above discussion, a lightweight DES implementation has been reported in [4]. Here, they have used a single substitution box, which are used eight times. The implemented design has been used for resource-limited applications that include, radio frequency identification (RFID) tags, wireless sensor nodes applications, etc. In another implementation [8], a JBits programming language based implementation of the DES algorithm for Xilinx Virtex field-programmable gate array (FPGA) has been described. In this implementation, sixteen rounds of the DES algorithm have been fully unrolled. A performance analysis of DES, 3DES, Blowfish and AES encryption/decryption algorithms have been provided in [9].

In this paper, we have proposed a simple and efficient VLSI architecture for computation of DES algorithm [10,11]. The proposed architecture requires nineteen clock cycles to encrypt a plaintext into ciphertext. The decryption process is identical to encryption operation and it completes the decryption process in nineteen clock cycles. The key generation process is realized in a combinational datapath and it provides all the required sixteen round keys to the encryption/decryption block in the first cycle of the clock. By this, it makes the decryption block to start the decryption process by using the last round key, which is mandatory as per the DES decryption algorithm. To implement an S-Box, five multiplexers (MUXs) are used. Out of the five MUXs, four MUXs are of 4-bit, 16-to-1 MUXs and one 4-bit, 4-to-1 MUX. To make the design work in pipelined mode the inputs and outputs are registered.

978-1-5090-0037-1/16 $31.00 © 2016 IEEE

Rest of this paper is organized as follows: in Section II, an overview of the DES algorithm has been provided. Section III has been used to present the proposed architecture for the DES encryption/decryption. This section has also been use to cover detailed architectural design of the various basic building blocks of the proposed architecture. Section IV has been used to provide FPGA-based implementation of the proposed architecture and its experimental results. Finally, conclusions are drawn in Section V.

II. THE DATA ENCRYPTION STANDARD (DES) ALGORITHM

The DES algorithm is a symmetric block cipher. It is an iterated block cipher algorithm, which is realized on permutation, XOR and substitution operations. All these operations are sequential in nature and iterates in sixteen internal rounds. The algorithm is based on the principles of *Feistel* cipher structure [2]. Here, a round function F consists of expansion/permutation, XOR, and substitution operations.

Here, a 64-bit plaintext is divided into, two, 32-bit halves, L_0 (left) and R_0 (right). These two halves pass through *sixteen* rounds of internal processing and then combine to produce the 64-bit ciphertext. Each round i has inputs L_{i-1} and R_{i-1} derived from the previous round. In addition, each round function F, requires a 48-bit unique subkey K_i, which is generated from the 64-bit input key K by the round key generation process. The overall processing at each round is summarized as:

$$L_i = R_{i-1} \tag{1}$$

$$R_i = L_{i-1} \oplus F\left(R_{i-1}, K_i\right) \tag{2}$$

The 32-bit R input is first expanded to 48-bits by using permutation plus expansion. The output is then XORed with round key K_i. This 48-bit output passes through a substitution table (S-Box) that produces a 32-bit output, which is permuted as per the DES algorithm [2]. The key generation process starts with first permutation choice, which transforms the 64-bit input key into 56-bit permutated output.

This output is treated as two 28-bits data C_0 and D_0. At each round, C_{i-1} and D_{i-1} are separately and circularly left-shifted by 1 or 2 bits which serve as input to the next round [1]. The shifted data is then provided to the second permutation choice, which produces a 48-bit output that serves as input to the function $F\left(R_{i-1}, K_i\right)$ as required in (2). In next section, we propose an architecture for the DES encryption/decryption engine.

III. A VLSI ARCHITECTURE FOR THE DES ALGORITHM

A top-level view of proposed architecture for the DES algorithm is shown in Fig. 1. Here, both the encryption and decryption operations use the same set of hardware building blocks. The three main components of the DES encryption/decryption computing scheme are: key generation, controller and a encryption/decryption engine.

The key generation block is use to take 64-bit input key and it generates sixteen, 48-bit round keys for the sixteen individual rounds. The complete round key generation process is covered in Subsection A.

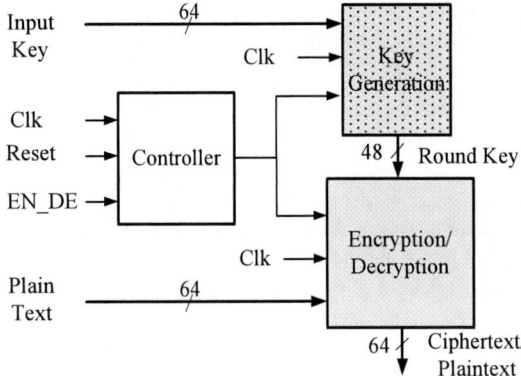

Fig. 1. A top-level block disgarm for DES encryption/decryption.

The controller is designed to generate various required control signals for controlling the key generation process and the encryption/decryption engine. The detailed scheme for the controller is given in Subsection B. Subsection C is used to discuss the architecture for a substitution box (S-Box).

A detailed view of the proposed VLSI architecture for encryption/decryption using DES algorithm is shown in Fig. 2. Here, the datapath of this architecture consists of a set of multiplexers, registers, permutation/expansion and substitution operations. The permutation/ expansion operation is a simple bit-transposition operation, which requires only simple data routing. The substitution operation requires eight different substitution boxes [2]. To implement an S-Box, we have proposed a multiplexer based design, which is explained in Subsection C.

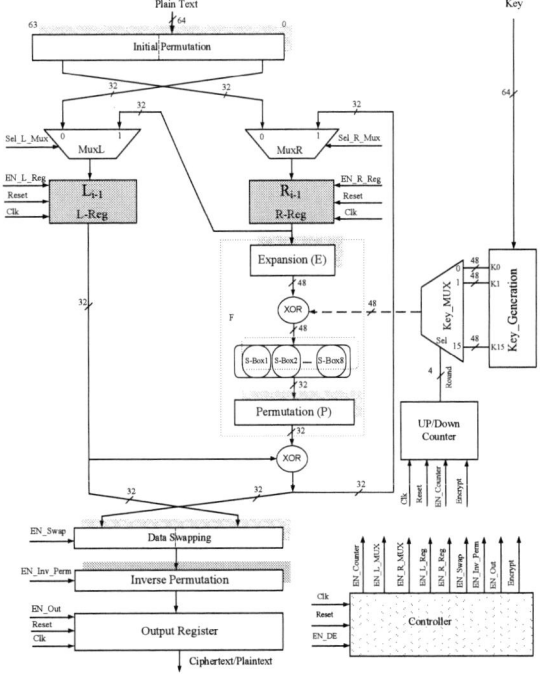

Fig. 2. A VLSI architecture for DES encryption/decryption computation.

978-1-5090-0037-1/16 $31.00 © 2016 IEEE

The main building blocks of the proposed architecture have been arranged in different subsections, which are described below.

A. Round Keys Generation Process

As per the DES algorithm needs, the architecture shown in Fig. 2 requires a round key in each rounds. The steps for the round key generation process are shown in Fig. 3. The round key generation process is use to generate 48-bit unique round keys for the each individual rounds.

In the DES algorithm, sixteen rounds are required to generate the sixteen round keys. To start the round key generation process, first a 64-bit input key is provided to the permutation choice-1 (permutation-1) block [1,2].The permutation-1 block permutes the input key into 48-bit data, which are provided to two 28-bit blocks (C and D). After that, the two 28-bit data are circularly left shifted by one or two bits, which depends upon the number of round as per the DES key generation algorithm [2]. The encryption/decryption operation is managed by the controller, which is explained below.

B. Controller for the Encryption/Decryption Operation

A controller has been designed for providing necessary control signals to the proposed architecture of Fig. 2. The proposed controller has six different states, which are shown in a form of finite state machine (FSM) in Fig. 4. As shown in the this figure, when the asynchronous reset input is at logic '1' value, the machine resides in the *Idle* state and it generates two control signals, *Sel_L_Mux and Sel_R_Mux*. Both of these control signals are kept at logic '0' level. After generating the above control signals, the machine waits for an external *En_De* input signal, which is used to select the encryption/decryption operation. When the *En_De* signal is at logic '1', the machine enters into the *Encrypt* state otherwise the next state would be a *Decrypt* state.

When the machine resides in the *Encrypt* state, two control signals *En_Counter* and *En_Encrypt* both at logic '1' level are generated. Whereas, in the *Decrypt* state, the *En_Counter* control signal is at logic '1' and *En_Encrypt* control signal has logic '0' level. When the signal *En_Encrypt* is at logic '1', the counter shown in Fig. 2 is works as an up counter (counts from 0 to 15). This counter has been used to select the round keys in increasing order. Similarly, when the *En_Encrypt* control signal is at logic '0' the counter works as a down counter and it selects the round keys in the decreasing order (from 15 down to 0).

The *Encrypt* state is followed by the *Encrypt_Rounds* state where the machine completes its sixteen encryption rounds, as governed by the DES algorithm [2]. In the *Encrypt_Rounds*, the generated control signals are, En_Counter, En_Encrypt, Sel_L_Mux, Sel_R_Mux, En_L_Reg and En_R_Reg. All of these generated control signals are kept at logic '1' level. Similar to this state, after the *Decrypt* state, the machine enters into the *Decrypt_Rounds* state and it use to generate the similar signals as with the *Encrypt_Rounds,* except the *En_Encrypt* signal, which is at logic '0' level.

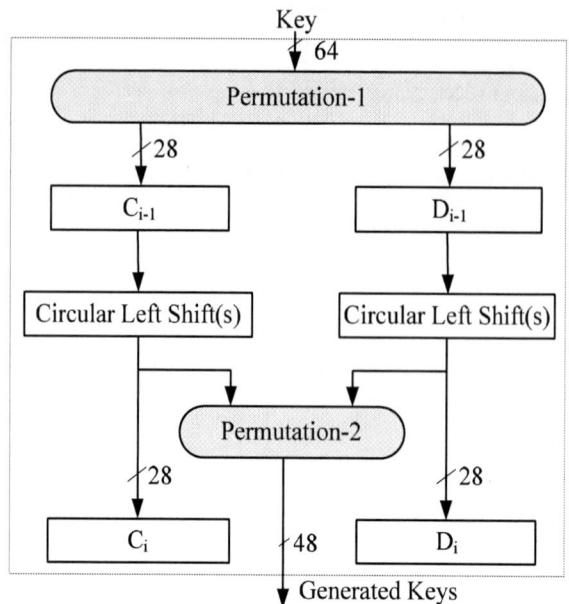

Fig. 3. Round key generation scheme.

In the *Decrypt_Rounds* state, the machine completes sixteen decryption rounds, which are required as per the DES algorithm [2]. The next state of both, the *Encrypt_Rounds* state and that of *Decrypt_Rounds* state is followed by a same state and it is called *Result* state. In the *Result* state, the generated control signals are: *En_Swap*, *En_Inv_Perm* and *En_Out*. All of these generated control signals have logic '1' level.

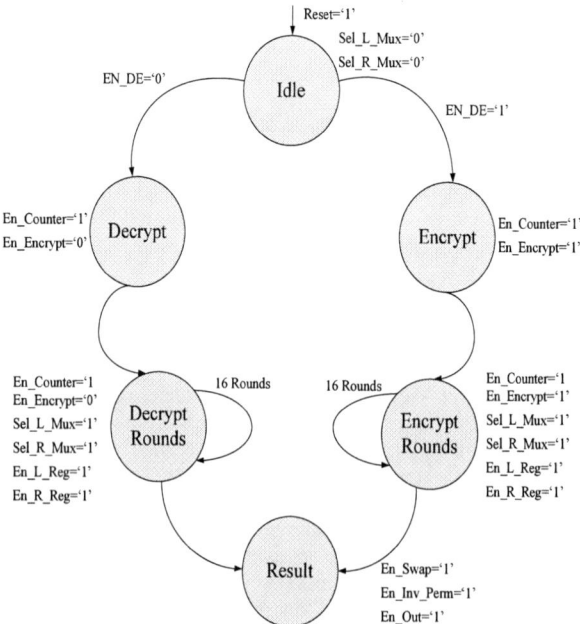

Fig. 4. Control signal generation using a FSM.

978-1-5090-0037-1/16 $31.00 © 2016 IEEE

C. Architecture for an S-Box

The substitution operation consists of a set of eight substitution boxes (S-Boxes). Each S-Box accepts 6-bit input and it provides 4-bits output. The S-Box transformation operation is defined in [2]. The S-Boxes can be designed by using only a set of five multiplexers (MUXs). Using this approach, a proposed architecture for the realization of an S-Box using the MUX-based approach is shown in Fig. 5.

Here, to implement the S-Box, five MUXs have been used. The first four MUXs (*MUX_S00*, *MUX_S01*, *MUX_S10*, *MUX_S11*) are of 4-bit, 16-to-1 MUX. These MUXs are used to provide four rows of the S-Box[2]. The same select lines drive the chain of four MUXs and it is used to select the content of different individual rows. The select lines of these MUXs arrive from the middle four bits of the input to the S-Box as per the S-Box selection rule [2].

The right MUX (MUX_SBox) is a 4-bit, 4-to-1 MUX. Here, the first-bit (MSB), S-Box_IN[5] and last-bit (LSB) S-Box_IN[0], of the input to the MUX_SBox, form a 2-bit binary number. These two bits are used as select lines for this multiplexer. This multiplexer is used to select one of the four outputs of the left four MUXs and provides 4-bit output signal SBOX_Out.

The design of the proposed S-Box is very simple and regular in nature. The architecture for the S-Box is highly regular and has very simple routing connections. By this, to design a set of eight S-Boxes (as shown in Fig. 2.), the above S-Box has been utilized eight times with their corresponding substitution values, which are provided by the DES algorithm and also discussed in [1].

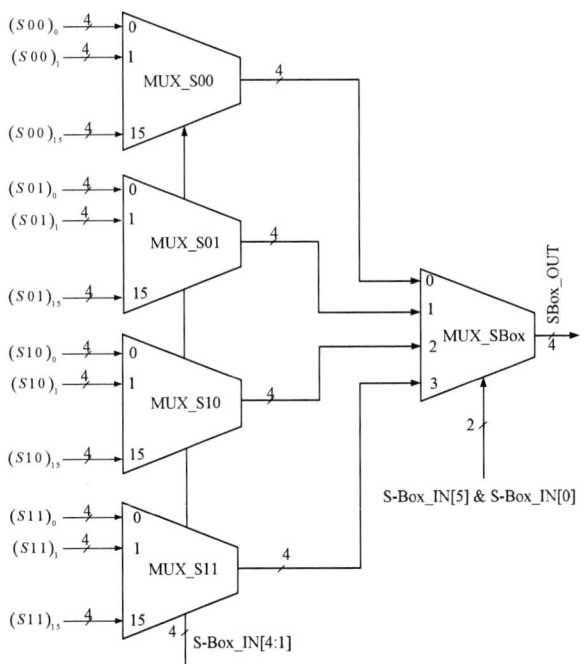

Fig. 5. Realization of an S-Box of DES algorithm.

D. The Encryption/Decryption Operation

The proposed architecture works for both encryptions as well as for decryptions, which depends upon the input signal '*EN_DE*'. When the signal *EN_DE* is at logic '1', the controller resides in the Encrypt state and it asserts *En_Encrypt* to logic high. The architecture performs the encryption operation. Similarly, when *EN_DE* is at logic low in the *Decrypt* state, the architecture performs decryption operation.

As shown in the proposed architecture as in Fig. 2, and the controller shown in Fig. 4, the *Sel_L_Mux* signal is used as a selection line for the *MuxL* and the *Sel_R_Mux* signal is used for input selection of *MuxR*. Initially, in *Idle* state, both of these select signals are at logic '0'. These selection lines are used for the selection of initial permuted inputs and routing them to the *L_Reg* and *R_Reg* registers respectively.

The architectures require a 48-bit XOR gate for signal '*E_XOR_K*'. One of the input signals of the gate comes from the expansion permutation operation (E) and the other signal is one of the sixteen, 48-bit round keys. The output of the XOR gate is provided to the substitution boxes (S-Boxes). The 32-bit output of the eight S-Boxes is then permutated (P). As per (2), the permutated output is then XORed with the left 32-bits of the initial permutated 32-bits, which have been selected by the *MuxL*. The output of the operation is then stored in the register *R-Reg* as per (1).

The above operations are executed sixteen times and it is controlled by the various control signals generated by the controller shown in Fig. 4. After completion of all the rounds, registered output is available after nineteen clock cycles. Depending on the encryption/decryption operation, which is controlled by *EN_DE* signal, the generated output is ciphertext or plaintext.

IV. EXPERIMENTAL RESULTS

A *ModelSim* simulation result of the proposed design is shown in Fig. 6. As depicted in the figure, the encryption/decryption operation requires nineteen clock cycles. In the operation, the *En_DE* signal is at logic '1' for the encryption. The signal has been set at logic '0' for performing the decryption operation. The design has been thoroughly verified by providing sample vectors in the form of uniform random numbers for plaintext and cipher keys.

The proposed architecture completes each of the encryption/decryption rounds in one clock cycle. Further, to make an estimate of the complete hardware resources need a fully loop-unrolled DES algorithm implementation [8] is done.

The utilization of the macro statistics is shown in Table I. The implemented design has been compared with the implementation of [8].

TABLE I.   MACRO-LEVEL COMAPRISION.

S. No.	Macro Name	Loop Unrolled Design [8]	Proposed Design
1.	Multiplexers	640	44
2.	XORs	32	2
3.	Counter	0	1
4.	Registers	3	3
5.	Comparators	0	2

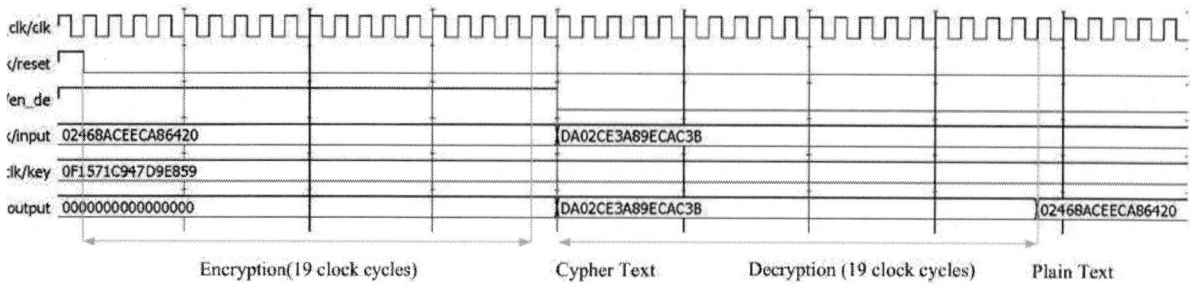

Fig. 6. A Modelsim simulation result of the design.

As shown in Table I, in comparison to the loop-unrolled DES implementation given in [8], the proposed design utilizes 14.55 times lesser multiplexers and 16 times lesser XOR gates. Additionally, the proposed design requires one 4-bit counter and two comparators.

The proposed architecture, shown in Fig. 2, is implemented in the VHDL design language, and synthesized using Xilinx ISE 14.4 for Virtex-5 xc5vfx70t FPGA device. The FPGA device utilization summary of the proposed architecture is given in Table II.

TABLE II.   FPGA DEVICE UTILIZATION SUMMARY.

Elements	Device Utilization	Utilization (%)
Slice LUTs	478/44800	1.07 %
Slice Registers	138/44800	0.31 %
Bonded IOBs	187/640	29.22 %
BUFG	1/32	3.13 %

The FPGA device utilization of the synthesized design shows that it utilizes, 1.07 % FPGA slices, 29.22 % Bounded IOBs and 3.13 % of total BUFG elements. A synthesized RTL schematic view of the proposed architecture is shown in Fig. 7.

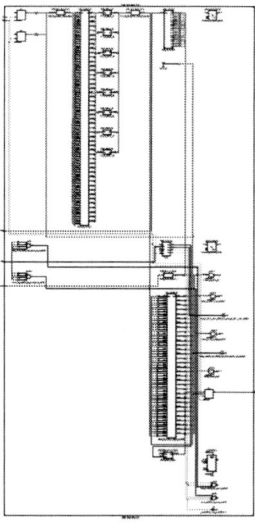

Fig. 7.   FPGA implemention of the proposed architecture.

V.   CONCLUSION

In this paper, we have proposed an efficient VLSI architecture for data encryption standard (DES) algorithm based encryption/decryption. As per the requirements of encryption/decryption operation, the same set of architecture can be used to perform both encryption as well as the decryption. The substitution operation (S-Box) needed in the DES algorithm has been implemented by multiplexer-based architecture. The proposed architecture is very regular and it requires very low amount of hardware resources, therefore it can be efficiently utilized in lightweight cryptography applications. The design has been modeled in the VHDL language and it has been synthesized for Xilinx Virtex-5 xc5vfx70t FPGA device.

REFERENCES

[1] W. Stallings, Cryptography and Network Security Principles and Practice, 5th ed. Prentice Hall, 2011.

[2] S. Vaudenay, A Classical Introduction to Cryptography: Applications for Communications Security. Springer Science & Business Media, 2006.

[3] T. Eisenbarth, S. Kumar, C. Paar, A. Poschmann, and L. Uhsadel, "A survey of lightweight-cryptography implementations," IEEE Design & Test of Computers, vol. 24, no. 6, 2007, pp. 522-533.

[4] G. Leander, C. Paar, A. Poschmann, and K. Schramm, "New Lightweight DES Variants," in Fast Software Encryption (FSE 2007), A. Biryukov, Ed. Springer Berlin Heidelberg: LNCS 4593, 2007, pp. 196-210.

[5] E. Biham and A. Shamir, Differential Cryptanalysis of the Data Encryption Standard. Springer Science & Business Media, 2012.

[6] S. Kelly. (2006, Dec.) Security implications of using the data encryption standard (DES). [Online]. https://tools.ietf.org/html/rfc4772

[7] J. Daemen and V. Rijmen, The design of Rijndael: AES-the advanced encryption standard. New York, USA: Springer Science & Business Media, 2013.

[8] C. Patterson, "High performance DES encryption in Virtex FPGAs using JBits," in IEEE Symposium on Field-Programmable Custom Computing Machines, Napa Valley, CA, 2000, pp. 113-121.

[9] O. P. Verma, R. Agarwal, D. Dafouti, and S. Tyagi, "Peformance analysis of data encryption algorithms," in 3rd Int'l Conf. on Electronics Computer Technology (ICECT), vol. 5, Kanyakumari, 8-10 Apr. 2011, pp. 399-403.

[10] M. E. Smid and D. K. Branstad, "Data encryption standard: past and future," Proc. of the IEEE, vol. 76, no. 5, pp. 550-559, 1988.

[11] S. Landau, "Standing the test of time: The data encryption standard," Notices of the AMS, vol. 47, no. 3, Mar. 2000, pp. 341-349.

Implementation of RNS and LNS based addition and subtraction Units for cryptography

Ch.satish kumar
School of Electronics Engineering
VIT University Chennai campus
Chennai, India
chintalapudi.satish2014@vit.ac.in

Prathiba A.
School of Electronics Engineering
VIT University Chennai campus
Chennai, India
prathiba.2@vit.ac.in

Kanchana Bhaskaran V S
School of Electronics Engineering
VIT University Chennai campus
Chennai, India
kanchana.vs@vit.ac.in

Abstract— **The need for fast data processing and reducing the power dissipation of digital signal processing(DSP) algorithms and Cryptographic algorithms have provoked the development of efficient hardware implementations of residue number system (RNS) and logarithmic number system(LNS) arithmetic. This paper describes the implementation of adder and subtractor units by using RNS and LNS arithmetic. Addition and subtraction units are major and basic operations in public key cryptographic algorithms like Elliptic Curve Cryptography (ECC). In cryptography the use of finite fields plays a major role which is time consuming, there is a need for fast and efficient implementations.**

Keywords— *Residue number system(RNS), logarithmic number system(LNS), adder, subtractor, power dissipation, digital signal processing(DSP), Cryptography.*

I. INTRODUCTION

Power dissipation and delay have emerged into an instrumental design optimization objective due to expanding demand for portable electronic devices. The residue number system (RNS) and logarithmic number system (LNS) are the two techniques for reducing power dissipation and to boost the speed of the device. The prime choice of number system can lessen the number of operations, strength of the operations and activity of the data by which computational load can be reduced substantially reduces the power dissipation and delay, these optimizations made attempt to put in hardware.

Addition and subtraction are the primitive operations in cryptography and DSP algorithms, they play a crucial role in public key cryptography algorithms like Elliptic Curve Cryptography (ECC), and in some applications of cryptography we use finite field arithmetic on large numbers or polynomials, the finite fields operations that need to be implemented in ECC are time consuming and the operands are large integers. Thus, an alternative based on RNS and LNS can be used for such applications.

Section II discusses the Residue Number System and its usage in cryptography, section II-A discusses about the implementation of forward conversion of conventional number system to Residue Number system, section II-B discusses about the basic modulo arithmetic operations and section II-C discusses about the Reverse conversion of residues to conventional arithmetic. Section III discusses about the

floating point arithmetic and representation of real number in the floating point number system. Section IV discusses about the logarithmic number system. Section V shows the simulation results of RNS and LNS arithmetic. Section VI concludes.

II. RESIDUE NUMBER SYSTEM

Mobile computing platforms must be secure, fast, compact and power efficient. While public key cryptography methods provide the frame work secure communications, their algorithms tend to be computationally complex and hence need some form of hardware acceleration [1].

The RNS promises efficient arithmetic by replacing the long integers used in public key cryptography with sets of smaller independent numbers. Doing so allows flexibility and reduces carry propagation delays. Residue number systems (RNS) were invented by the third-century Chinese scholar Sun Tzu. The residue number system is a non-weighted number system, it is capable of parallel processing, carry free operation and high speed [2].

Residue Number System (RNS), which originates from the Chinese Remainder Theorem, offers a promising future in VLSI because of its carry-free operations in addition, subtraction and multiplication. This property of RNS is very helpful to reduce the complexity of calculation in many applications. A residue number system represents a large integer using a set of smaller integers, called residues. But the area overhead, cost and speed not only depend on this word length, but also the selection of moduli, which is a very crucial step for residue system [3].

In 1950's, RNS was implemented by computer scientists, to put them in to use of fast arithmetic and fault tolerant applications. The design of RNS arithmetic falls under three steps.

- A. Forward conversion
- B. Modulo channel
- C. Reverse conversion

A. Forward conversion:

The forward conversion is a technique in which we can convert Conventional representation of an integer to N-tuples of Residues [4] [5]. (All the equations are referenced from [2])

$$X \xrightarrow{RNS} \{x_1, x_2, \ldots\ldots\ldots, x_N\} \qquad (1)$$

where $x_i = \langle X \rangle m_i$, $\langle \ \rangle m_i$ denotes the mod m_i operation and m_i is the member of co-prime integers $B = \{m_1, m_2, \ldots\ldots, m_N\}$ called moduli set, the $\langle X \rangle m_i$, operation gives the integer remainder of the integer X when divided by m_i, the design of modular arithmetic falls under two categories. On one hand, they are implemented for flexibility, in this case we are allowed to take any moduli set. On the other hand, due to the demand for architectural simplicity and best performance, the RNS arithmetic are designed for particular set of moduli (for e.g. 2^n-1, 2^n, 2^n+1)[6].

The choice of moduli set should satisfy the some conditions, the members of moduli set should be

   i.   relatively prime.
   ii.   the members of the moduli should be as small as possible so that it takes minimum computational time.
   iii.   the product of the moduli should be large enough in order to meet the desired dynamic range, the speed and cost also depends on chosen moduli.

B. Modulo channel:

The basic arithmetic operation is carried out in this block, a larger computation can be decomposed into a smaller

TABLE. I Residues for various moduli.

N	Relatively Prime Moduli			Relatively Non-Prime Moduli		
	$m_1=2$	$m_2=3$	$m_3=5$	$m_1=2$	$m_2=4$	$m_3=6$
0	0	0	0	0	0	0
1	1	1	1	1	1	1
2	0	2	2	0	2	2
3	1	0	3	1	3	3
4	0	1	4	0	0	4
5	1	2	0	1	1	5
6	0	0	1	0	2	0
7	1	1	2	1	3	1
8	0	2	3	0	0	2
9	1	0	4	1	1	3
10	0	1	0	0	2	4
11	1	2	1	1	3	5
12	0	0	2	0	0	0
13	1	1	3	1	1	1
14	0	2	4	0	2	2
15	1	0	0	1	3	3

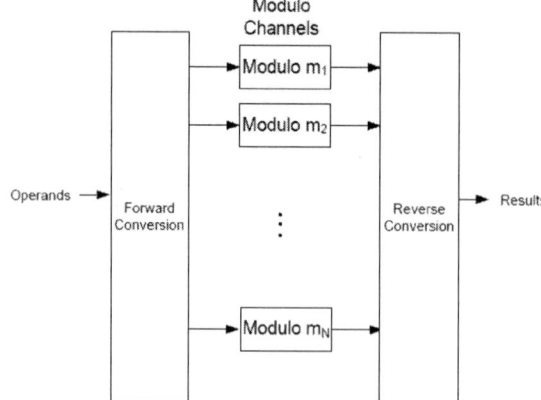

Figure.1 General Structure for RNS Processor.

computations that can be accomplished as independent and parallel for each residue without any carry propagation between the residues as shown in the Figure.1 [7].

The RNS is of interest because basic arithmetic operations are accomplished in digit parallel and carry free manner. The arithmetic on pair of residues is represented as

$$z = \langle x_i\ O\ y_i \rangle m_i \qquad (2)$$

where "*O*" represents the basic operations like addition, subtraction, multiplication etc.

$$X \xrightarrow{RNS} \{x_1, x_2 \ldots\ldots\ldots\ldots, x_N\} \qquad (3)$$

$$Y \xrightarrow{RNS} \{y_1, y_2 \ldots\ldots\ldots\ldots, y_N\} \qquad (4)$$

The RNS images of Z=X+Y is obtained as follows

$$Z \xrightarrow{RNS} \{\langle x_1\ O\ y_1 \rangle m_1, \langle x_2\ O\ y_2 \rangle m_2 \ldots\ldots, \langle x_N\ O\ y_N \rangle m_N\} \qquad (5)$$

C. Reverse conversion:

Inverse conversion of RNS residues to its conventional format makes use of Chinese remainder theorem (CRT) or mixed radix conversion. The Chinese remainder theorem converts the RNS basis to its conventional form. It is one of the more difficult and major RNS operation, the implementation of Chinese remainder theorem is as follows

$$X = \langle \sum_{k=1}^{N} \bar{m}_i \langle \bar{m}_i^{-1} * x_i \rangle m_i \rangle M \qquad (6)$$

where $\bar{m}_i = \frac{M}{m_i}$, $M = \prod_{i=1}^{N} m_i$ and $\bar{m}_i^{-1}$ is the multiplicative inverse of the $\bar{m}_i$ of modulo m_i i.e., an integer such that $\langle \bar{m}_i * \bar{m}_i^{-1} \rangle = 1$.

By using Euler's theorem we can calculate the multiplicative inverse of a number ($\overline{m}_i$) as

$$\overline{m}_i^{-1} = \langle \overline{m}_i^{(m_i-2)} \rangle m_i \qquad (7)$$

III. FLOATING POINT NUMBER SYSTEM

Digital signal processing algorithms need to be done in real-time, for that they require large dynamic range of numbers. Due to the demand for large dynamic range and fast performance, the floating point or logarithmic number systems came in to existence [8].

There are two ways of representation of floating point numbers, single precision representation and double precision representation. Every floating point number has three fields sign bit, exponent field and mantissa field.

For single precision number there are 8 exponent bits, 23 unsigned significant bits and one sign bit total 32 bits, the range of the exponent E is 0 to 255, for double precision number there are 11 exponent bits, 52 unsigned significant bits and one sign bit total 64 bits, the floating point representation is shown in Figure. 2.

The representation of floating point number is as follows

$$F = -1^s * 1.f * 2^E \qquad (8)$$

$$E = E^{true} + bias \qquad (9)$$

The most common value for bias is $2^{(e-1)} - 1$ where 'e' is the number of bits in the exponent.

To meet the specified dynamic range by floating point number system with cost of lower precision and the increased complexity over the fixed point representation. Logarithmic number system (LNS) also provides the required dynamic range of numbers with precision and advantages over floating point number system for certain applications.

IV. LOGARITHMIC NUMBER SYSTEM

Logarithmic number system (LNS) was first introduced into computer systems for processing of low-precision FFT in 1970s. Due to its similar representation range and better relative error behavior, LNS has long been considered as an alternative to the floating-point (FLP) representation. The logarithmic number system is the substitute of the floating point number system providing same precision and required dynamic range of numbers. (All the equations are referenced from [2])

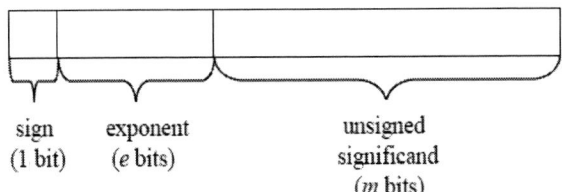

Figure.2 Floating point representation.

Sign Bit	Fixed-point Logarithmic Value	
	Integer: m bits	Fractional: f bits
S	M	F

Figure.3 Logarithmic number system representation.

Unlike the floating point numbers defined by IEEE 754 standard, there is no commonly accepted standard for LNS numbers. In this paper, we use the signed-logarithmic representation format similar to which consists of a sign bit and a fixed-point number to record the logarithmic value, shown as in Figure.3 [9].

Its value is given by $(-1)^S \times 2^{M.F}$, which provides a similar representation range to FLP numbers with m-bit exponent, f-bit significant and one sign bit. Note that the first bit of the fixed-point logarithmic value is also a sign bit, while the other bits indicate the absolute magnitude.

In this arithmetic the basic operations are performed on the logarithmic value of operand, instead of doing on operand itself, LNS converts some complex functions makes easy like

$$\log A * B = \log A + \log B \qquad (10)$$

$$\log \frac{A}{B} = \log A - \log B \qquad (11)$$

$$\log A^B = B \log A \qquad (12)$$

On the other hand implementation of addition and subtraction by using LNS becomes very complex, it makes use of lookup table for its implementation [13].

$$\log_2(A + B) = a + \log_2(1 + 2^{b-a}) \qquad (13)$$

$$\log_2(A - B) = a + \log_2(1 + 2^{b-a}) \qquad (14)$$

where $\qquad a = \log_2(A), b = \log_2(B)$

The implementation of Logarithmic number arithmetic can fall in three steps

1. Logarithmic number system maps floating point in to a triplet as follows

$$X \xrightarrow{LNS} \{z, s, x = \log_b|X|\} \qquad (15)$$

Where X is represented as $(1.\text{mantissa})*2^{exp}$

$$\log_2(1.\text{XXXX} * 2^{exp}) = \log_2 1.\text{XXXX} + \log_2 2^{exp}$$

$$= \log_2 1.\text{XXXX} + exp \qquad (16)$$

where $\log_2 1.\text{XXXX}$ is in the range of (0, 1] and 'exp' is an integer, for $\log_2 1.\text{XXXX}$ we need to store the values in LUT.

2. Perform operation on the logarithmic values.

2016 International Conference on VLSI Systems, Architectures, Technology and Applications (VLSI-SATA)

3. Reverse conversion of triplet in to its conventional form.

$$(z, s, x = \log_b |X|) \xrightarrow{LNS^{-1}} X : X = (1 - z)(-1)^s b^x \quad (17)$$

where 'z' denotes the zero flag, 's' denotes the sign of the number X and 'x' denotes the logarithm of a number.

The operation like multiplication and division are become easy in logarithmic arithmetic and the implementation of addition and subtraction becomes complex as discussed below.

The organization of LNS adder with adder, subtractor, multiplexer and lookup table is as shown in Figure. 4.

The organization of the LNS subtractor [10] is same as the LNS addition but in the LUT we need to store the $\log(1 - 2^{-|x-y|})$ non linear function for subtraction, the stored values of non linear function are used to compute the addition or subtraction operation, the complexity of the LNS operation lies in the implementation of the LUT, the choice of base effect the bit activity and power dissipation, the base 2 has less bit activity compared to other bases so less power dissipation.

The functions in addition ($\log(1 + 2^{-|x-y|})$) and subtraction ($\log(1 - 2^{-|x-y|})$) represents the exponential behavior due to implementation complexity of LUT in the hardware. The main complexity lies in the implementation of the LUTs for storing the values of the functions $\log(1 + 2^{-|x-y|})$ and $\log(1 - 2^{-|x-y|})$.

V. SIMULATION RESULTS

The simulation results of basic operations such as adder and subtractor in Residue number system and Logarithmic number system is as shown below.

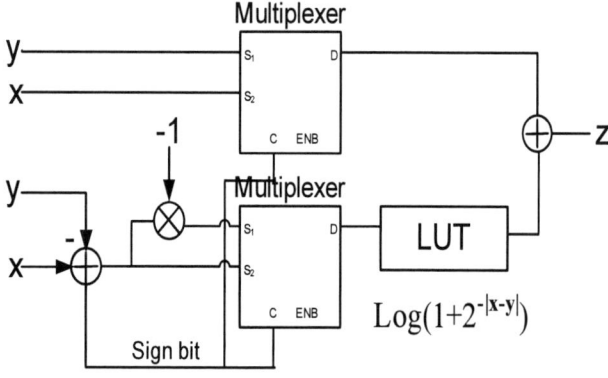

Figure.4 Organization of LNS adder.

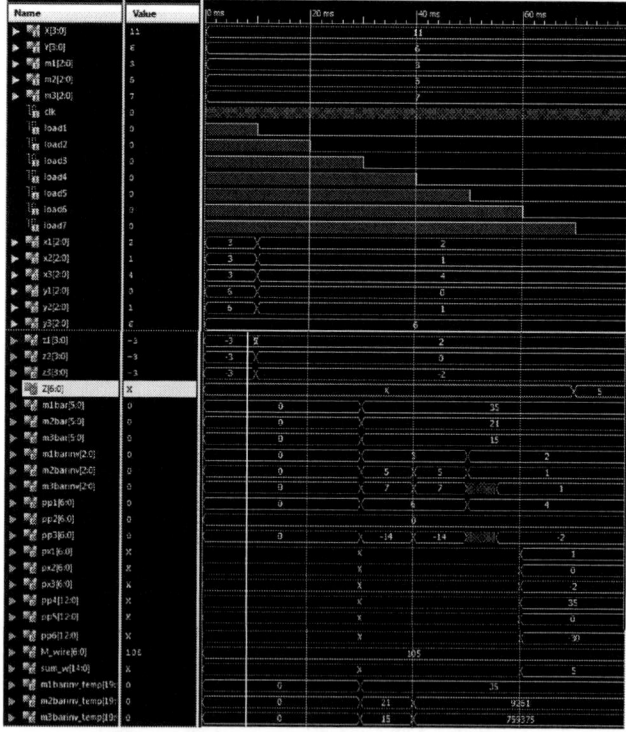

Figure. 5 Simulation result of the RNS Adder.

Figure. 6 Simulation result of the RNS Subtractor.

The simulation result shown in the above figure.5&6 denotes the RNS addition and subtraction operation on 11 and 6 and gives the output 17 and 5(shown in pink color).

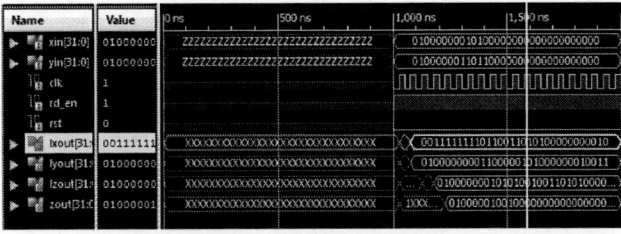

Figure. 7 Simulation result of the LNS Adder.

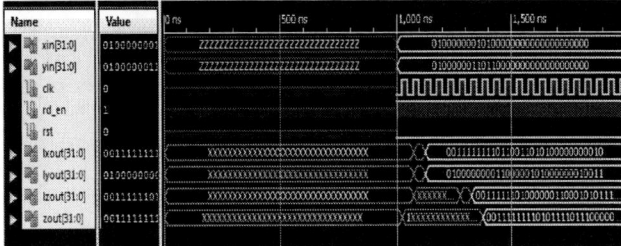

Figure. 8 Simulation result of the LNS Subtractor.

The simulation result shown in the above figure.7&8 denotes the LNS addition and subtraction operation which operates on 11 and 6 and gives the output 17 and 5 in floating point format.

VI. CONCLUSION

Recent advancement in the technology and portable devices demands the need for fast and low power consumption devices which is addressed by the two techniques residue number system and logarithmic number system, by which bit activity can be reduced and parallel processing can also be achieved. Power dissipation can also depend on some factors such as number system, dynamic range, efficiency of the number system. Due to huge application in several fields such as Cryptography, Digital Signal processing these techniques becomes very popular.

REFERENCES

[1] Zhining Lim, *An RNS Enabled Micro processor for Public Key Cryptography,* 2010.

[2] Dimitrios Soudris, *Designing CMOS circuits for low power,* 2002.

[3] Jean-claude Bajard, Nicolas Meloni and Thomas Plantard. "Efficient RNS for Cryptography."*World Congress: Scientific ComputationApplied Mathematics and Simulation,* Jul 2005, Paris (France), 2005.

[4] Mohammad Esmaeildoust, Shirin Rezaei, Marzieh Gerami and Keivan Navi, "On the Design of RNS Bases for Modular Multiplication, " *INTERNATIONAL JOURNAL OF NETWORK SECURITY, VOL.*16, NO.2, PP.118-128, MAR. 2014.

[5] Amos Omondi, Benjamin Premkumar, *Residue Number System theory and implementation* ,vol.2. Imperial college press, 2007.

[6] R.A. Patel, M. Benaissa, N. Powell, s.boussakta. "A New Low Power High-Speed Adder for RNS, " *Signal Processing Systems, 2004.* SIPS 2004.

[7] Melanie Dugdale,"VLSI Implementation of Residue Adders Based on Binary Adders, " *IEEE TRANSACTIONS ON CIRCUITS AND SYSTEMS-11: ANALOG AND DIGITAL SIGNAL PROCESSING, VOL. 39,* NO. 5, MAY 1992.

[8] Michael Haselman, "A Comparision of Floating Point and Logarithmic Number Systems on FPGAs." *Field-Programmable Custom Computing Machines, 2005.* FCCM 2005. 13th Annual IEEE Symposium .

[9] V. Paliouras and T.stouraitis, *Low Power Properties of the Logarithmic Number System,* 2001.

[10] K.Jhansson, O.Gustafsson, and L. Wnhammar, "Implementation of Elementary Functions for Logarithmic Number system," *IET COMPUTERS AND DIGITAL TECHNIQUES, VOL.*2,NO. 4, PP.295-304.

Real Time Watermarking of Grayscale Images using Integer DWT Transform

Sakthivel. S.M.
Asst. Prof (Sr.G), SENSE
VIT University, Chennai Campus
Chennai, India
sakthivel.sm@vit.ac.in

Dr. Ravi Sankar. A
Professor, SENSE
VIT University, Chennai Campus
Chennai, India
ravisankar.a@vit.ac.in

Abstract — In this paper, 2D integer wavelet transform based watermarking is carried out for the grayscale image with its VLSI architectural implementations. In the 2D integer wavelet transformation the lifting scheme is adopted and the watermarking operation is carried out in the LL2 frequency sub-bands. The entire watermark embedding process and extraction process are modeled in MATLAB and analyzed against the signal processing attacks like compression, salt & pepper noise, rotation and Intensity transformation attacks. Finally the same algorithm is modeled using Verilog HDL and implemented using ALTERA QUARTUS-II.

Keywords—DWT-Watermarking,LiftingScheme,RealTime-wateramrking,Transform Domain.

I. INTRODUCTION

Due to the rapid growth of computer technologies the usage of digital media has become more and more popular, which leads to the process of data copying and editing with help of available advanced software packages and tools[1]. One of the best ways to protect the multimedia content is digital watermarking. In the watermarking process the secret data watermark is embedded visibly or invisibly inside the multimedia content [2]. The multimedia data can be in any of the forms as image, audio and video [3]. Generally the watermarking process is carried out either in spatial or transform domain in three phases as watermark insertion (embedding), verification (verify the presence of watermark) and watermark separation (extortion)[4]. The spatial domain embedding process will insert the secret signature by the direct manipulation pixel intensities at redundant spaces whereas the transform domain based techniques will alter the frequency components of multimedia content for secret data insertion. The insertion of the secret signature at the frequency components yields a high robustness as against the signal processing attacks as compared with spatial domain embedding. There exists different transform domain embedding techniques like Fast Fourier transform(FFT), Discrete Cosine transform(DCT) and Discrete Wavelet transform(DWT)[5] to carry out the watermarking process effectively. Inspite of its high robustness nature during the watermarking process all the above stated embedding techniques involves high computational complexity [6]. In this paper the DWT based approach in Integer to Integer lifting Scheme is used to carry out the watermarking process. The

reason for choosing the DWT approach is due to its excellent spatial localization and multiresolution characteristics to hide the data in an invisible manner. In this DWT approach, the Integer lifting scheme is incorporated to transform the host signal into several frequency components which involves less complex operations as compared with other complex wavelets like Harr and Daubechies.

In this paper a grayscale image is considered as a host and a secret data. During the watermarking process the integer lifting wavelet transform method is carried out with its VLSI implementation. All the software based watermarking algorithms requires a specific CPU or GPU to run the mnemonics of the algorithm from a rom or ram whereas the hardware implementation using Field programmable Gate Array (FPGA) and Application Specific Integrated Circuit (ASIC) has a customized circuitry to carry out the watermarking process [7].The presence of custom circuit elements leads to the area, power and timing optimization [8]. Moreover the software based approach is quite easy to hack using advanced software tools and packages when compared with hardware watermarking approach [9,10]. The added advantage of hardware based approach over the software implementation is that the watermarking can happen in real time.

In this paper the Integer wavelet transform in forward lifting and backward lifting mode is incorporated to carry out the watermarking process. First the entire watermarking and extraction scheme is modeled using MATLAB and analyzed against the various signal processing attacks like Salt & pepper noise, intensity transformation, rotation along with JPEG compression. Then the same procedure of Lifting scheme is modeled using Verilog HDL with its VLSI architecture and implemented using ALTERA QUARTUS-II software.

II. LIFTING SCHEME OF 2D INTEGER DWT

The lifting scheme of DWT analyzes the filtering operation as a series of polyphase matrix operations, which helps in the implementation perspective using Verilog HDL. The general block scheme of DWT is shown in the Fig-1 with their corresponding subband filters for high pass and low pass operation.

The first part of the block diagram represents the Forward integer lifting process with a high pass and low pass filter

978-1-5090-0037-1/16 $31.00 © 2016 IEEE

bank. The second half of the block diagram represents the backward lifting operation with the same series of filter banks. Here the signal a (n) is the average signal and the d(n) is the difference signal which results from the forward lifting scheme operation. The advantage of the lifting scheme is that the backward lifting is realized by the reverse operation of the forward lifting.

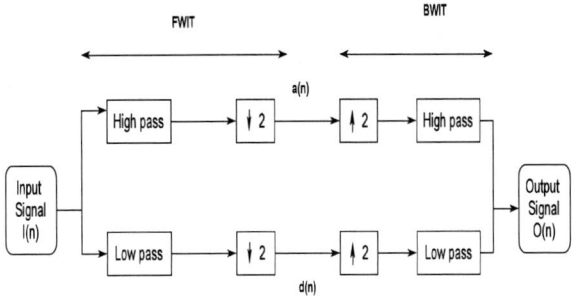

Fig- 1 DWT Filter Bank

The average and difference signals are calculated using the following equations 1 & 2

$$d_i(n) = a_{i-1}^{odd}(n) - a_1 * [(a_{i-1}^{even}(n-1) + a_{i-1}^{even}(n)] \qquad - 1$$
$$a_i(n) = a_{i-1}^{even}(n-1) + a_2 * [(d_i(n) + d_i(n-1)] \qquad - 2$$

Here the values of a_1 and a_2 are considered as ½ and ¼ for the lifting scheme of this Integer DWT operation. This entire operation of the DWT in forward mode and backward mode is happening in the split, predict and update phases as in the Fig-2.

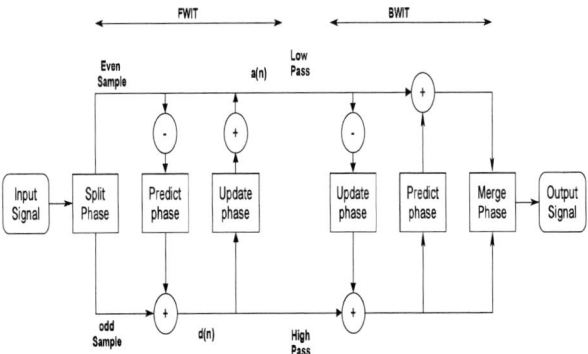

Fig - 2. Forward & Backward Lifting Scheme.

In the split phase of the Forward DWT operation the input signal is split into even and odd component. Then in the predict and update phases the average and difference signal components are calculated using the equations for $d_i(n)$ and $a_i(n)$. Finally the lowpass and highpass signal components are obtained from the forward lifting mode. Similarly in the backward lifting the output signal is reconstructed using the same update and predict concept in reverse manner and finally merged to get the output equation. The equations for reconstructing the odd and even components of the signal in the backward lifting mode operation is given below

$$a_{i-1}^{even}(n) = a_i(n) - a_2 * [d_i(n-1) + d_i(n)] \qquad - 3$$
$$a_{i-1}^{odd}(n-1) = d_i(n) + a_1 * [a_{i-1}^{even}(n-1) + a_{i-1}^{even}(n)] \qquad - 4$$

The entire forward and lifting scheme of operation is discussed for the 1-D DWT processing of the signal. For getting a the 2D DWT of an image signal the 1D DWT has to be performed row wise and column wise of the image signal. It means that the 2D DWT is an extended version of two 1D DWT in row wise and column wise manner. Here also the values of a_1 and a_2 are considered as ½ and ¼ for the lifting scheme of this Integer DWT operation

III. PROPOSED WATERMARKING

Generally in the DWT mode watermark embedding and extraction process, the LL subbands are chosen for embedding the secret image, as it has high energy compaction and these frequency components are perceptible to human eyes. In this entire procedure the two levels of DWT decomposition is carried out and the LL2 frequency component is selected for embedding. The process of watermark embedding and extraction is explained in a detailed manner in the following subsections.

A. Watermark Embedding

1. Read the host and watermark grayscale image.
2. Perform the 2D Integer DWT of the host image for two levels using 2D integer lifting Scheme.
3. Resize the watermark to size of the LL2 subband frequency component.
4. Choose the normalization factor n as 0.01 for normalizing the secret data.
5. Then carry out the watermarking process using the equation
   LL2 of W (i,j) = LL2[X(i,j)] + n * [S(i,j)]
   Here the W (i,j) is the watermarked image, X (i,j) is the input image and S(i,j) is the Watermark image.
6. Finally perform the inverse 2D Integer DWT to reconstruct the watermarked image.

B. Watermark Extraction

1. Read the watermarked image
2. Perform the 2D integer DWT on this watermarked image for two levels of decomposition
3. Choose the LL2 frequency subband
4. Also select the same normalization factor for extraction n as 0.01.
5. Then extract the watermark data as
   S (i,j) = [LL2 of W(i,j) – LL2 of X(i,j)]* (1/n)
6. Round of the value after normalization and thus give the secret data.

The entire process of watermark embedding and extraction are pictorially represented in the Fig-3 & 4 respectively.

978-1-5090-0037-1/16 $31.00 © 2016 IEEE

2016 International Conference on VLSI Systems, Architectures, Technology and Applications (VLSI-SATA)

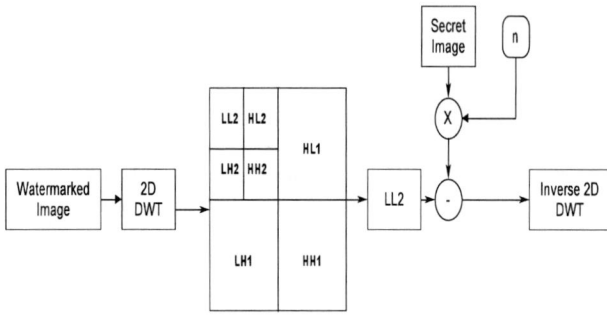

Fig - 3. Watermark Embedding.

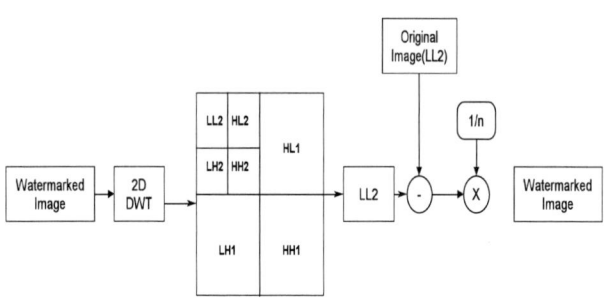

Fig - 4. Watermark Extraction.

IV. DATAPATH ARCHITECTURE FOR WATERMARK EMBDDING AND EXTARCTION

This section presents the hardware architecture for the 2D integer lifting scheme of watermark embedding and extraction. The corresponding datapath architectures are explained in the following subsections

A. Data Path for Watermark Embedding

The data path architecture of the watermark embedding module consists of three main units DWT processor, Watermarking Unit and IDWT processor.

The data path unit of DWT processor unit consists of two image RAMS one with a size of host image and other with a size of secret image to store it. Then the DWT processor module consists of two 1-D DWT module and a 1D FSM controller. The 1D- DWT module is used to predict the difference signal component and average signal component. The same 1D-DWT module is used twice to carry out the 2D-DWT operations once in row wise and then in the column wise mode. The entire operation of the DWT operation is controlled by a FSM controller inside a DWT processor with control signal command 1D start and 2D start. A transpose memory is also used to store the intermediate co-efficient during lifting dwt operation.

The data path unit of the watermarking unit consists of an adder, multiplier and FSM controller to carry out the

watermarking process. Here the LL2 sub-band components are added using adder with the multiplied versions of the secret data and fed into the IDWT processor unit.

The IDWT processor consists of two 1D IDWT module, which is used to predict the odd and even signal components. The same 1D IDWT module is used twice inside the IDWT processor with control signal command 1DI start and 2DI start. The entire operation of the IDWT is controlled by the IDWT FSM controller with the corresponding commands. The entire datapath architecture of watermark embedding is pictorially represented in the Fig-5.

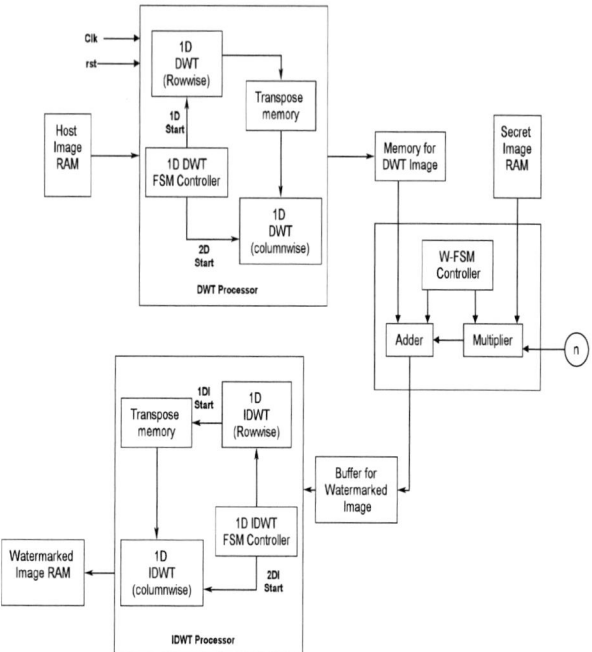

Fig – 5 Datapath Architecture for Watermark Embedding.

B. Data Path for Watermark Extraction

The datapath architecture of the watermark extraction module consists of two modules DWT processor and a Extraction unit.

The datapath unit of DWT processor consists of the same components as in the watermark embedding stage as 1D-DWT lifting module and a FSM controller. The 2D DWT operation is carried out for the watermarked image and the corresponding LL2 sub-band component is fed into the extraction unit.

The extraction unit consists of a subtractor, multiplier and FSM controller. The watermarked image is subtracted from the original LL2 component of original image and multiplied by a normalized factor to retrieve the secret data. The entire operation of the extraction is controlled by the FSM controller with the corresponding command signals. The entire datapath architecture of watermark extraction is pictorially represented in the Fig-6.

978-1-5090-0037-1/16 $31.00 © 2016 IEEE   172

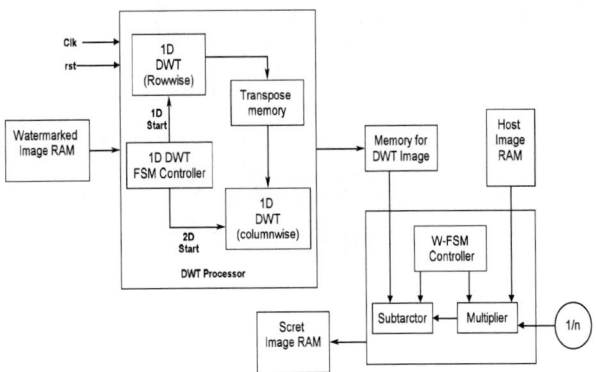

Fig – 6 Datapath Architecture for Watermark Extraction.

C. Data Path for 1D DWT & IDWT unit

The module 1D-DWT is used to predict the difference (High - pass) and the average (Low-pass) component of the input signal row-wise and column-wise. For the calculation of the difference signal d_i and average signal a_i the corresponding data paths in the Fig-7 & 8 are used respectively.

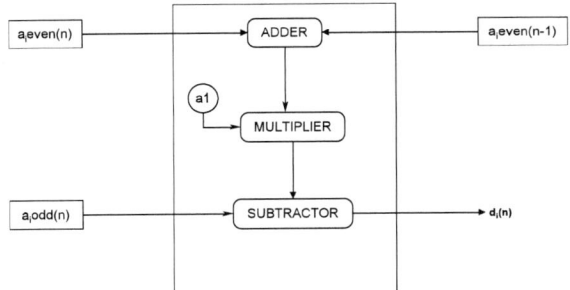

Fig – 7 Datapath for predicting the difference component d_i.

In the 1D-DWT calculation the difference d_i component is calculated using the equation- 1. Hence this calculation requires an adder, multiplier and then subtractor to predict the difference (highpass) value. Similarly the average component a_i is calculated using the equation-2. Hence for this calculation once again an adder and multiplier circuit is incorporated to predict the average (lowpass) value.

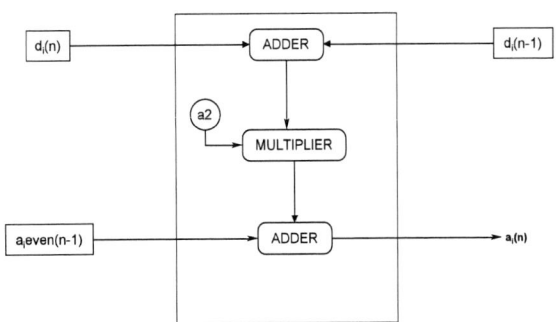

Fig – 8 Datapath for predicting the average component a_i .

Similarly in the IDWT calculation same adder, multiplier and subtractor circuit is used according to the equations 3& 4. Here the prediction of even and odd signal components is carried out with the same components as in DWT calculation.

D. FSM controller and VLSI Flow

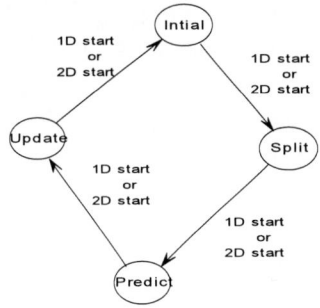

Fig – 9 FSM Controller State Machine for 1D-DWT Operation.

The data path of FSM controller for the 1D-DWT is shown in the Fig-9. The entire datapath of watermark embedding and extraction is driven by the commands of the corresponding FSM. Here the FSM is having four states as Initial/merge split, predict and update operation. In the split phase, even and odd components are separated and in the predict phase the difference signal component is calculated using the corresponding equation and data path component. Then in the update phase the average signal component is calculated. All these operations could happen inside the DWT processor only if the command ready signal 1D or 2D start is high. Similarly the FSM for IDWT has four stages to carry out the inverse DWT operation.

For the chip level implementation of the embedding and extraction corresponding data path modules are modeled in Verilog HDL. The host image and secret data is initialized as Hexa decimal values and stored in their corresponding RAMS. The modeled code is functionally verified by simulation and implemented as a real time chip using the below VLSI Design flow mechanism in Fig-10.

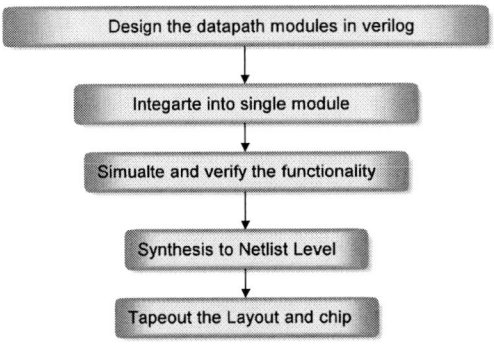

Fig – 10 VLSI Flow

V. RESULTS AND DISCUSSIONS

This section gives the simulation results of 2D-DWT integer wavelet transform watermarking using MATLAB and ALTERA QUARTUS-II software. In the MATLAB implementation a 512 X 512 lenna image in Fig-11(a) is considered as the host and the cameraman image of size 128 X 128 in Fig-11(b) as a secret image. The entire watermarking and extraction principle are modeled in MATLAB and simulated.

Fig- 11(a) Host Image

Fig-11(b) Watermark Image

After the Integer DWT based wateramarking implemention in MATLAB, the algorithm is analyzed against the different signal processing attacks with the paramaetric values of Peak Sinal to Noise(PSNR), Normalized Cross Corrleation Co-efficient (NCC), Normalized Absolute Error(NAE) and Strctural Content(SC).

The Watermarked image is subjected to the intensity variation attacks like intensity transformation, sharpening, Blurring and Contrast adjustment. It is infered that for all the attacks the PSNR value stands above 50DB which clearly indicates that the retived image is good in quality.Also by observing the values of NCC, NAE and SC for the intensity variation operation shows that the original and wateramark are similar to eachother as it values are closer to the factor one.

Then the watermarked image is analyzed against the rotation and compression signal processing attacks. Even for the rotation of the 2 degree the extracted watermark is similar to the original watermark which is reflected by the higher PSNR value greater than 49DB. For the compression operation also the PSNR value stands in the range of 49DB to 51 DB which ensures that the original and extracted watermark is similar. The values of the NCC, NAE and SC for the above attacks are closer to the factor one, which justifies the above stated condition. For the entire signal processing attack analysis the values of PSNR, NCC, NAE and SC are given in the Table-I.

Finally the watermarking and extraction algorithm are modeled in Verilog HDL and the functionally is verified using Modelsim simulator. For the image reconstruction process the Hexa-Decimal image values from the image RAM are wrote into a text file after the Verilog HDL simulation and the image is reconstructed using MATLAB. Then results of the MATLAB and Verilog Simulation are compared and inferred that both the simulation results are not differing in their pixel intensity values.

Table-I Analysis with different types of Attacks

Attacks		Extracted Watermark	PSNR	NCC	NAE	SC
Without Attack			51.23	.997	.998	.994
Pepper salt Noise			50.91	.935	.978	.964
Blurring			51.11	.981	.973	.993
Sharpening			51.21	.967	.928	.954
Intensity Transformation			50.21	.957	.918	.943
Image Contrast			51.01	.971	.988	.959
Rotation (in degree)	1		49.92	.987	.929	.968
	2		49.81	.957	.928	.961
JPEG Compression	5		51.42	.976	.956	.977
	10		51.34	.931	.934	.963
	15		49.97	.943	.921	.952
	20		49.67	.979	.911	.946

Table –II Device Utilization Summary

Device Utilization Summary for Cyclone-II EP2C5AF256A7			
Logic Utilization	Used	Available	Utilization
No.of.Slice Registers	534	4608	12%
No.of.Bonded IOBs	68	158	43%
Total Combinational Function Used	1240	4608	27%
Maximum Fan Out	462		
Avearge Fan Out	2.81		
No.of.Slice LUTs Used	1288		

After the functional verification, the entire Verilog code for watermarking and extraction unit is synthesized using ALTERA QAURTUS-II tool using the family **Cyclone-II EP2C5AF256A7**. The device utilization summary for the hardware watermarking and extraction model is given in the Table-II. The Entire watermarking and extraction procedure of this DWT and IDWT consumes 534 Logic Slice Registers, 68 IOBs, 1240 logical functions and 1288 Logical LUTs.

For the effective comparison of the VERILOG based approach with MATLAB implementation, a comparative study has been made based on the timing. It is inferred that the MATLAB implementation of watermarking and extraction using the Integer DWT requires 1.365s to carry out process on a Intel® core ™ i3-2350M processor with 8GB internal RAM and 2.3GHz operating speed. Whereas the VERILOG based implementation requires only 467 µS for the same watermarking and extraction with a 100MHz clock. The above comparison has been made for the image size of 512 X 512 Host image and 128 X 128 Secret (watermark) image. After this analysis the power analysis and timing analysis of the entire watermarking and extraction module has been carried out using the Power Play Power Analyzer tool and Time Quest Timing analysis tool in ALTERA QUARTUS software. The entire unit consumes 37.18mW of power and from the timing analysis report it is inferred that there is no critical path in the design. This means that the entire logic is meeting the timing requirements with the clock during the design.

VI. CONCLUSION

In this work an Integer DWT based robust watermarking and extraction has implemented in real time with its architectural specifications. As the watermarking and extraction is carried out using a FPGA chip, it can be incorporated as Co-Processor for watermarking with any multimedia devices. Due to the reconfigurable nature of the FPGA chip this algorithm can be modified for any complex protocols also. Hence in this paper an effective robust invisible watermarking and extraction process has been designed to carry out the watermarking process in real time using ALTERA tools.

VII. REFERENCES

[1] A.M. Eskicioglu, E.J. Delp, An overview of multimedia content protection in consumer electronics devices, Elsevier Signal Processing : Image Communication 16 (2001) 681-699.

[2] C.T. Li, Digital fragile wateramrking scheme for authentication of JPEG images, in: Proceedings – Vision, Image and Signal processing, vol. 151(6), 2004, pp. 460-466.

[3] M.Hussain,M.Hussain,A survey of Image Stegnography Techniques, International Jouranl of Advance science and technology, Vol.54.May 2013

[4] M. Chaumont and W. Puech, "DCT-Based Data Hiding Method To Embed the Color Information in a JPEG Grey Level Image", 14th European Signal Processing Conference (EUSIPCO 2006), Florence, Italy,

[5] S.P. Mohanty, N. Ranganathan, K. Balakrishnan, "A dual voltage-frequency VLSI chip for image watermarking in DCT domain", IEEE Transactions on Circuits and Systems II (TCAS-II) 53 (5) (2006),pp. 394–398.

[6] Anand Darji, Dr. A.N.Chandorkar, Dr. S.N.Merchant, Vipul Mistry, "VLSI Architecture of DWT based Watermark Encoder for Secure Still Digital Camera Design", Third International Conference on Emerging Trends in Engineering and Technology,2010, pp. 760-764.

[7] Cox I., Kilian J., Leighton T. and Shamoon T., Secure Spread Spectrum Watermarking for Multimedia, IEEE Transactions on Image Processing, Vol. 6, No. 12, December 1997

[8] Saraju P. Mohanty, N. Ranganathan and Ravi K. Namballa, "VLSI Implementation of Visible Watermarking for a Secure Digital Still Camera Design", IEEE Proceedings of the 17th International Conference on VLSI Design (VLSID'04)

[9] S.P. Mohanty, N. Ranganathan, K. Balakrishnan, "A dual voltage-frequency VLSI chip for image watermarking in DCT domain", IEEE Transactions on Circuits and Systems II (TCAS-II) 53 (5) (2006),pp. 394–398.

[10] Cheng Mingzi, " A Combined DWT and DCT Watermarking Scheme Optimized Using Genetic Algorithm", Journal of Multimedia, Vol-8, No.3, June 2013, pp. 299-304

Ultra Low Power Capacitive Power Management Unit in $0.18\mu m$ CMOS

Sanjay K. Kasodniya
[1]VLSI & Embedded Systems
Research Group, DAIICT
Gandhinagar-382007, India
[2]Space Applications Centre,
ISRO, Ahmedabad-380015, India
Email: ksanjay@sac.isro.gov.in

Biswajit Mishra
[1]VLSI & Embedded Systems
Research Group, DAIICT
Gandhinagar-382007, India
Email: biswajit_mishra@daiict.ac.in

Nilesh M. Desai
[2]Space Applications Centre,
ISRO, Ahmedabad-380015, India
Email: nmdesai@sac.isro.gov.in

Abstract—**This paper presents the design and simulation of a two stage power management circuit implemented in $0.18\mu m$ CMOS that operates from very low voltages starting from $460mV$ and higher up to a maximum of $800mV$. The proposed capacitive power management unit consumes very low power of $11\mu W$ @ $500mV$ sufficient to be operated from tiny photovoltaic cells, dimensions of few mm^2. In addition to the lower power consumption, the proposed circuit does not need any off chip components; ideal for ultra low power wireless sensor nodes.**

Index Terms– Capacitive, Power Management Unit, Energy harvester, CMOS, $0.18\mu m$.

I. INTRODUCTION

Energy Autonomy in battery less systems and wireless sensor nodes is gaining a lot of attention from researchers [1] working on ultra low power systems. Energy harvesting techniques is the key to attain energy autonomy for these electronic systems. There have been numerous efforts discussing AC (piezo, vibration) and DC (solar, thermal) energy harvesters that produce useful quantities of energy driving low power electronic systems [2]. Mostly the low voltage output from these harvesters is not sufficient to drive the electronics as is, and require up converting. Dedicated power conversion and power management circuit is of utmost importance in such harvesting circuits [3]. The key to such power management unit (PMU) is low power consumption, high efficiency and fewer number of off chip components. Both inductive [4] and capacitive methods [5], [6] exist and offer elegant solutions for these energy harvesting electronic circuits. The works described in [7-12] also emphasize the importance of capacitive energy harvesting methods. For example, in [7], [8], [9] authors have demonstrated the power management unit that operates at very low voltage (120 mV) and very low power (sub-μWatt) that can be useful for several applications including wireless sensor nodes. Authors in [10], [11] discuss their long efforts on energy harvesting circuits and their application. In this paper, a capacitive based method is adopted for the proposed PMU, as on chip MOS capacitors offer systems to have lesser number of off chip components and therefore yields to smaller form factor realisations.

In this paper, a capacitive power management unit for a DC energy harvester such as a photovoltaic (PV) is proposed. It is assumed that the input will have a minimal voltage requirement of approximately $460mV$ and can go up to $800mV$, typical output from a PV cell. This is our initial attempt to design a novel PMU based on standard $0.18\mu m$ CMOS models to be used for applications (WSN, payload sensors) that require energy autonomy.

This paper is organised as follows: in Section II, the operating principle of the proposed PMU is discussed. In Section III, the operating principle of the voltage doubler along with the circuit is discussed. In this section, response of the first stage of the PMU is provided. In Section IV, the circuit and the results of the five stage ring oscillator is discussed. The individual building blocks of the PMU in voltage monitor, voltage reference circuit is discussed in Section V. The performance of the reference circuit along with the simulation results of the op-amp and the comparator is discussed in this section. The results along with the key performance metric of the proposed PMU are discussed in Section VI. Finally in Section VII, conclusion and future work is discussed.

II. OPERATING PRINCIPLE

The proposed PMU circuit is shown in Fig 1. It has two stages of voltage doublers, each one approximately doubling the input voltage. Output of the 1st stage voltage doubler will charge the capacitor C_{out1} approximately to twice the input voltage V_{in} while the output of 2nd stage voltage doubler will charge the capacitor C_{out2} to V_{out} *i.e.* approximately twice the V_{DD1}. The 2nd stage voltage doubler is connected to a voltage monitor that monitors the voltage across C_{out2} and regulates. If the voltage goes beyond the regulation limit, the monitor disconnects the path from input V_{DD2} to C_{out2}. This voltage monitor is based on a low-voltage V_{ref} and two comparators that compares V_{ref} with the voltage at output capacitance C_{out2}. If the voltage (V_p) at node P is more than the reference voltage, the voltage-monitor will open the transmission gate, otherwise it lets the input voltage (V_{DD2}) charge the output capacitance C_{out2} to the desired voltage.

978-1-5090-0037-1/16 $31.00 © 2016 IEEE

2016 International Conference on VLSI Systems, Architectures, Technology and Applications (VLSI-SATA)

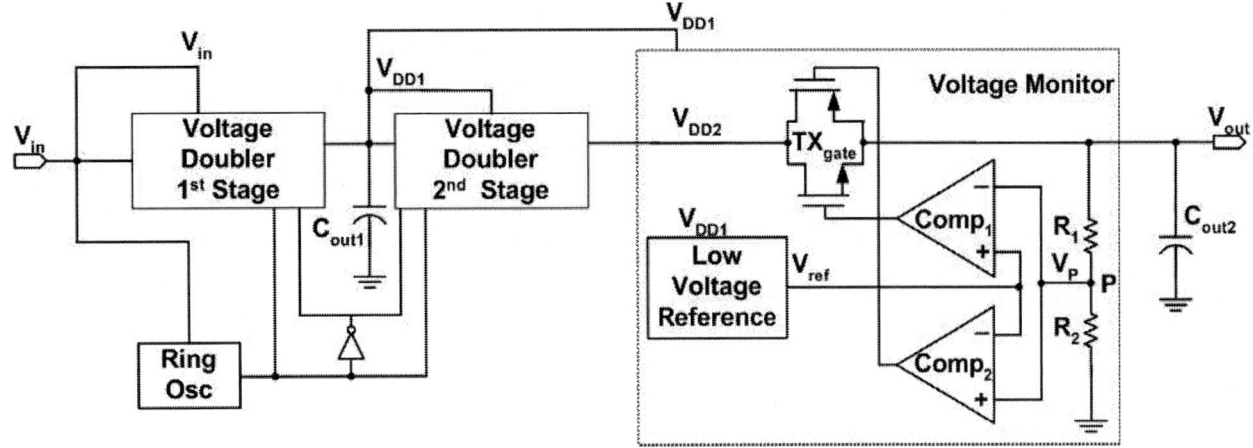

Fig. 1: Proposed PMU Circuit

III. Voltage Doubler

The voltage doubler circuit is shown in Fig 2 [12]. As shown in the Fig 2, the output voltage is approximately charged to twice the input voltage V_{in} (in this case the supply). The voltage doubling requires two clock phases ϕ and ϕ_b. In Fig 2, the basic operation of the voltage doubler is discussed. During clock phase ϕ, switches S_1 and S_3 are closed (S_2, S_4 open) so that the capacitor C is charged to the supply voltage V_{in}. In the next clock phase ϕ_b, the switches S_1 and S_3 are opened while S_2 and S_4 are closed. The bottom plate of the capacitor assumes a potential of V_{in}, while the capacitor maintains its charge ($Q = CV_{in}$) from the previous phase. Where C is the intermediate capacitance . This results in doubling the input voltage during ϕ_b, it can be seen that the charge:

$$Q = C(V_{out} - V_{in}) = CV_{in} \tag{1}$$

$$V_{out} = 2V_{in} \tag{2}$$

In order to accommodate a load, output capacitance C_{out} is placed at the output. Equation (1) and (2) can be modified to (3) and (4) to get the output voltage [12] as :

$$Q = (C + C_{out})V_{out} - CV_{in} = CV_{in} \tag{3}$$

$$V_{out} = \frac{C}{C + C_{out}}.2.V_{in} \tag{4}$$

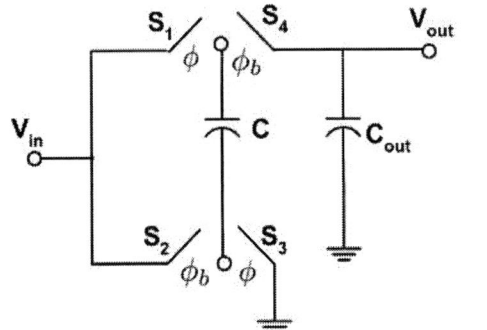

Fig. 2: Voltage Doubler— Principle of Operation

In this proposed work, one stage Dickson charge pump circuit is used to implement the doubler circuit as shown in Fig 3. The voltage doubler has inverter and transmission-gate switches (marked $S_1 - S_4$). The inverters are for phase reversal of the clock to generate ϕ_b from ϕ and the transmission gate switches are referenced $S_1 - S_4$, exactly as in Fig 2.

The simulation result of voltage doubler circuit with an input voltage range of $200mV - 500mV$ is shown in Fig 4. As can be seen, the doubler circuit starts working approximately

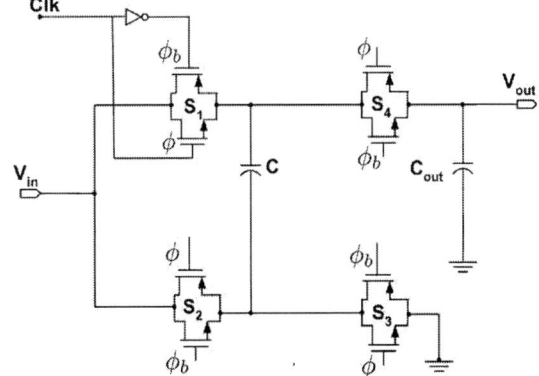

Fig. 3: Voltage Doubler Circuit

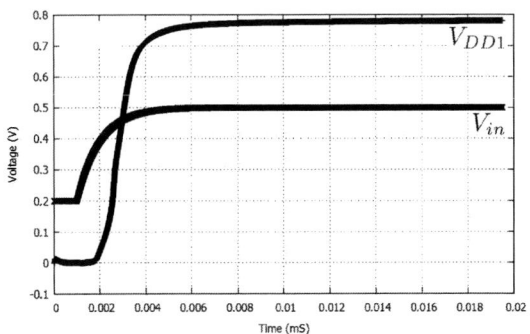

Fig. 4: Simulation of Voltage Doubler (V_{in} varying from $200mV - 500mV$, V_{out} up-to $780mV$)

978-1-5090-0037-1/16 $31.00 © 2016 IEEE

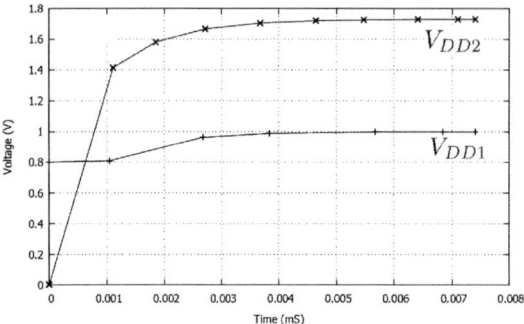

Fig. 5: Simulation of Voltage Doubler (V_{in} varying from 780mV to 1V, V_{out} Varying from 1.5V to 1.75V)

at an input voltage of $460mV$. We can observe the gain beyond this voltage where output reaching to $780mV$ at an input voltage at $500mV$. With a second set of input voltage changing from $780mV$ to approximately $1V$ we observe that output of the second stage voltage doubler charges the C_{out} from $1.5V$ to $1.8V$ and is shown in Fig 5.

IV. RING OSCILLATOR

For the voltage doubler to operate, a clock circuit is necessary that provides both the phases. Single stage voltage doubler discussed in this paper required frequency of $900kHz$[12]. This is achieved with a 5 stage RC ring oscillator circuit and is shown in Fig 6. Here number of stages are kept minimum (at five) and RC delay is used to achieve the frequency of $900kHz$. The oscillator works with $R = 560K\Omega$ and $C = 0.2pF$ and generates a clock frequency of $900kHz$ at $500mV$. At this voltage, the current requirement is very low, consuming $12nA$ and hence is adopted for the PMU. The simulation results for the ring oscillator are shown in Fig 7. Both the stages of the voltage doubler circuit is being fed from the same ring oscillator.

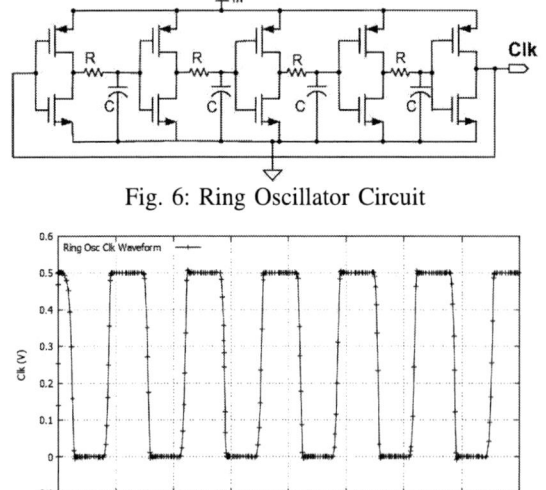

Fig. 6: Ring Oscillator Circuit

Fig. 7: Simulation Results of the Oscillator Circuit @ $900kHz$

V. VOLTAGE MONITOR

The voltage monitor consist of a low voltage reference (V_{ref}), two comparators, transmission gate (TX_{gate}) and a resistor network as shown in Fig 1. The voltage monitor provides a voltage within two specific levels at the output. In the proposed design, the higher level and the lower level is fixed at $2.0V$ and $1.5V$ respectively. This voltage level is maintained by the two comparators ($Comp_1$ and $Comp_2$ and the TX_{gate}. Both the comparators use a stable V_{ref} and compares a part of output voltage V_p through the ratioed logic using R_1 and R_2. The output of the comparators drive the TX_{gate} that ensures the V_{out} to be within $1.5V$ to $2.0V$. For example, if the $V_p < V_{ref}$, output of $Comp_1$ will be high (NMOS will turn ON), and output of $Comp_2$ will be low (PMOS will turn ON). This will provide a direct path from the input (V_{DD2}) to output (V_{out}). In case of $V_p > V_{ref}$, both the NMOS and PMOS are off, thereby no charging at C_{out2}. As V_p changes, repetition of the above cycle occur. The individual circuits of low voltage reference, Op-Amp and comparator are discussed in the following sub-sections.

A. Low Voltage Reference (V_{ref})

The reference circuit consists of an op-amp, BJTs Q_1 to Q_3, and MOSFETs M_a to M_e as shown in Fig 8. The reference circuit works on principle of PTAT (Proportional To Absolute Temperature) and CTAT (Complimentary To Absolute Temperature), where the V_{BE} of Q_1 is inversely proportional to the temperature that exhibits CTAT where

$$V_{BE@Q_1} = V_T ln(\frac{I_{C@Q_1}}{I_S}) \qquad (5)$$

while PTAT is exhibited by $\triangle V_{BE}$.

$$\triangle V_{BE} = (V_{BE@Q2} - V_{BE@Q1}) \qquad (6)$$

$$\triangle V_{BE} = V_T ln(\frac{I_{C@Q_2}}{I_S} \cdot \frac{I_S}{I_{C@Q_1}}) = V_T ln(m) \qquad (7)$$

Where ($I_{c@Q_2} = m \times I_{c@Q_1}$) (area of Q_2 is m times the area of Q_1), V_{BE} is base to emitter voltage, I_C is collector current, I_S is saturation current of a BJT and V_T is thermal voltage. These two temperature-coefficients cancel each other to give nominally zero temperature-coefficient.

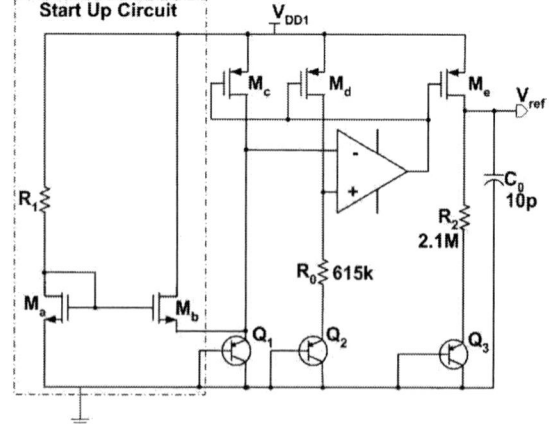

Fig. 8: Low Voltage Reference Circuit

978-1-5090-0037-1/16 $31.00 © 2016 IEEE

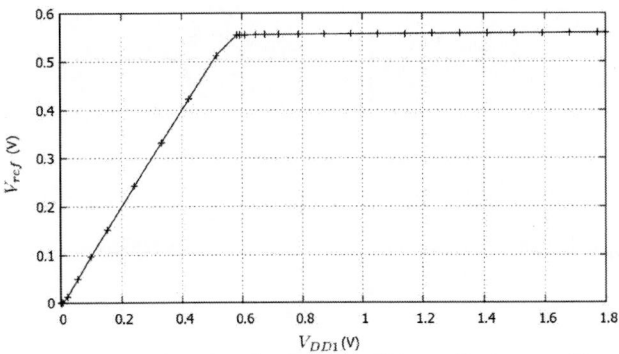

Fig. 9: Simulation Results for Low Voltage Reference V_{ref}

The V_{ref} circuit consumes current of approximately $7\mu A$. The simulation results of the low voltage V_{ref} is shown in Fig 9. As can be seen in Fig 9, the crossover of V_{in} and V_{ref} is at $600mV$. The effect of PTAT and CTAT can be observed from a stable V_{ref}.

B. Op-Amp

The reference circuit uses a low voltage op-amp, circuit shown in Fig 10. It is a two stage op-amp with RC compensation. Input stage is NMOS differential pair and output stage is a common source amplifier. The op-amp provides gain of $23dB$ at $500mV$ supply. The gain and phase plot are shown in Fig 11, which shows UGB at $150kHz$ and Phase-Margin of 70°.

C. Comparator

The comparator circuit is shown in Fig 12. This circuit is a modified version of the low voltage op-amp discussed earlier. In the voltage monitor circuit, the comparator acts when the output voltage on capacitor goes beyond $2V$, then it will generate switching signals for transmission gate. This will in effect disconnect the path from input to output and ensure the voltage to ripple within a range- say $1.5V$ to $2V$ in our proposed PMU. The comparator circuit takes current of around $6\mu A$.

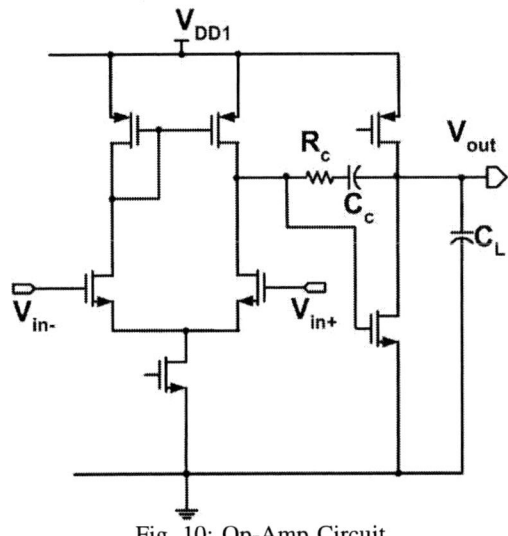

Fig. 10: Op-Amp Circuit

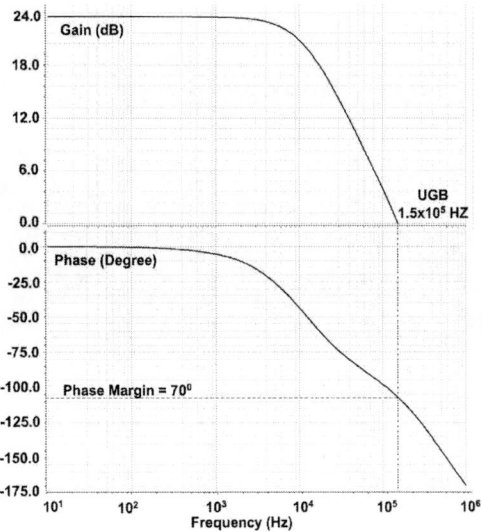

Fig. 11: Gain and Phase plot of Op-Amp

VI. RESULTS AND DISCUSSION

To validate the feasibility of proposed PMU, circuit simulations are performed in Cadence Virtuoso (resolution set at $200\mu s$ for PMU simulation) with the input at $800mV$ (typically an output of a solar cell at indoor lighting conditions at 200 lux gives an output of $400mV$) for the first voltage doubler. As can be seen in Fig 13, the output from the voltage doubler stage varies from $0V$ to $2.5V$ (marked as (A) in the Fig 13). However, most of the standard CMOS circuit operates at $1.8V$. The voltage monitor ensures that the voltage from V_{DD2} (stage 2 doubler) that can fluctuate due to varying environmental conditions, stabilizes somewhere close to $1.8V$. In our proposed PMU, V_{out} ripples between $1.5V$ to $2.0V$ (marked (B) in Fig 13).

Fig 14, shows the zoom-in of Fig 13 at start-up condition. Assuming V_{DD2} is available, the response of V_{out} is shown here. As can be seen, V_{out} is initially below $1.5V$ due to the changing nature of V_{DD2}. Also, the limits in which V_{out} is switching is limited to within $0.5V$. Therefore, V_{out} magnitude

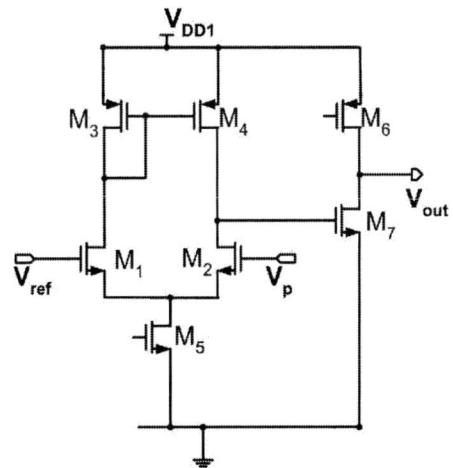

Fig. 12: Comparator Circuit

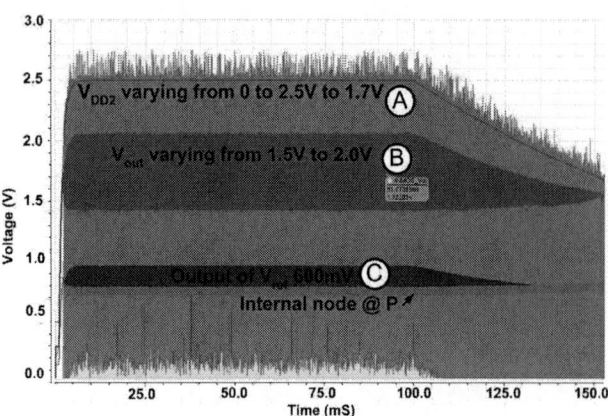

Fig. 13: Output of the PMU

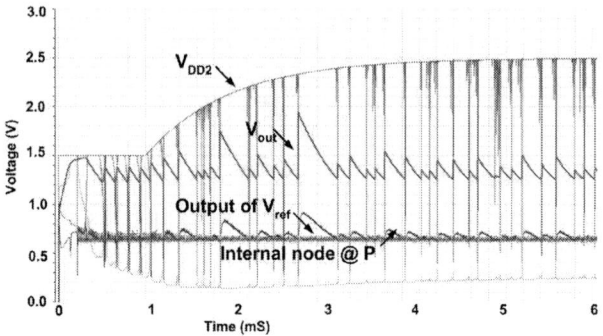

Fig. 14: Start-up Condition of PMU

is between $1.3V$ to $1.8V$ during the initial phase. After some time ($10ms$), V_{DD2} reaches the final value of $2.5V$, at which time, the V_{out} is switching between the range of $1.5V$ to $2.0V$ (see Fig 13).

The output of the V_{ref} is at $600mV$ (marked Ⓒ). The current consumption at different voltages of the individual blocks of the PMU is provided in Table 1. The voltage reference and the comparator consume the most current and is the subject of our future work to optimize it to nano-ampere consumption. The total power consumption of our proposed PMU is $22\mu A$ at $500mV$ ($11\mu W$).

VII. CONCLUSION AND FUTURE WORK

A capacitive power management unit is proposed that can work with DC energy harvesters like photovoltaic and thermo-electric generators. The PMU can work from very low voltages of $460mV$ and can output a stable voltage output for the circuits to operate reliably. The proposed capacitive PMU circuit has an added advantage of operating from cold start without the need of any additional circuit for jump-start compared to similar circuits in [5] and [6] , where they need jump start voltages to start the PMU. Quantification of the harvesting efficiency of the harvesters along with the peak efficiency is important for PMUs and is left as future work. The proposed PMU consumes $11\mu W$ at an input voltage of $500mV$. The power consumed is comparable to the power cited in [5], [7] and [8]. The voltage monitor consumes the

TABLE I: Current and Voltage of different Sub-blocks

Individual Blocks	Voltage(V)	Current(A)
Voltage Reference	$800mV - 1.5V$	$7\mu A - 20\mu A$
Op-Amp	$800mV$	$1\mu A$
Comparator	$800mV - 1.5V$	$6\mu A - 20\mu A$
Ring Oscillator	$500mV$	$12nA$
Voltage Doubler 1	$500mV$	$1\mu A$
Voltage Doubler 2	$800mV$	$2\mu A$
PMU	@$500mV$	$22\mu A$

greater part of this power and is the scope of our future work.

REFERENCES

[1] Sean Dieter Tebje Kelly, Nagender Kumar Suryadevara and Subhas Chandra Mukhopadhyay, "Towards the implementation of IoT for Environmental Condition Monitoring in Homes," *IEEE Sensors Journal*, vol. 13, no. 10, Oct. 2013, pp. 3846-3853.

[2] Toshishige Shimamura, Mamoru Ugajin, Kenji Suzuki, Kazuyoshi Ono, Norio Sato, Kei Kuwabara, Hiroki Morimura and Shinichiro Mutoh, "Nano-Watt Power Management and Vibration Sensing on a Dust Size Batteryless Sensor Node for Ambient Intelligence Applications," *IEEE ISSCC* 2010, pp. 504-506.

[3] Mamoru Ugajin, Toshishige Shimamura, Shinichiro Mutoh and Mitsuru Harada, "Design and Performance of a Sub Nano Ampere Two Stage Power Management Circuit in 0.35 um CMOS for Dust-size Sensor Nodes," *IEICE Transaction on Electronics*, vol. E94-C, no. 7, July 2011, pp. 1206-1211.

[4] E.J. Carlson, K. Strunz and B.P.Otis, "A 20 mV Input Boost Converter With Efficient Digital Control for Thermoelectric Energy Harvesting," *IEEE Journal of Solid-State Circuits*, vol. 45, no. 4, April 2010, pp. 741-750.

[5] Inge Doms, Patrick Merken, Chris Van Hoof and Robert P. Mertens, "Capacitive power Management Circuit for Micropower Thermoelectric Generators With a 1.4 μA Controller," *IEEE Journal of Solid-State Circuits*, vol. 44, no. 10, Oct 2009, pp. 2824-2833.

[6] Yi-Chun Shih and B.P. Otis, "An Inductorless DC-DC Converter for Energy Harvesting With a 1.2μW Bandgap-Referenced Output Controller," *IEEE Transactions on Circuits and Systems II: Express Briefs*, vol. 58, no. 12, Dec 2011, pp. 832-836.

[7] Biswajit Mishra, Cyril Botteron, Gabriele Tasselli, Chritian Robert and Pierre-Andre Farine, "A Sub-μA Power Management Circuit in 0.18μm CMOS for Energy Harvesters," *DATE*, 2013, pp. 1-6.

[8] Biswajit Mishra, Cyril Botteron and Pierre-Andre Farine, "A 120mV startup circuit based on charge pump for energy harvesting circuits," *IEICE Electron Express*, vol. 8, no. 11, 2011, pp. 830-834.

[9] Po-Hung Chen, Koichi Ishida, Xin Zhang, Yasuyuki Okuma, Yoshikatsu Ryu, Makoto Takamiya and Takayasu Sakurai, "A 120-mV Input, Fully Integrated Dual-Mode Charge Pump in 65-nm CMOS for Thermoelectric Energy Harvester," *IEEE Design Automation Conference, ASP-DAC*, 2012, pp. 469-470.

[10] Suyoung Bang, Yoonmyung Lee, Inhee Lee, Yejoong Kim, Gyouho Kim, David Blaauw and Dennis Sylvester, "A Fully Integrated Switched-Capacitor Based PMU with Adaptive Energy Harvesting Technique for Ultra-Low Power Sensing Applications," *IEEE Intrnational Symposium on Circuits & Systems (ISCAS)*, 2013, pp. 709-712.

[11] Wanyeong Jung, Sechang Oh, Suyoung Bang, Yoonmyung Lee, Zhiyoong Foo, Gyouho Kim, Yiqun Zhang, Dennis Sylvester and David Blaauw, "An Ultra-Low Power Fully Integrated Energy Harvester Based on Self-Oscillating Switched-Capacitor Voltage Doubler," *IEEE Journal of Solid-State Circuits*, vol. 49, no. 12, 2014, pp. 2800-2811.

[12] Feng Pan and Tapan Samaddar, *"Charge Pump Circuit Design"* McGraw-Hill, 2006.

[13] Luca Magnelli, Francesco A. Amoroso, Felice Crupi, Gregorio Cappuccino and Giuseppe Iannaccone, "Design of a 7 nW, 0.5 V subthreshold complementary metal oxide semiconductor operational amplifier," *International Journal of Circuit Theory and Applications*, 2012.

SWITCHING BASED EVALUATION OF SUBSTRATE CURRENT IN LIGHTLY AND HEAVILY DOPED CMOS AT 45nm

Sanjay Sharma, R.P.Yadav, Vijay Janyani

Department of Electronics and Communication Engineering, M.N.I.T., Jaipur

Abstract: Integration of analog and digital circuits is a vital design issue in mixed signal circuits. Switching activity at the digital end widely affects the analog circuitry. This paper presents the generation and variation of substrate current due to the switching activity in a digital circuit and also how the substrate current varies with the switching frequency. The circuit under test is a CMOS inverter at 45nm technology node. Circuit is made up of virtually fabricated NMOS and PMOS devices, the devices are made using ATHENA process simulator and the circuit is then implemented and evaluated in MixedMode. Transient simulation is performed and then substrate current is plotted with respect to two different input pulses for lightly doped and heavily doped substrate CMOS inverter. As the switching frequency increases the substrate current increases, this is applicable for both the lightly and heavily doped CMOS.

Keywords: **Substrate current, lightly and heavily doped CMOS, ATLAS.**

I INTRODUCTION

Substrate coupling is a major hindrance in mixed signal circuits [1]. Particularly at radio frequency (RF) this becomes a vital issue, as the device size shrinks substrate coupling affects the devices on the common substrate. Feature miniaturization has been mainly responsible for dramatically reducing the distance between high-frequency noise sources and sensitive devices, the substrate being a major carrier of this type of spurious signals. In digital circuits switching activity introduces substrate noise [2].Digital logic gates inject current into the substrate as they charge and discharge their loads. This type of noise is caused by the sudden discharge of electrons or holes during a transition. It is often known as switching noise. The switching noise is two to three times greater than the device noise [3]. In order to characterize substrate noise, its generation is a major issue. Switching activity at the circuit level is an important parameter to predict the substrate current [4]. The objective of this paper is to evaluate this very switching noise, which finally ends up as substrate noise. In mixed-signal circuit design, monolithic CMOS process remains the choice of implementation technology for its superior device packing density and reduced fabrication cost. CMOS logic gates inject current into the substrate as they charge and discharge their loads [5]. Single CMOS switching is considered in this paper, while in a mixed signal environment multiple switching takes place simultaneously. Thus the substrate current generated through the switching activity is a serious problem [6]-[9]. With a decrease in feature size and increase in switching frequency,

the substrate noise evolved by digital switching also increases. Therefore in nanometer process substrate noise analysis becomes important [10]. Thus this paper evaluates the substrate current variation as a function of switching frequency for CMOS at 45nm process. In section II, process, device and circuit simulation are described. Results are analyzed in section III, section IV concludes the paper.

II PROCESS, DEVICE & CIRCUIT SIMULATION

CMOS inverter circuit is made with MixedMode tool from ATLAS. Two CMOS inverter circuits with heavily and lightly doped substrate are integrated in MixedMode. Four devices are virtually fabricated with ATHENA at 45nm technology node. These four devices are lightly and heavily doped NMOS and PMOS respectively. Impact ionization at the device level is mainly responsible for the substrate current [11]-[13]. Process specification sheets for NMOS and PMOS are given below in Table 1 and Table 2 respectively:

Table 1 NMOS process sheet

Process	Lightly doped NMOS	Heavily doped NMOS
Initial substrate	P-Type-1e15	P-Type-1e18
P well implant	Boron dose1e12/cm^2	Boron dose1e12/cm^2
Gate oxide thickness	1nm	1nm
V_t implant	Boron=1.5e13	Boron=1.5e13
Poly deposition	80nm	80nm
S/D implant	Arsenic=1e15	Arsenic=1e15
Halo implant	Boron 5e13 Energy 25 Angle 30^0 full rotation	Boron 5e13 Energy 25 Angle 30^0 full rotation
S/D implant (deep)	3e15,7.5Kev	3e15,7.5kev
RT Annealing	750-800 nitro for 1 min.	750-800 nitro for 1 min.
Metal deposition	Al-10nm	Al-10nm

Table 2 PMOS process sheet

Process	Lightly doped PMOS	Heavily doped PMOS
Initial substrate	n-type-1e15	n-Type-1e18
n well implant	Phosphorous dose=7e13/ cm^2	Phosphorous dose=7e13/ cm^2
Gate oxide thickness	1nm	1nm
V_t implant	Arsenic 5e12	Arsenic 5e12
Poly deposition	80nm	80nm
S/D implant	Boron 1.5e14	Boron 1.5e14
Halo implant	Arsenic 1.5e13 Energy 20 Angle 35^0 full rotation	Arsenic 1.5e13 Energy 20 Angle 35^0 full rotation
S/D implant (deep)	1.5e15, 3Kev	1.5e15, 3Kev
RT Annealing	750-800 nitro for 1 min.	750-800 nitro for 1 min.
Metal deposition	Al-10nm	Al-10nm

Device characteristics are verified by ATLAS device simulator. The results of process simulator are used as input for device simulator and thus device characteristics are evaluated. D.C. analysis is performed for all the four devices. The extracted device parameters for lightly and heavily doped NMOS are given in Table 3:

Table 3 Device parameters (NMOS)

Parameters	Lightly doped NMOS	Heavily doped NMOS
V_t sat	0.21164 V	0.224777 V
V_t lin	0.253492 V	0.261598 V
SS sat	0.0777042 V/dec	0.0784734 V/dec
SS lin	0.0792496 V/dec	0.0798562 V/dec
DIBL	0.036393 V/V	0.0320183 V/V
I_{on}	0.00213986 A	0.00188337 A
I_{off}	3.79865e-09 A	2.69667e-09 A

The above and below extracted sheets confirm devices are fully functional. Similarly for lightly and heavily doped PMOS the extracted device parameters are given in Table 4:

Table 4 Device parameters (PMOS)

Parameters	Lightly doped PMOS	Heavily doped PMOS
V_t sat	-0.18089 V	-0.26236 V
V_t lin	-0.277837 V	-0.348957 V
SS sat	0.0988542 V/dec	0.0706196 V/dec
SS lin	0.0722356 V/dec	0.0707719 V/dec
DIBL	-0.0843017 V/V	-0.0753017 V/V
I_{on}	-0.000818387 A	-0.000747102 A
I_{off}	-1.43462e-07 A	-4.74127e-10 A

III RESULTS

Transient simulation is performed by applying two different input pulses of rise and fall times of 1ps and 10ps to the designed CMOS inverter at a GHz frequency. Input is applied to two different types of CMOS i.e., lightly and heavily doped. The rise and fall time of both the pulses are same. The duration of transient simulation is 600ps.
The netlist defined in MixedMode is:

vin 1 0 0 PULSE 0 1.2 10ps 10ps 10ps 140ps 300ps
an 2=drain 1=gate 0=source 4=substrate infile=NMOS.str
width=1.
ap 2=drain 1=gate 3=source 4=substrate infile=PMOS.str
width=3.0
vcc 3 0 1.2

The above defined netlist is for pulse for 10ps similarly we can define the same for 1ps pulse. Delay for the 10ps pulse is more than the 1ps pulse so switching is faster in 1ps pulse than in 10ps pulse. The substrate needs to be made common to both driver (NMOS) and load (PMOS) so that to study the impact of switching on circuit's substrate. If the substrate is grounded impact evaluation will be almost zero, which is applicable in the case of single CMOS circuit, but for mixed environment

where separate supplies are used for analog and digital entities, the device's individual contribution to the substrate current has to be evaluated. Also by grounding the common substrate the parasitic and capacitances are ignored, which can be dangerous for mixed signal design at radio frequency (RF). In Fig. 1 and Fig. 2, the plot of Substrate current verses Transient time is shown. Fig. 1 evaluates substrate current for lightly and heavily doped CMOS when the pulse input is given with 1ps rise and fall time respectively, for pulse period. The plot clearly depicts that substrate current in heavily doped CMOS has greater peaks than lightly doped CMOS.

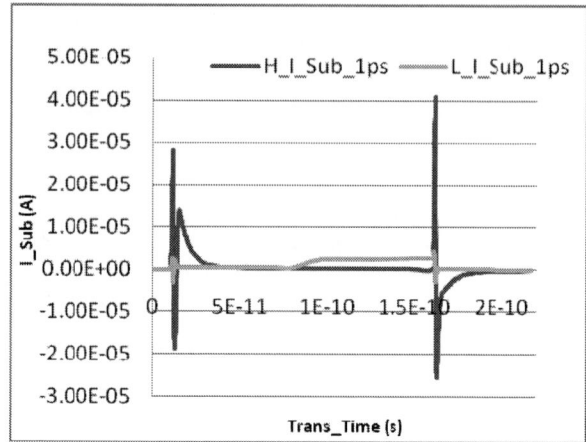

Fig. 1 Substrate current Vs Transient time for 1ps pulse for lightly and heavily doped CMOS

Similar characteristics are evaluated in Fig. 2 only the delay is increased from 1ps to 10ps. When the delay is 1ps, switching is faster so while comparing Fig. 1 and Fig. 2, we notice that substrate current is more pronounced when transient time is 1ps. Thus there will be more substrate coupling with an increased switching activity.

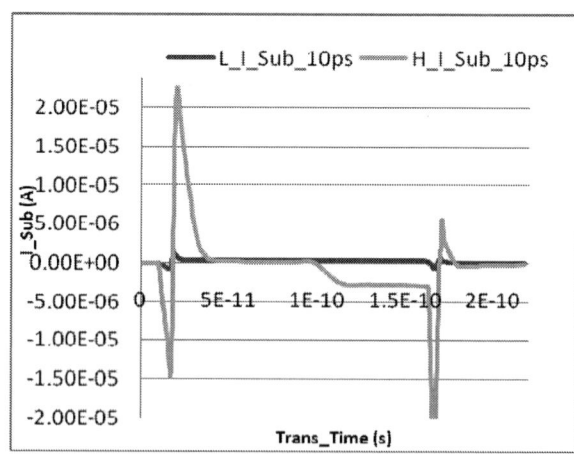

Fig. 2 Substrate current Vs Transient time for 10ps pulse for lightly and heavily doped CMOS

The increase in substrate current with an increase in switching is shown in Fig. 3 and Fig. 4. In these figures lightly and heavily doped CMOS are taken separately and respectively two different input pulses (1ps and 10ps) are applied to each of them. In Fig. 3 substrate current in lightly doped CMOS increases when the delay decreases from 10ps to 1ps, similar plot is there in Fig. 4 which describes the behavior of heavily doped CMOS for same set of input pulse. With an increase in pulse delay, substrate current decreases thus verifying that at higher switching frequency, substrate coupling increases [14]-[17].

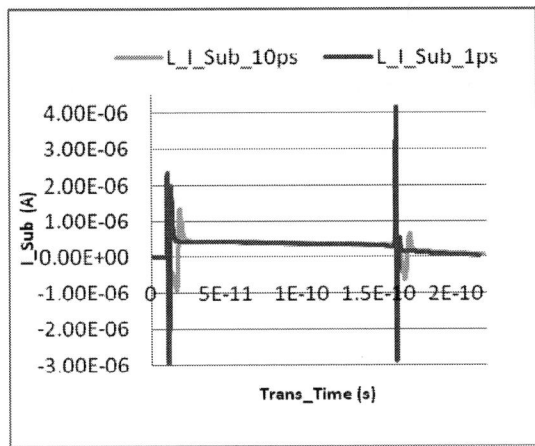

Fig. 3 Substrate current variation at 1ps and 10ps pulse for lightly doped CMOS

Out of Fig. 3 and Fig. 4, substrate peaks are more dominant in heavily doped CMOS for both 1ps and 10ps simulations.

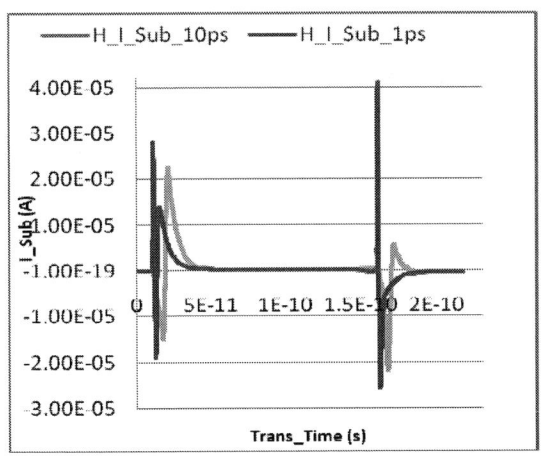

Fig. 4 Substrate current variation at 1ps and 10ps pulse for heavily doped CMOS

IV CONCLUSION:

Substrate current in heavily doped CMOS is more than by an order in comparison to lightly doped CMOS. As the switching frequency increases the substrate current also increases, thus generating more substrate noise and this noise at RF interferes with the sensitive analog circuit. There is decrease in substrate current when lightly doped NMOS and PMOS devices are used in inverter. Thus, at RF in mixed signal design we have to use lightly doped substrate devices if we want to reduce the impact of substrate current on the circuits that follow digital circuits (inverter).

REFERENCES

[1] A. Afzali-Kusha, M. Nagata, N. Verghese, and D. Allstot, "Substrate Noise Coupling in SoC Design: Modeling, Avoidance, and Validation," Proceedings of the IEEE, vol. 94, no. 12, pp. 2109–2138, Dec. 2006.

[2] P. Erratico, "The evolution of analog/digital interface in submicron IC's," in Proc. IEEE-CAS Region 8 Workshop Analog Mixed IC Des., Baveno, Italy, Sep. 1997, pp. 1–7.

[3] M. Shoji, Theory of CMOS Digital Circuits and Circuit Failures. Princeton University Press, 1992.

[4] Ming Shen; Mikkelsen, J.H., "An analytical model for spectral peak frequency prediction of substrate noise in CMOS substrates," in NORCHIP, 2013 , vol., no., pp.1-4, 11-12 Nov. 2013

[5] M. Badaroglu, P. Wambacq, G. Van der Plas, S. Donnay, G. Gielen, and H. De Man, "Evolution of substrate noise generation mechanisms with CMOS technology scaling," IEEE Trans. Circuits Syst. I, Reg. Papers, vol. 53, no. 2, pp. 296–305, Feb. 2006.

[6] Substrate Noise Coupling in Mixed-Signal ASICs, S. Donnay and G.Gielen, Eds. Boston, MA: Kluwer,2003.

[7] R. Gharpurey and E. Charbon, "Substrate coupling: Modeling, simulation and design perspectives," in Proc. Int. Symp. Quality Electron.Des. (ISQED), San Jose, CA, Mar. 2004, pp. 283–290.

[8] B. Owens, S. Adluri, P. Birrer, R. Shreeve, S. K. Arunachalam, K. Mayaram, and T. S. Fiez, "Simulation and measurement of supply and substrate noise in mixed-signal ICs," IEEE J. Solid-State Circuits, vol.40, no. 2, pp. 382–391, Feb. 2005.

[9] A. Afzali-Kusha, M. Nagata, N. K. Verghese, and D. J. Allstot, "Substrate noise coupling in SoC design: Modeling, avoidance, and validation," Proc. IEEE, vol. 94, no. 12, pp. 2109–2138, Dec. 2006

[10] E. Charbon R. Gharpurey, P. Miliozzi R. G. Meyer, A. Sangiovanni-Vincentelli, "Substrate Noise, Analysis and Optimization for IC Design," Kluwer Academic Publishers, 2003.

[11] X. Aragones, J. Gonzalez, and Rubio, Analysis and Solutions for SwitchingNoise Coupling in Mixed-Signal ICs. Kluwer Academic Publishers, Boston, 1 edition, 1999.

[12] P. Miliozzi, L. Carloni, E. Charbon, and A. Sangiovanni-Vincentelli, "SUBWAVE: a methodology for modeling digital substrate noise injection in mixed-signal ICs". In Proc. of IEEE on Custom Integrated Circuits Conference, pages 385–388, May 5-8, 1996.

[13] Sadiku, M.N.O.; Issa, E.M.; Attia, J.O.; Momoh, O.D., "Substrate coupling in mixed signal integrated circuits," in Southeastcon, 2011 Proceedings of IEEE , vol., no., pp.401-404, 17-20 March 2011

[14] C. Soens et al., "RF Performance Degradation Due to Coupling of Digital Swithing Noise in Lightly Doped Substrates," Proc. Southwest Symposium on Mixed-Signal Design, 2003, pp. 127-132.

[15] A. Nardi, H. Zeng, J. L. Garrett, L. Daniel, A. L. Sangiovanni-Vincentelli, "A Methodology for the Computation of an Upper Bound on Noise Current Spectrum of CMOS Switching Activity," Proc. IEEE ICCAD '03, pp. 778, 2003.

[16] J. F. Osorio, L. Elvira, F. Martorell, J.L. González, X. Aragonès, "Substrate Noise Macro-Modeling of Digital Cores," Proc. XVIII of Conference on Design of Circuits and Integrated Systems, DCIS '03, 2003, 75-80.

[17] W. Rhee, K. Jenkins, J. Liobe, and H. Ainspan, "Experimental analysis of substrate noise effect on pll performance," IEEE Trans. Circuits Syst. II, Exp. Briefs, vol. 55, no. 7, pp. 638–642, Jul. 2008.

OptMem: Dark-Silicon Aware Low Latency Hybrid Memory Design

†Salman Onsori, ‡Arghavan Asad
†‡Computer Engineering Department
†Bilkent University
Ankara, Turkey
salmanonsori@bilkent.edu.tr, ar_asad@comp.iust.ac.ir

*KaamranRaahemifar,‡Mahmood Fathy
‡Iran University of Science and Technology, Tehran, Iran
*Electrical and Computer Engineering Department
*Ryerson University, Ontario, Canada
kraahemi@ee.ryerson.ca, mahfathy@iust.ac.ir

Abstract—In this article, we present a convex optimization model to design a three dimension (3D)stacked hybrid memory system to improve performance in the dark silicon era. Our convex model optimizes numbers and placement of static random access memory (SRAM) and spin-transfer torque magnetic random-access memory(STT-RAM) memories on the memory layer to exploit advantages of both technologies. Power consumption that is the main challenge in the dark silicon era is represented as a main constraint in this work and it is satisfied by the detailed optimization model in order to design a dark silicon aware 3D Chip-Multiprocessor (CMP). Experimental results show that the proposed architecture improves the energy consumption and performanceof the 3D CMPabout 25.8% and 12.9% on averagecompared to the Baseline memory design.

Keywords—Dark silicon, Non-Volatile Memory (NVM), Hybrid memory architecture, Embedded Chip-Multiprocessor (eCMP), Convex optimization, uncore components, 3D integration.

I. INTRODUCTION

In nowadays multicore architectures, energy efficiency becomes the primary concern during system design. Especially, energy consumption is a primary constraint in embedded system design since many of them are generally limited by battery lifetime. Main memory and cache subsystemsconsume a significant portion of overall energy in memory-intensive embedded applications. Due to the exponential contribution of leakage power in total power consumption in nanoscale era, leakage power can be a major driver of dark silicon in future multicore systems. However, leakage power also constitutes a major fraction of power consumption of memory modules [1]. Consequently, architecting new classes of memory systems with the minimum leakage power is essential for embedded systems.

One of the newest challenges in multicore design is the management ofdark silicon [2-4]. The rise of utilization wall due to thermal and power budgets restricts active components and results in a large region of dark silicon. Among the on-chip components, the cores anduncore components consume most of the power.Uncore components such as memory and on-chip network play a significant role in consuming large portion of power. Power management of these uncore components can be critical to maximize design performance in the dark silicon era. Thus, in addition to the embedded system requirements, dark silicon constraint forces designers to reduce energy consumption.

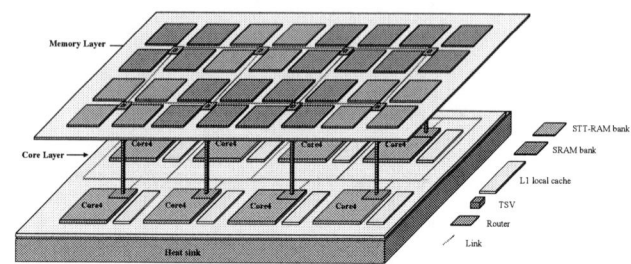

Fig. 1. 3D architecture of the proposed design

Spin Transfer Torque RAM (STT-RAM) as a promising candidate of non-volatile memories (NVMs) are considered to be attractive replacement for traditional SRAM memories due to their favorable characteristics such as highdensity, non-volatility and near-zero leakage power [5,6]. Nevertheless, they sufferfrom a longer write latency, limited write endurance and higher write energy consumption when compared to the traditional SRAM memory technology. In order to overcome the mentioned disadvantages of both memory technologies and benefit from their positive features, we use SRAM and STT-RAM as two different types of memory banks in the proposed memory architecture. This heterogeneous memory design is the best design possibility because it benefits from both memory technologies. Several research have also exploredmicro-architectural heterogeneity to combat the dark silicon problem [4,7].Fig. 1 shows an overview of the proposed architecture using an example of a 16 core homogeneous CMP in the lower layer and hybrid memory architecture in upper layer. In the proposed heterogeneous memory system, STT-RAM memory banks are incorporated with SRAM memory banks.

In this paper, we propose a convex optimization based approach for designing the heterogeneous memory system in order to maximize performance of the three dimensional(3D) CMP with respect to the peak power budget which is the main constraint in the dark silicon era. The proposed model maps applications/threads with more dependency and communication intensity closer to each other while at the same time it finds optimal distance of these applications/threads to each memory banks in order to reduce latency of the3D CMP design. More specifically, the proposed convex model optimally chooses efficient number and placement of SRAM and STT-RAM memory banks on the memory layer, and maps applications/threads on cores in the core layer.

2016 International Conference on VLSI Systems, Architectures, Technology and Applications (VLSI-SATA)

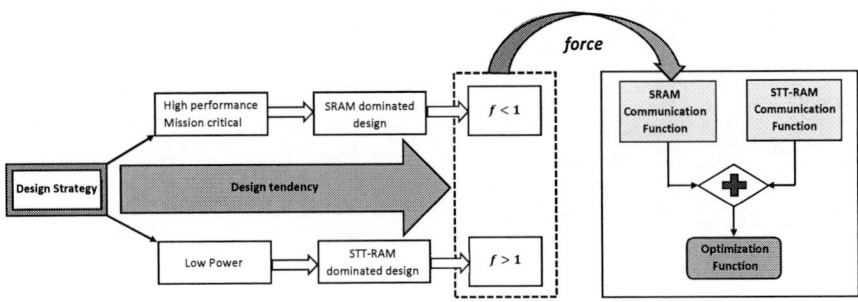

Fig. 2. Impact of force coefficient on the design strategy.

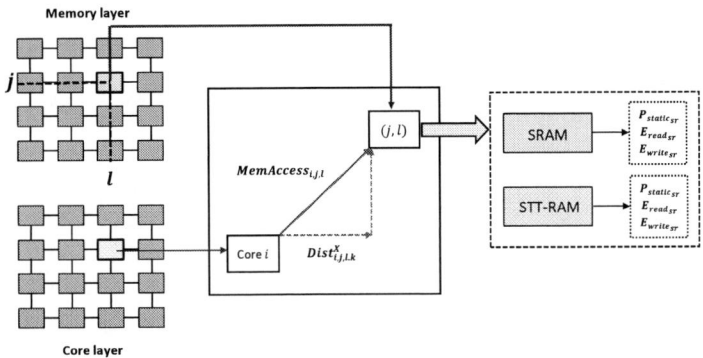

Fig. 3. Define parameters of a sample core and a memory bank and the communication between them.

The rest of this paper is organized as follows. Section II describes the optimization problem and its formulation. In Section III, excremental evaluation is presented. Finally, Section IV concludes the paper.

II. OPTIMIZATION PROBLEM AND FORMULATION

In this section, we propose a convex optimization model for achieving the following goals: 1) satisfying the dark silicon constraint with exploiting emerging technologies like NVM memories and 3D integration technology, 2) designing a hybrid memory architecture in order to use maximum advantages of SRAM and STT-RAM memories 3) efficient arrangement of memory banks with different technologies in the memory layer to decrease communication overhead on the CMP.

In order to achieve these goals, the proposed objective function is presented as follow:

$$Minimum \quad Overal_{com} = f.CommLatency_{SRAM} + CommLatency_{STTRAM} \quad (1)$$

In Equation (1), the overall communication is comprised of SRAM and STT-RAM communication functions. f is called a *force* coefficient that is used to show priority of memory layer to be designed by SRAM or STT-RAM banks.

In other words, f is a factor which can change impact of STT-RAM or SRAM communication cost in the overall cost function, $Overal_{comm}$. For example, if the goal is designing a mission critical embedded system and it should be highly reliable, SRAM banks are chosen with the cost of additional energy consumption but it is acceptable due to functionality of the device. On the other hand, if the reliability is not that much important or our goal is simply reduce the power consumption, force is changed to design a memory with higher possibility for choosing STT-RAM banks. However, there should be a tradeoff for these decisions since we need to satisfy power constraint of the desirable design. These design strategies and their impact on the optimization function is shown in Fig. 2. Also, it should be mentioned that each of SRAM and STT-RAM communication costs are combination of costs in x and y dimensions. These functions are introduced in the following paragraphs.

$CommLatency_{SRAM}^X$ is the latency of accessing to SRAM banks by cores in dimension x:

$$CommLatency_{SRAM}^X$$
$$= \sum_{i=1}^{C} \sum_{j=0}^{D_X-1} \sum_{l=0}^{D_Y-1} \sum_{k=1}^{D_X-1} MB_{j,l}^{SR}$$
$$\times (MemAccess_{i,j,l,r} \times Dist_{i,j,l,k}^X \times k$$
$$+ MemAccess_{i,j,l,w} \times Dist_{i,j,l,k}^X \times k) \quad (2)$$

In Equation (2), j and l are x and y dimension of a memory bank in the second layer. C is number of cores and dimension of the chips for x and y coordinates are D_X and D_Y, respectively. $Dist_{i,j,l,k}^X$ is a binary variable and is set to 1 if the distance between cores i and a memory bank in position (j, l) is equal to k for x-dimension. $MemAccess_{i,j,l,r}$ is number of read accesses of core i to a SRAM bank in position (j, l). Also, $MemAccess_{i,j,l,w}$ is number of write accesses of core i to SRAM bank in position (j, l). In addition, Fig. 3 demonstrates two dimensional communications of a sample core and a memory bank. Similarly, $CommLatency_{SRAM}^Y$ is defined like $CommLatency_{SRAM}^X$ just for dimension y.

$CommLatency_{SRAM}^Y$

$$= \sum_{i=1}^{C} \sum_{j=0}^{D_X-1} \sum_{l=0}^{D_Y-1} \sum_{k=1}^{D_X-1} MB_{j,l}^{SR}$$
$$\times (MemAccess_{i,j,l,r} \times Dist_{i,j,l,k}^Y \times k$$
$$+ MemAccess_{i,j,l,w} \times Dist_{i,j,l,k}^Y \times k) \quad (3)$$

Equation (2) is comprised of four summations. It finds different commination costs between cores and SRAM banks based on possible distances between cores and SRAM banks in x coordinate. With minimizing objective function $Overal_{com}$, we also will achieve minimum distances (k) in $CommLatency_{SRAM}^X$. This procedure again is done for $CommLatency_{SRAM}^Y$ in the y coordinate. Therefore, we can have the best placements for SRAM banks to have minimum communication latencies between cores and these banks.

$CommLatency_{STTRAM}^X$ is another cost function that is the communication cost for accessing to STT-RAM banks by cores in dimension x. More specifically,

$CommLatency_{STTRAM}^X$

$$= \sum_{i=1}^{C} \sum_{j=0}^{D_X-1} \sum_{l=0}^{D_Y-1} \sum_{k=1}^{D_X-1} MB_{j,l}^{ST}$$
$$\times (MemAccess_{i,j,l,r} \times R_{Cost} \times Dist_{i,j,l,k}^X \times k$$
$$+ MemAccess_{i,j,l,w} \times W_{Cost} \times Dist_{i,j,l,k}^X$$
$$\times k) \quad (4)$$

Equation (4) models communication costs of cores with STT-RAM banks. Conceptually, this equation is similar to Equation (2) but it has additional parameters to model STT-RAM memory instead of SRAM. Two parameters namely R_{Cost} and W_{Cost} are defined in this part which are the cost of reading and writing to a STT-RAM normalized with SRAM memory. $CommLatency_{STTRAM}^Y$ are defined as same as $CommLatency_{STTRAM}^X$ for dimension y.

$CommLatency_{STTRAM}^Y$

$$= \sum_{i=1}^{C} \sum_{j=0}^{D_X-1} \sum_{l=0}^{D_Y-1} \sum_{k=1}^{D_X-1} MB_{j,l}^{ST}$$
$$\times (MemAccess_{i,j,l,r} \times R_{Cost} \times Dist_{i,j,l,k}^Y \times k$$
$$+ MemAccess_{i,j,l,w} \times W_{Cost} \times Dist_{i,j,l,k}^Y$$
$$\times k) \quad (5)$$

Note that, as there are only two layers in this work, communication cost in z dimension is assumed to be same for all cores and banks; therefore we do not consider it in the model.

In order to design the memory layer with SRAM and STT-RAM banks, there should be some constraints to architect the system in order to achieve the optimal placement and positions of SRAM and STT-RAM banks. These constraints are defined at Equation (6) and (7).

$$MB_{i,j}^{ST} + MB_{i,j}^{SR} = 1 \quad \forall i, j \quad (6)$$

$$\sum_{i=0}^{D_X-1} \sum_{j=0}^{D_Y-1} (MB_{i,j}^{ST} + MB_{i,j}^{SR}) = C \quad (7)$$

$MB_{i,j}^{ST}$ is a binary variable and is set to 1 if the existing memory bank in (i, j) is a STT-RAM bank. Similarly, $MB_{i,j}^{SR}$ is a binary variable and is set to 1 if the existing memory bank in (i, j) dimension is a SRAM banks. Equation (6) allows only assignment of one SRAM or STT-RAM bank to a single coordinate. In addition, sum of used STT-RAM and SRAM banks in second layer is equal to C that is defined in Equation (7).

The total power consumption of the proposed memory architecture during run time period of the mapped workload must be less than the available power budget. More specifically,

$$P_{Total} = (Power_s + Power_d) \leq P_{budget} \quad (8)$$

Equation (8) is the dark silicon constraint for the proposed memory architecture. Power consumption is the main constraint of the dark silicon era and uncore components such as on-chip memories are responsible for significant amount of power consumption [1]. On the other hand, satisfying power budget, P_{budget}, which in the dark silicon era is well-known to Thermal Design Power budget (TDP), is a main factor of the proposed model. Therefore, the achieved memory architecture based on the proposed model mitigates the dark silicon challenge by reducing power of the memory system as one of the most important uncore components. Focusing on uncore components architectures as a solution to combat dark silicon is unexplored in these days [1].

Since this optimization approach is solved at design time and static power dissipation depends on temperature, we consider pessimistic worst-case scenario and calculate P_{Static}^{SR} and P_{Static}^{ST} at maximum temperature limit.

$$Power_s = \sum_{i=0}^{D_X-1} \sum_{j=0}^{D_Y-1} (MB_{i,j}^{SR} \times P_{Static}^{SR} + MB_{i,j}^{ST} \times P_{Static}^{ST}) \quad (9)$$

Equation (9)finds static power of the hybrid memory by summing static power consumption of each SRAM and STT-RAM bank.

In Equation (10), P_{read}^{SR}, P_{write}^{SR}, P_{read}^{ST} and P_{write}^{ST} indicate average dynamic power consumed by the SRAM and STT-RAM banks per read and write access, respectively. $P_{dynamic}$ as the dynamic power consumption of the proposed hybrid memory system is calculated as bellow:

$$Power_d = \sum_{i=0}^{D_X-1} \sum_{j=0}^{D_Y-1} \sum_{c=1}^{C} (MB_{i,j}^{SR} \times (MemAccess_{i,j,c,r} \times P_{read}^{SR}$$
$$+ MemAccess_{i,j,c,w} \times P_{write}^{SR}) + MB_{i,j}^{ST}$$
$$\times (MemAccess_{i,j,c,r} \times P_{read}^{ST}$$
$$+ MemAccess_{i,j,c,w} \times P_{write}^{ST})) \quad (10)$$

To summarize, objective function $Overal_{com}$ is minimized under constraints (2) through (10). We only mentioned main constraints and their related variables in this section for brevity.

III. EXPRIMENTAL EVALUATION

A. Experimental Setup

We use GEM5[8] as a full system simulator to implement memories and cores. To simulate accurate behaviour of the 3D CMP design and its NoC architecture, we integrated GEM5 with a NoC simulator[9]. In addition, to calculate power consumption of the design, mention platform is integrated with McPAT [10]. The cache capacities and energy consumptions of SRAM and STT-RAM are estimated from CACTI [11] and NVSIM [12], respectively. The simulation platform of the work is shown in Fig. 4. Also, detailsof the memory parameters and the baseline system configurationwhichwe used in our experiments for SRAM and STT-RAM banks are shown in Table I and Table II, respectively.

We use multithreaded workloads for performing our experiments. The multithreaded applications with small working sets are selected from the PARSEC benchmark suit [13]. In our setup, programs in a given workload are randomly mapped to cores to avoid a specific OS policy. For the experimental evaluation, P_{budget} and T_{max} are considered $100W$ and $80\ ^{\circ}C$, respectively. Furthermore, we use CVX [14]to model the proposed convex optimization problem and solve it.

TABLE I.   DIFFERENT MEMORY TECHNOLOGIES COMPARISON AT32NM

Technology	Area	ReadLatency	WriteLatency	LeakagePower at80 °C	ReadEnergy	WriteEnergy
1MB SRAM	$3.03mm^2$	0.702ns	0.702ns	444.6mW	0.168nJ	0.168nJ
4MB STT-RAM	$3.39mm^2$	0.880ns	10.67ns	190.5mW	0.278nJ	0.765nJ

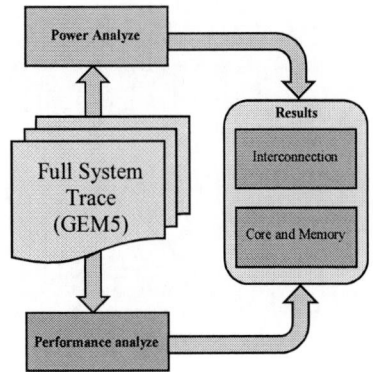

Fig. 4.   Simulation platform of the work.

TABLE II.   SPECIFICATION OF THE BASELINE EMBEDDED CMP CONFIGURATION

Component	Description
Number of Cores	16, 4× 4 mesh
Core Configuration	Alpha21164, 3GHz, area 3.5mm², 32nm
Private Cache per each Core	SRAM, 4 way, 32B line, size 32KB per core
On-chip Memory	Hybrid-fix: 8MB SRAM (8 banks , each 1MB) and 32MB STT-RAM (8 banks , each 4MB)
Network Router	2-stage wormhole switched, virtual channel flow control, 2 VCs per port, a buffer with depth of 4 flits per each VC, 5 flits buffer depth, 8 flits per data packet, 1 flit per address packet, each flit is set to be 16-byte long

B. Experimental Result

In this section, we evaluate our proposed 3D CMP with stacked memory in two different cases: 1) the CMP with hybrid stacked memory with same number of SRAM and STT-RAM banks in which STT-RAM banks are on the left and SRAM banks are on the right part of the memory layer (Hybrid-fix), 2) CMP with the proposed hybrid stacked memory on the core layer.

Fig.5 shows the results of the normalized energy consumption of the proposed method with respect to Hybrid-fix. As shown in this figure, the proposed design reduces energy consumption by about 25.8% on average compared to Hybrid-fix design.

Fig. 6 compares the normalized performance results. As shown in this figure, the proposed design improves IPC as a best parameter shows performance of the system up to 12.9% (4.87% on average) compared to the Hybrid-fix baseline design.

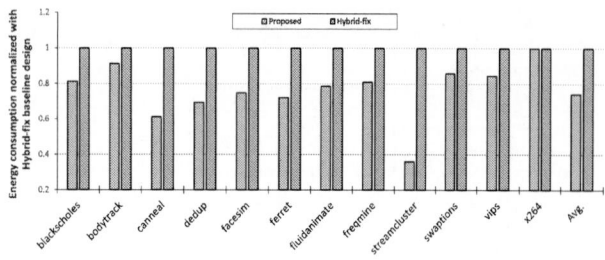

Fig. 5. Normalized energy consumption of the proposed design with respect to Hybrid-fix.

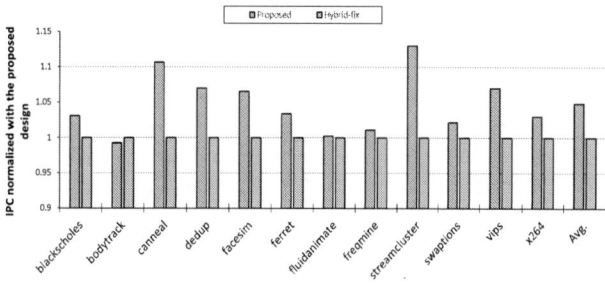

Fig. 6. Normalized performance comparison of each application with respect to the proposed design.

IV. CONCLUSION

In this work, we proposed a model to design an optimal heterogeneous memory system with using SRAM and STT-RAM memory banks. Our proposed optimization model finds optimal number and placement of different memory banks to satisfy peak power budget which is the main challenge in the dark silicon era. Experimental results show that the proposed architecture improves the energy consumption and performance of the 3D CMP on average about 25.8% and 12.9% respectively, compared to the Baseline memory design.

REFERENCES

[1] H. Cheng, et al. "Core vs. Uncore: The Heart of Darkness," Design Automation Conference(DAC), USA, 2015.

[2] H. Esmaeilzadeh, E. Blem, R. S. Amant, K. Sankaralingam, and D. Burger. "Dark silicon and the end of multicore scaling," In Computer Architecture (ISCA), pp. 365-376, 2011

[3] J. Henkel, H. Khdr, S. Pagani, and M. Shafique. "New trends in dark silicon." In Design Automation Conference (DAC), pp. 1-6, 2015.

[4] J. Allred et al. "Designing for dark silicon: a methodological perspective on energy efficient systems," In ISLPED, 2012.

[5] A. K. Mishra, T. Austin, X. Dong, G. Sun, Y. Xie, N. Vijaykrishnan and C. R. Das, "Architecting on-chip interconnects for stacked 3D STT-RAM caches in CMPs," ISCA, pp. 69–80, 2011.

[6] J. Ahn, S. Yoo, and K. Choi, "Dasca: Dead write prediction assisted stt-ram cache architecture," International Symposium on High Performance Computer Architecture (HPCA), 2014.

[7] Y. Turakhia et al. "Hades: Architectural synthesis for heterogeneous dark silicon chip multi-processors," In DAC, 2013.

[8] N. Binkert, B. Beckmann, G. Black, S. K. Reinhardt, A. Saidi, A. Basu, J. Hestness et al. "The gem5 simulator." ACM SIGARCH Computer Architecture News 39, vol. 39, no. 2, May 2011.

[9] M. Palesi, S. Kumar and D. Patti, "Noxim: Network-on-chip simulator," http://noxim.sourceforge.net, 2010.

[10] S. Li, J. H. Ahn, R. D. Strong, J. B. Brockman, D. M. Tullsen, and N. P. Jouppi, "McPAT: an integrated power, area, and timing modeling framework for multicore and manycore architectures," In Annual IEEE/ACM International Symposium on MICRO-42, pp. 469-480, 2009.

[11] N. Muralimanohar, R. Balasubramonian and N. P. Jouppi, "CACTI 6.0: A tool to model large caches," HP Laboratories, Technical Report, 2009.

[12] X. Dong, C. Xu, N. Jouppi, and Y. Xie, "NVSim: A Circuit-Level Performance, Energy, and Area Model for Emerging Non-volatile Memory," In Emerging Memory Technologies Springer, pp. 15-50, New York, 2012.

[13] M. Gebhart, Gebhart, Mark, Joel Hestness, Ehsan Fatehi, Paul Gratz, and Stephen W. Keckler. "Running PARSEC 2.1 on M5." University of Texas at Austin, Department of Computer Science, Technical Report, 2009.

[14] M. Grant, S. Boyd and Y. Ye, "CVX: Matlab software for disciplined convex programming," Available at www.stanford.edu/ boyd/cvx/.

Ultra Low Power 12-Bit SAR ADC for Wireless Sensing Applications

Raja Hari Gudlavalleti, Subash Chandra Bose

IC Design Group, CSIR-Central Electronics Engineering Research Institute,
Pilani, Rajasthan, India
{rajahari|subash}@ceeri.ernet.in

Abstract—This paper presents a 12-bit SA-ADC for portable low power wireless sensor systems. The proposed SA-ADC operates for rail-to-rail input range and achieves low power consumption. Split capacitor array based DAC and a novel charge-integration based dynamic comparator are used for low power consumption of the ADC. Measured DNL and INL are -0.59/0.67 LSB and -1.2/1.33 LSB respectively. At sampling rate of 100-kS/s with 1.8-V supply, the ADC consumes only 2-μW power and achieves a SNDR of 64.42-dB, SFDR of 71.2-dB resulting in an FoM of 14-fJ/Conversion-step. The ADC core occupies an area of 0.238-mm² and is fabricated in AMS 0.35-μm CMOS technology.

Keywords—ADC; low power; successive approximation register

I. INTRODUCTION

Portable sensor systems [1] typically require an analog-to-digital converter (ADC) of 10-12 bits resolution and bandwidth of upto 100-kHz. However, wireless sensor systems operating with battery or ambient energy sources like solar, RF, vibration etc., demand extremely low power circuits and ADC is one of the important sub-blocks for signal conversion in these circuits. Typically, power consumption is expressed in energy per conversion step defined using figure of merit (FoM) given by

$$FoM = \frac{P}{2^{ENOB} \, F_S}$$

where, F_S is the sampling frequency, P is the power consumed and $ENOB$ is the effective number of bits. For medium resolution and sampling rates of upto few kHz, a successive approximation (SA) based ADC has been an appropriate architecture because of its low power consumption. The primary sources of power consumption in SA-ADCs are the comparator, capacitive reference digital-to-analog (DAC) array and digital control logic circuit. The power consumption of digital circuits scales with the advancement of the technology. The supply voltage reduction is the effective way of reducing power consumption since the power of digital circuits directly benefits from supply voltage reduction. Therefore, the SA-ADC in the present work is operated at 1.8v as compared to typical 3.3v for AMS 0.35μm technology. The low supply voltage makes the design of analog circuits part

more challenging. However, changes in circuit topology needs to be done to analog circuits to meet the challenges while lowering the supply voltage. Dual supply is another method of reducing power consumption at the cost of increased pins in which the analog circuits are operated at higher supply voltage and digital circuits at lower supply voltage. Instead of lowering the supply voltage, various circuit schemes [1-6] have been proposed to reduce the power consumption in ADCs by scaling of sampling rate, reducing the number of boosted switches, charge-based schemes and time-domain comparator. A SA-ADC can be implemented using fully differential, pseudo-differential and singled ended architectures. A fully-differential implementation of the ADC reduces the common-mode noise and signal distortion. A rail-to-rail operation would exclude the use of additional reference to the ADC that reduces the overall power consumption at system level.

This paper proposes a novel charge-integration based dynamic comparator for a fully-differential rail-to-rail 12-bit SA-ADC has been designed. The prototype design is fabricated in AMS 0.35μm CMOS process. The proposed ADC has FoM of 14-fJ/conversion-step for 1.8v supply and occupies an active area of 0.238-mm². The remainder of the paper is organized as follows. Section II discusses the circuit architecture of SA-ADC and implementation of various sub-blocks. Section III presents the measurement results of the proposed ADC. Finally, Section IV draws the conclusion.

II. CIRCUIT ARCHITECTURE

A. SA-ADC Architecture

The architecture of 12-bit SA-ADC is shown in Fig. 1. The fundamental building blocks of the ADC are the comparator, sample-and-hold (S/H) circuit, capacitor array, and digital control logic (also called SAR registers). A fully differential SA-ADC is chosen to suppress supply noise and reject common-mode noise. A charge-redistribution based DAC that serves as both S/H circuit and reference capacitor array is implemented. Since this ADC is fully differential, the operation of the two sides is complementary. For simplicity, only the positive side of the ADC operation is described. The operation of the ADC is as follows. In the reset phase, the charges on the MSB and LSB array of the DAC capacitor array are reset to common-mode voltage V_{cm} (shown in Fig. 5). In the sampling phase, the input signal V_{inp}, is sampled on to

2016 International Conference on VLSI Systems, Architectures, Technology and Applications (VLSI-SATA)

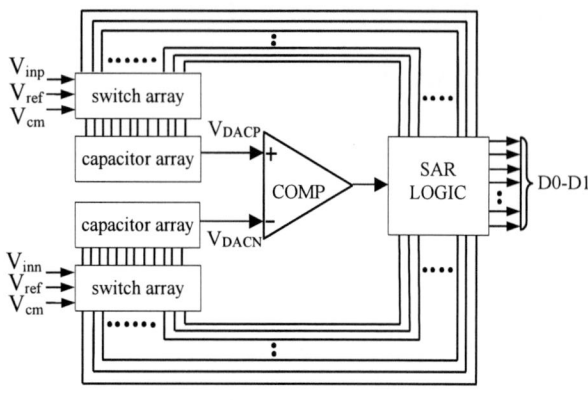

Fig. 1. Proposed SA-ADC

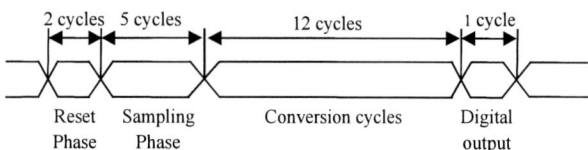

Fig. 2. Timing diagram of ADC generated by digital control logic

the bottom plates of the capacitor array and top plates are still held at V_{cm}. The reset and sampling phase are overlapped to optimize the sampling and reset time of the capacitor DAC. In the conversion phase, the largest capacitor is switched to V_{ref} and the other capacitors are switched to gnd. The comparator then performs the first comparison. If V_{inp} is higher than V_{inn}, the most significant bit (MSB) is 1 otherwise, it is 0 and the largest capacitor is reconnected to ground. Then, the second largest capacitor is switched to V_{ref}. The comparator does the comparison again based on the input voltages. The ADC repeats this procedure until the least significant bit (LSB) is decided. In the last cycle, the 12-digital output bits are stored into the output register. The SAR control logic generates the required timing signals. The timing diagram for the ADC is shown in Fig. 2. The ADC is synchronous to clock running at a frequency of upto 2MHz and also, this clock goes to SAR logic and comparator. The SAR logic is implemented with D-FFs and additional digital logic using full-custom approach to optimize area and power.

B. Comparator Architecture and Operation

A novel charge-integration based dynamic comparator is shown in Fig. 3. Initially, when the clock signal *clk* is low, capacitor C_{sa} is set to the reference voltage (here V_{dd}) using S1 switch. Differential input signals, V_{inp} and V_{inn} are applied to M1 and M2 respectively, that act as the constant current sinks

for given input signals. When clock signal goes high, the input voltage signals are integrated for a controlled time interval ΔT. During ΔT, the voltages V_p and V_n discharges the capacitors C_{sa}, at a constant rate and the discharge rate during charge-integration depends on the input voltage applied to the current sinks. The rate of discharge amplifies the differences in the input voltages and after ΔT instant, the voltages are compared, further amplified using pre-amplifier in the dynamic

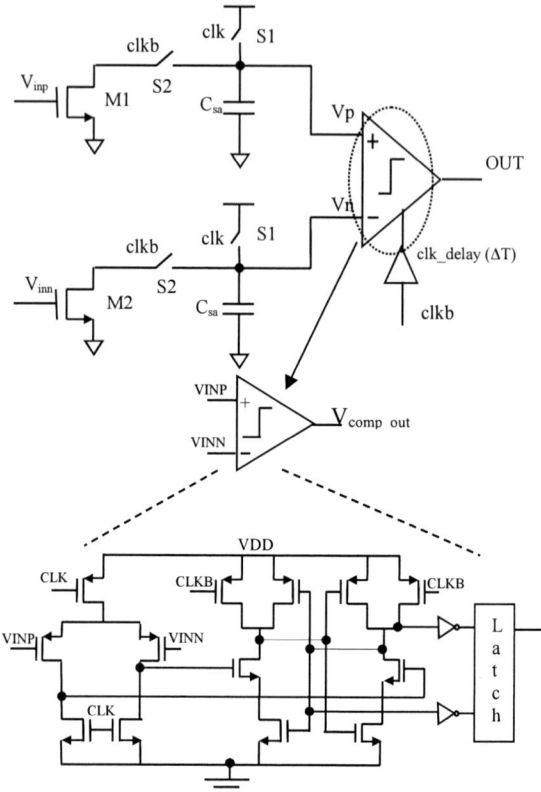

Fig. 3. Charge-integration based comparator

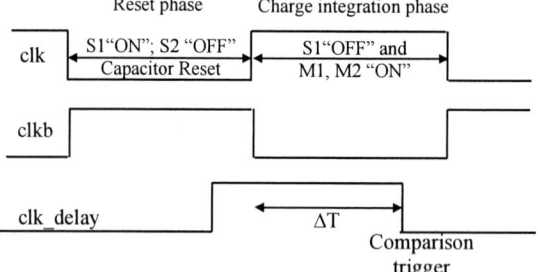

Fig. 4. Timing diagram of the comparator

comparator and finally latched to appropriate digital level. The ΔT shown in the Fig. 3, is implemented using controlled delay logic circuit that acts as the clock to the the dynamic comparator. The ΔT is obtained by delaying *clk* by more than half of its time period. The timing diagram of the comparator operation is shown in Fig. 4. The sizing of C_{sa} and transistors

978-1-5090-0037-1/16 $31.00 © 2016 IEEE   190

M1, M2 depend upon integration time. M1, M2 and both C_{sa} are matched using common-centroid layout. Post layout monte-carlo simulations for 1000 runs ensured 14-bit resolution for the ADC in the design. The individual comparator has been tested for a clock rate of upto 1MHz.

C. Digital to Analog Converter

The conventional binary-weighted capacitor DAC has good intrinsic matching. However, the total capacitance grows

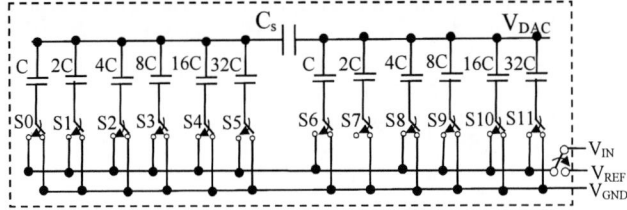

Fig. 5. Split capacitor based digital-to-analog converter

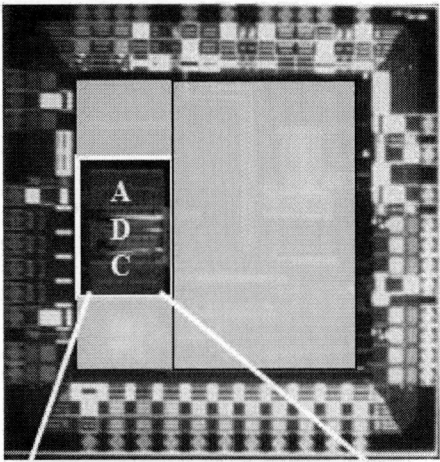

Fig. 6. Layout floor-plan of the capacitor array

exponentially with resolution resulting in the increase of total area and settling time. The total power consumption of the DAC is proportional to the total capacitance and switching frequency. It is shown that [6] a split capacitor based DAC has 37% switching power reduction compared to binary-weighted capacitor array. Hence, charge-redistribution DAC is implemented using a split-capacitive architecture with two 6-bit binary weighted capacitors each coupled by a split capacitor as showing in Fig. 5. The split capacitor C_s, separates the capacitor array into LSB and MSB arrays. Sizing of the unit capacitor plays an important role for better matching and kT/C noise performance of the capacitor DAC. Although increasing the capacitor size would improve matching and noise performance, an optimized value of 100fF has been chosen for unit cell of the poly2-poly1 capacitors. The routing and floor plan of the layout affects the performance of the DAC. Fig. 6 shows the layout strategy of the capacitor DAC. D in the figure indicates the dummy, C is the capacitor and its associated LSB or MSB unit value and CS is the split capacitor.

III. EXPERIMENTAL RESULTS

The SA-ADC is fabricated in AMS 0.35-μm CMOS technology. The die photograph and zoomed in-view layout of the fabricated ADC is shown in Fig. 7. The active core area of the ADC is 0.238-mm². The measured peak DNL and INL are -0.59/0.67 LSB and -1.2/1.33LSB respectively as shown in Fig. 8. Fig. 9 plots the FFT spectrum for an input signal frequency of 1-kHz. The measured SNDR and SFDR are 64.42 and 71.2-dB respectively. The resultant ENOB is 10.41bits. The total power consumption of the ADC is 2μW. The prototype achieves a FoM of 14-fJ/conversion-step. Table I summarizes the comparison to state-of-the-art ADCs.

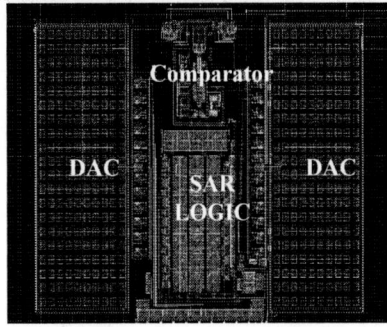

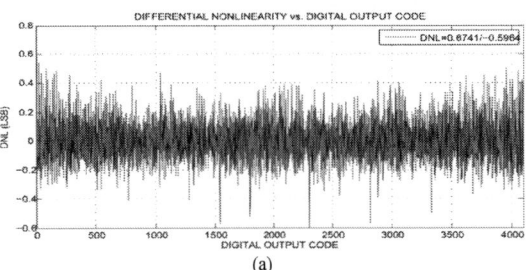

Fig. 7. Die photograph and zoomed-in layout view of the SA-ADC

(a)

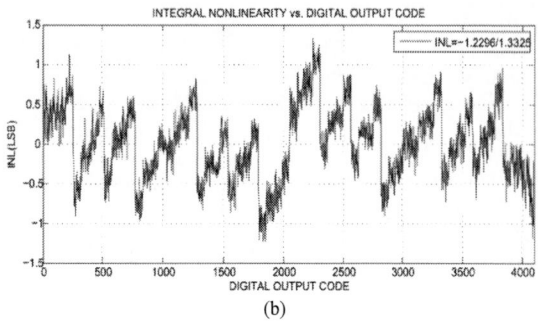

(b)

Fig. 8. Measured (a) DNL and (b) INL of ADC

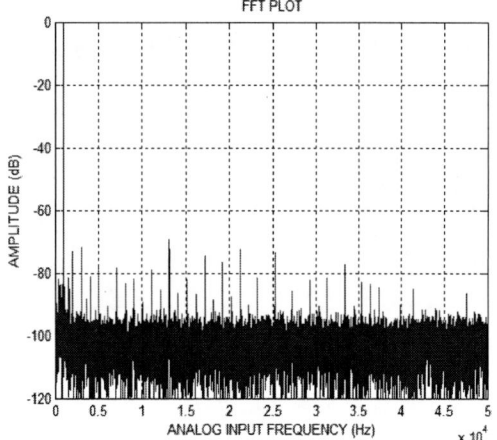

Fig. 9. Measured spectrum at 100-kS/s.

TABLE I. SUMMARY OF STATE-OF-ART 12-BIT ANALOG-TO-DIGITAL CONVERTERS

Specification	[1]	[7]	[8]	This work
Technology	0.18μm	0.18μm	0.13μm	0.35μm
Supply	1.8V	1.8V	1.2V	1.8V
Sampling	100kS/s	200kS/s	45 MS/s	100kS/s
Resolution	12-bit	12-bit	12-bit	12-bit
SNDR (dB)	65	60	70.1	64.42
SFDR (dB)	71	--	90.3	71.2
Power	25μW	72μW	3.0mW	2μW
FoM(fJ/step)	165	477	36.7	14

IV. CONCLUSION

This paper presents an ultra low power rail-to-rail fully differential 12-bit SA-ADC. The ADC consumes 2μW with 1.8v supply at a sampling rate of 100-kS/s having a FoM of 14fJ/conversion-step. The ADC achieves an ENOB of 10.41 bits. A novel charge-integrated based comparator with low offset and achieves upto 14-bit resolution having very low static power consumption is proposed. The split capacitor structure also saves capacitor area and power consumption. The proposed layout strategy for the split-capacitor DAC improves the performance of the ADC. The ADC core occupies an area of 0.238-mm^2 fabricated in AMS 0.35-μm CMOS technology. The experiment results demonstrate the power and performance of the ADC.

ACKNOWLEDGMENT

The authors would like thank CSIR-CEERI, Pilani for research support and DeitY/MCIT, New Delhi for chip fabrication.

REFERENCES

[1] N. Verma, A.P. Chandrakasan, "An ultra low energy 12-bit rate-resolution scalable SAR ADC for wireless sensor nodes," *IEEE J. Solid-State Circuits*, vol. 42, pp. 1196-1205, Jun. 2007.

[2] Z. Cao, S. Yan, and Y. Li, "A 32mW 1.25GS/s 6b 2 b/step SAR ADC in 0.13μm CMOS," in *IEEE ISSCC Dig. Tech. Papers*, pp. 542–543, Feb. 2008.

[3] Hao-Chio Hong, Guo-Ming Lee, "A 65-fJ/conversion-step 0.9-V 200-kS/s rail-to-rail 8-bit successive approximation ADC," *IEEE J. Solid-State Circuits*, vol. 42, pp. 2161-2168, Oct. 2007.

[4] P. Harpe, E. Cantatore and A. van Roermund, "A 10b/12b 40 kS/s SAR ADC with data-driven noise reduction achieving up to 10.1b ENOB at 2.2 fJ/Conversion-step", *IEEE J. Solid-State Circuits*, vol. 48, no. 12, pp. 3011 -3018, Nov. 2013.

[5] A. Agnes, E. Bonizzoni, P. Malcovati, and F. Maloberti, "A 9.4-ENOB 1V 3.8μW 100kS/s SAR ADC with time-domain comparator," *IEEE ISSCC Dig. Tech. Papers*, pp. 246-247, 2008.

[6] Yung-Hui Chung, Meng-Hsuan Wu, and Hung-Sung Li, "A 12-bit 8.47-fJ/Conversion-Step Capacitor Swapping SAR ADC in 110-nm CMOS," IEEE Trans. Circuits and Syst. I, Reg. Papers, vol. 62, no. 1, Jan. 2015.

[7] B. P. Ginsburg and A. P. Chandrakasan, "An energy-efficient charge recycling approach for a SAR converter with capacitive DAC," in *Proceedings of IEEE Int. Symp. Circuits and Systems* (ISSCAS '05), pp. 184–187, May 2005.

[8] Yang Siyu, Zhang Hui, Fu Wenhui, Yi Ting and Hong Zhiliang, "A low-power 12-bit 200kS/s SAR ADC with a differential time domain comparator," *Journal of Semiconductor*, vol.32, no.3, March 2011.

[9] W. Liu, P. Huang, and Y. Chiu, "A 12b 22.5/45MS/s 3.0mW 0.059mm^2 CMOS SAR ADC achieving over 90dB SFDR," in *IEEE ISSCC Dig. Tech. Papers*, pp. 80–381, Feb. 2010.

Distance Estimation and Direction Finding Using I2C Protocol for an Auto-navigation Platform

Rajesh Kannan Megalingam
Amrita School of Engineering
Amrita Vishwa Vidyapeetham
Kollam, Kerala, India
Email: megakannan@gmail.com

Jeeba M Varghese
Amrita School of Engineering
Amrita Vishwa Vidyapeetham
Kollam, Kerala, India
Email: jeeba91@gmail.com

Aarsha Anil S
Amrita School of Engineering
Amrita Vishwa Vidyapeetham
Kollam, Kerala, India
Email: aarsha.anil@gmail.com

Abstract—**This paper presents an auto navigation platform in Arduino which uses the i2c protocol to interface a digital compass and a rotation encoder to calculate distance travelled and direction with respect to the Earth's magnetic field. The digital compass IC used here is the HMC6352, while the rotation encoder is designed with the help of a MOC7811 coupler IC. The compass contains complete 2-axis sensors, analog, and digital electronics and also contains all the firmware for heading computation and calibration. The rotation encoder is an electromechanical device which obtains the angular position of the motor shaft it is connected to and then converts this position into some analog or digital value. This is mostly done using optoelectronic sensors which provide electric pulses in response to some stimulus. We use this property of the device to accurately guide our auto navigated device to pre-determined distances.**

Keywords- MOC7811, HMC6352, I2C protocol, Rotational encoder.

I. INTRODUCTION

Automobile navigation systems are a combination of diverse technologies used to guide a device to any location. Over the years a number of technologies have been introduced like GPS which help in auto-navigation, most of which works on the principle of position mapping. The first step is to identify the current location of the device and then correlate this data to navigate the device to any other location. But most of these technologies if not all have been concentrated on outdoor navigation of bigger automobiles while technological advancements in navigation platforms developed for indoor navigation which can be used for commercial applications has been few and far between. For instance in case of GPS, there is significant power loss in indoor systems due to signal attenuation, which leads to requirement of additional circuitry to setup the whole system. This is practically infeasible considering the cost to setup such a system for even the most basic of indoor navigation models. Alternatively other methods for positioning like Wi-Fi based positioning systems (WPS), Bluetooth, Magnetic positioning etc. have been on the rise. In this paper the method we use is Magnetic positioning. By using this method we can reduce the overall cost to implement such a system indoors. This is the most significant advantage of our system.

Here we use a digital compass IC HMC6352 to measure the orientation of the device with respect to the Earth's magnetic field. HMC6352 is the integrated compass module developed by Honeywell which works on Anisotropic Magneto-resistive (AMR) technology. The compass senses Earth's magnetic field (0.6 gauss) and provide the sensitivity for enhanced accuracy and performance. The compass has a Wheatstone bridge configuration that converts magnetic fields into a millivolt output and contains complete 2-axis sensors, analog and digital electronics and also other firmware for heading computation and calibration. This compass interfaced with the Arduino UNO is used to check the orientation of any device with respect to Earth's magnetic field. The communication between HMC6352 and Arduino uses the twin wire (TWI)or the I2C(Inter- integrated circuit) protocol in standard data rate mode of 100kbps, where the Arduino becomes the master and HMC6352 module becomes the slave.

The basic arrangement of an optical rotation encoder consists of a Light Emitting Diode (LED) and a light detector mounted exactly opposite each other. A patterned disc is placed between the LED and the light detector, while the disc itself is attached to the motor shaft. The disc contains alternate opaque and transparent sections. Once the disc starts rotating the opaque patterns block the light while the transparent sections lets it pass through. This results in a square wave pattern at the output of the detector. Now to calculate linear distance the formula used is

$$Distance = \frac{Wheel\ Circumference\ \times\ Counts}{Counts\ per\ Revolution} \tag{1}$$

In the following sections we discuss various aspects of the paper. In section II we discuss the basic motivation behind the development of such a platform and its real world applications. Section III lists the previous developments regarding the topic. In section IV, we explain the design and implementation of our hardware and software sections. Section V records the observations made during our experiments and finally we conclude our paper in section VI.

II. MOTIVATION

The aim of this paper is to design a cost effective, compact auto navigation platform using a digital compass and rotation encoder to measure direction and guide the device to a predetermined location. Most navigation systems utilize systems like GPS to guide devices through any path. But such systems lose their credibility in indoor conditions, because of signal attenuation and other defects like reflection losses from obstacles in a confined space. Hence need for cheap navigation systems, especially for indoor conditions have been on the rise. Our system uses simple rotation encoders and magneticcompass to guide the devices which are not only cheap but also provide highly accurate values since they are not affected by internal reflection losses. Also since range of signal send and receive is not as wide as GPS systems, attenuation losses are also negligible. With further development this design can be used to guide small scale devices like automated wheelchairs without using satellite positioning or other similar procedures followed by bigger automobiles in outdoor condition. This also reduces the energy consumed by the system as a whole.

III. RELATED WORKS

Auto navigation gaining more importance in medical and space applications has led to considerable advancements in development of navigation systems based on distance and orientation calibrations. The Arduino platform helps to interface the digital compass for the purpose of orientation awareness so that the wheel chair motor can be controlled according to the distance which has already been calculated using the rotational encoder. The digital compass HMC6352 is interfaced with Arduino and follows I2C protocol to do so. Inter Integrated Circuit protocol [2] helps to establish communication between a numbers of devices with only two wires in common i.e. Serial Data Line and Serial Clock Line. The I2C protocol should be multi slave with single master or multi master multi slave or multi master single slave.

As compared to other types of compass, the 1490 compass sensor by Dinsmore gives heading information in eight directions such as North, South ,East and West ,which are the fundamental points and the intermediate points of North east, Northwest, South east and South west. The compass is simple, cheaper and easy to construct but it shows very less number of heading directions and has slower response times. A tilt of greater than 12 degrees causes errors in heading output.[1] The E Gizmo compass is better than 1490 compass which has resolution of 0.5 degrees gives the values from 0 to 359.5, but it needs to be positioned correctly so that objects which disrupt magnetic field cannot causes faulty readings. The digital compass with R1655 sensors has a wider range of temperature tolerance (-40 to +85 degree Celsius) but its in-efficiency and inaccuracy makes it unsuitable for applications requiring higher accuracies. The Honeywell's HMC6352 used here has 2 axis magneto resistive sensors and needs only 2.5 - 5.2 V supply for battery operations having a compass accuracy of 2.5 to 3 degrees. HMC 6352 has various applications in fields like Consumer electronics, Hand held devices like mobile phones, General Compassing Applications, Integration with GPS in case of outdoor navigation technologies, Vehicle direction and telematics and Satellite positioning systems.

Here HMC6352 and a rotational encoder are interfaced with Arduino Uno which has an embedded ATmega328 microcontroller chip. The Arduino has inbuilt Wire library which helps in the I2C protocol communication [2] with the compass. The compass has SDA and SCL lines which can be connected to Arduino. There are various types of auto-navigation based positioning systems such as Wi-Fi (WPS), Bluetooth based systems etc. In indoor conditions, positioning using Wi-Fi signals is easier to implement using Android smart phones with specialized applications for localization[3]. But this method is consumes more energy and installation charges are high. In WPS, the Wi-Fi signal strength of each location is calculated with reference to the pre-selected points and tabulated. This is known to be the fingerprint of that location. During the positioning the current fingerprint is compared with the previously obtained ones and the position is located[3].In Bluetooth positioning, Bluetooth hotspots are used to avail information about the device, for instance the hotspot zone in which it is present. These devices are more accurate than WPS, but requires expensive geometrical computations[3] and are much more expensive.

IV. DESIGNAND IMPLEMENTATION

Our design utilizes an Arduino UNO board which is interfaced with the HMC6352 compass and a rotation encoder. The basic block diagram of the arrangement is shown in Fig 1. The microcontroller in Arduino UNO is ATmega328, which controls the actions of the compass and motor driver according to information provided by the rotation encoder. The whole arrangement can be fitted into any indoor navigation device. In this case we use a wheelchair with two 320W, 24V motors with 4600RPM. These motors are driven by two Hercules 6-36V, 16A motor drivers with PWM control. An Encoder disc is connected to the motor shaft, which is placed between the LED and light detector sections of the MOC7811 encoder IC. The output from the rotation encoder is fed back to the microcontroller. Additionally the digital compass IC HMC6352 is directly interfaced to the Arduino board via I2C interface lines.

A. ATMega Microcontroller

The microcontroller embedded within the Arduino UNO board is ATmega328. It is an 8-bit microcontroller developed by Atmel which follows the RISC architecture and supports serial communication protocols like USART, SPI and TWI (otherwise known as I2C). Here we need to use the I2C communication protocol to interface the digital compass with the microcontroller. Microcontrollers provides overall control by performing functions like send and receive data from the HMC6352, control the distance travelled by the device by utilizing data from the rotation encoder, measure and control direction in which device needs to travel using inputs from compass IC and control speed of device via motor drivers.

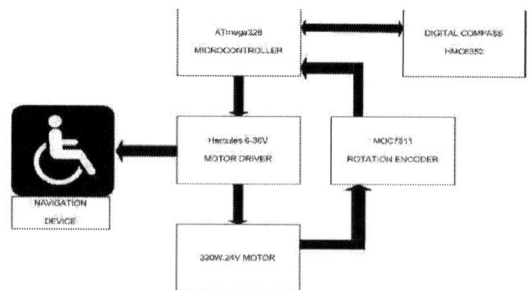

Fig. 1. Block diagram

2016 International Conference on VLSI Systems, Architectures, Technology and Applications (VLSI-SATA)

B. Rotation Encoder

In our design we can program the ARDUINO UNO to use the rotation encoder output to move our device to set distances, by counting the number of pulses per time. The circuit arrangement is shown in Fig2. A patterned disc connected to the motor shaft is placed between the diode and detector sections as shown in Fig3. When the disc rotates a square pulse is generated according to the patterns on the disc. From this we obtain the number of pulses, which in turn can be used to calculate the linear distance as given below. Hence to move the device through a given distance, say 1m, we have to program the Arduino to stop moving once the required number of pulses is obtained from the rotation encoder.

C. HMC6352 Digital Compass

The compass as mentioned before follows the I2C protocol and is directly interfaced with the microcontroller via SDA and SCL lines as shown in Fig4. The supply voltage provided by the Arduino is a nominal voltage of 3.3V. In I2C protocol, there are only 2 wires for the communication between the devices i.e. SDA (Serial Data Line) and the SCL (Serial Clock Line). The master Arduino controls the communication by sending the clock pulses through SCL. The default value of the SCL and SDA line are logic high. When there is a transition of high to low occurs in SDA line during SCL with logic high shows the start condition which initiates the start of the communication. I2C communication is byte oriented, where the address of the compass module is send by the Arduino which will be the hexadecimal value 43 initiating the read from the compass module and hexadecimal value 42 for write operations as default. So the 8-bit data send through the SDA line following the start bit consist of a 7-bit address and a read or write bit. The logic high in 8th bit of transmission initiates the read operation which would give us heading values in return. The 9th bit on the SDA is the Acknowledgement bit if it is logic high otherwise if it is a logic low shows a not acknowledgement signal (NACK) which determines whether the master slave communication is achieved or not. The transition of low to high in SDA with SCL in logic high initiates the stop condition terminating the communication.

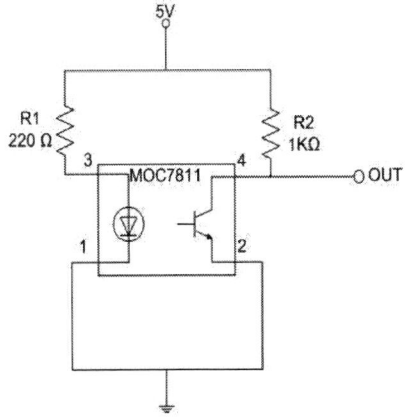

Fig. 2. Circuit diagram for detecting number of pulses using MOC7811 encoder IC

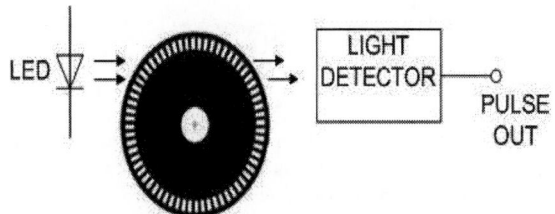

Fig. 3. Schematic of encoder disc placed between led and light detector section of MOC7811

D. Other Components

Test runs were conducted with 320W, 24V wheelchair motors which were driven by Hercules 6-36V, 16Amps motor drivers. The speed of the motors can be controlled by the drivers through a PWM input, maximum allowable frequency of which is 10 KHz. The speed required for our test device was 1.2km/hr. To control the motor we need three inputs to the motor from the driver. Two inputs to control the direction of rotation and one PWM input to control speed as shown in Fig5

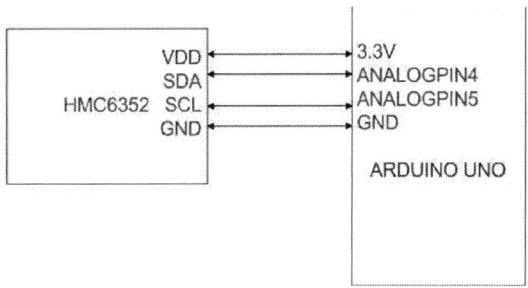

Fig. 4. Port diagram of HMC6352 and Arduino Uno

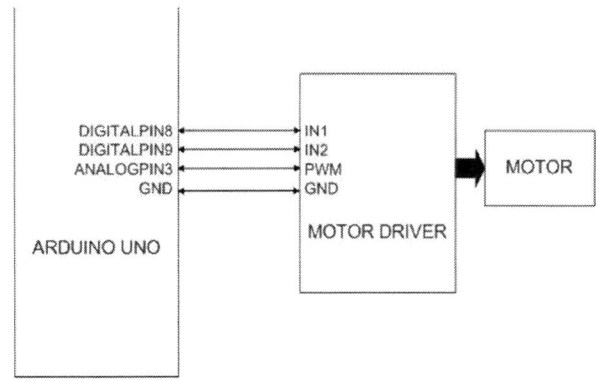

Fig. 5. Port diagram of Arduino Uno and Hercules motor driver

V. Experimental Results

A. Rotational Encoder with Arduino UNO

Test runs were conducted by using a wheelchair connected with two 320W, 24V motors driven by two Hercules 6-36V, 16A motor drivers. These motor drivers are controlled by the Arduino UNO board which in turn has a rotation encoder and digital compass interfaced to it. We designed a rotation encoder with diameter 6.5cms, with four leaves for initial testing. The accuracy of tests were high and were recorded as shown below.

The table 1 shows the readings obtained when Arduino Uno is interfaced with Hercules motor driver using the rotational encoder. We have discerned how to navigate any device to set distances via the rotation encoder. In Table 1, the expected distance to be travelled is compared with the obtained readings. From the table, it is found that the error percentage varies from a maximum of 1 to some negligible values(given in Table 1). Thus the accuracy obtained using the rotational encoder is higher than other navigation systems. The negligible amounts of error percentages can be ignored by careful calculations. Thus the rotational encoder based distance calculation is much more efficient.

TABLE I.   ROTATION ENCODER READINGS WITH ARDUINO UNO

No. of tries	Expected distance in cm	Distance travelled in cm	Error Percentage
1	100	101	1
2	150	149	-0.6667
3	100	99.5	-0.5
4	150	149.5	-0.333
5	160	161	0.625
6	130	129.5	-0.3846
7	200	201	0.5
8	180	179	-0.556
9	50	50.5	1
10	80	79	-1.25

TABLE II.   DIGITAL COMPASS READINGS WITH ARDUINO UNO

No. of tries	Expected direction	Degrees obtained	Error percentage
1	NORTH	359.2	-0.133
2	NORTH	359.1	-0.25
3	NORTH	359.6	-0.111
4	NORTH	0.8	0.2999
5	NORTH	1.9	0.599
6	SOUTH	179.8	-0.3889
7	SOUTH	181.6	0.889
8	SOUTH	179.8	-0.3889
9	SOUTH	180.56	0.889
10	SOUTH	180.1	0.3111
11	EAST	89.5	0.05556
12	EAST	91.8	2
13	EAST	89.9	-0.111
14	EAST	90.81	0.9
15	EAST	89.69	-0.3444
16	WEST	271.2	0.444
17	WEST	270.8	0.2563
18	WEST	269.3	-0.259
19	WEST	270.3	0.111
20	WEST	269.79	-0.778

B. HMC6352 with Arduino UNO

The digital compass HMC6352 is interfaced with the Arduino Uno to set orientation in any navigation device. The heading values obtained from the digital compass is converted to degrees in Arduino. The readings are taken in accordance with the cardinal reference points, north as 0 or 360 degrees, South as 180 degrees, East as 90 degrees and West as 270 degrees. The turning of device in any of these directions has to show the exact readings as above. The maximum error percentage obtained is from 2 to some negligible values (given in Table 2) which can be ignored for further calculations. Thus the orientation calculation using HMC6352 is simpler, cheaper and more accurate.

VI. Conclusion

Experiment results show that by using rotation encoder MOC7811 magnetic compass and HMC6352 we can effectively design a navigation platform which is cost effective, consumes much less energy and is not considerably affected by noise as compared to existing technologies. The precision of compass and encoder readings are fairly high with very low error percentage. With further development, like adding features like location mapping and obstacle detection and avoidance, our system can be commercially used in indoor navigation and manufacture of fully automated devices like auto-navigated wheelchair would be possible, which would be affordable without any sacrifices in terms of accuracy.

Acknowledgment

We wish to thank Almighty God who gave us the opportunity to successfully complete this venture. The authors wish to thank Amrita Vishwa Vidyapeetham University, Kollam Campus, Kerala for providing us support and funding to carry out this project. We also wish to thank the Humanitarian Technology Lab of ECE Dept., for aiding us in this endeavor.

References

[1] C. K. Agubor, G. N. Ezeh, M. Olubiwe, O.C. Nosiri, "Design and Implementation of a Simple HMC6352 2-Axis-MRDigital Compass" , International Journal of Emerging Technology and Advanced Engineering Website: www.ijetae.com (ISSN 2250-2459, ISO 9001:2008 Certified Journal, Volume 5, Issue 3, March 2015)

[2] Mr. J. J Patel, Prof B. H. Soni, "Design And Implementation Of I2C Bus Controller Using Verilog", ISSN: 0975 6779— Nov 12 To Oct 13 — Volume 02, Issue – 02

[3] Yutaka Arakawa , Yuki Sonoda , Koki Tomoshige , Shigeaki Tagashira and Akira Fukuda, "Implementation of WiFi/Bluetooth-based Smart Nar-row Field Communication", 2014 Seventh International Conference on Mobile Computing and Ubiquitous Networking (ICMU)

[4] Hong-Shik Kim and Jong-Suk Choi , "Advanced indoor localization using ultrasonic sensor and digital compass", International Conference on Control, Automation and Systems 2008 Oct. 14-17,2008 in COEX, Seoul,Korea

2016 International Conference on VLSI Systems, Architectures, Technology and Applications (VLSI-SATA)

Mathematical Modeling and Analysis of New Modified Glitch Free Adiabatic Inverter Circuit with Trapezoidal Power Supply

Alak Majumder, Rahul Kaushik

Department of Electronics and Communication Engineering
National institute of Technology Arunachal Pradesh
Yupia, India - 791112
majumder.alak@gmail.com, rahulkaushik415@gmail.com

Abstract—**The growing demand of low power electronics equipments has forced the research community to think of some methods by which energy of a circuit can be recycled and then adiabatic logic was born. The literature has experienced a numerous number of adiabatic circuits which faces a lot of complication in terms of glitch, noise and huge no. of transistors employed leading to the increase in device area. In this work, we have presented a new model of GFCAL circuit removing the diodes from its original architectures. The mathematical modeling of the energy of inverter circuit employing the new logic has been presented considering a Trapezoidal power supply considering it could be a good option for the adiabatic circuit as it will charge the load capacitor optimally giving lesser energy dissipation in ON path resistance. The simulation of the circuit is done in 180 nm process technology and compared the result with conventional CMOS.**

Keywords—Adiabatic circuit; GFCAL; power; energy; glitch

I. INTRODUCTION

With the advancement of technology no. of cores integrated in a single chip is increasing thereby leading to huge power loss which has become a threat for the low power circuit designer. There is considerable work going on in the low power design application throughout the world [1]. In the recent past, to reduce the circuit energy without degrading the circuit performance people has come up with an idea to recycle the energy stored in the load capacitance by sending it back to the main supply which is named Adiabatic logic [2, 3,4]. The word 'Adiabatic' comes from the fundamental of thermodynamic process in which no exchange of energy happens with the environments leading to the concept of no energy dissipation in the circuit. In VLSI circuits, the charge transfer between two nodes is considered as process and various techniques can be applied to minimize the energy during that process [5, 6]. The fully adiabatic process is an ideal phenomenon and may be achieved with very low switching speed. But, in real life application, energy dissipation based on charge transfer is based on adiabatic and non-adiabatic components which violets the zero energy computation statement [7].

In our investigation, we have proposed a new modified version of Glitch Free & Cascadable Adiabatic Logic (GFCAL) [8, 9] replacing the diodes available in the older circuits by the MOS transistors. Also, we are using Trapezoidal pulse as the power supply of the circuits. The mathematical modeling of the circuit is done to compute the total energy dissipation in a single cycle of the main supply. Measurement of dissipated power and energy is carried out through the circuit simulation using 180 nm technology file. It needs single phase power and supply and input and may be utilized in Cascadable and Hierarchical circuits.

II. OPERATION OF PROPOSED GFCAL INVERTER

Fig.1 shows the diode based Glitch Free and Cascadable Adiabatic Logic (GFCAL) structure. The modified structure of GFCAL is shown in fig. 2 where diodes D1 and D2 are replaced by M4 and M3 respectively.M1 & M3 creates the charging path providing current flow from supply to load capacitor (C1) and M2 & M4 creates the discharging path to have the reverse current flow towards supply which leads to recycling of energy.

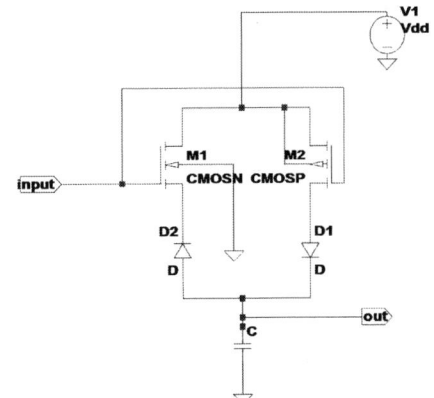

Fig .1 GFCAL inverter circuit

978-1-5090-0037-1/16 $31.00 © 2016 IEEE

2016 International Conference on VLSI Systems, Architectures, Technology and Applications (VLSI-SATA)

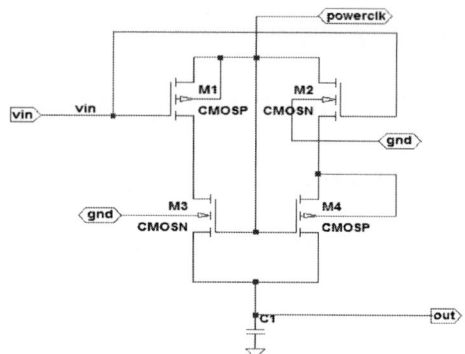

Fig.2: New modified GFCAL based Inverter Circuit

The operation of the proposed circuit will follow the following cases -

A. 1st Case: Logic 0 input with load capacitor uncharged

In this case, transistor M1 will be ON and when the trapezoidal power supply will be in transition from 0 to V_{DD}, at some point of time M3 will be ON creating a path to current flow from the supply to the load capacitor to generate logic 1 at the output.

B. 2nd Case: Logic 1 input with load capacitor uncharged

In this case, M2 gets ON and during the transition of power supply from V_{DD} to 0, somewhere M4 becomes ON leading to a current flow towards the supply leaving logic 0 at the output. But, a small amount of leakage may interfere to charge the capacitor during the transition of supply from 0 to V_{DD} as M3 gets ON in that phase.

C. 3rd Case: Logic 0 input with load capacitor charged

In this case, M1 will be On & M3 also gets ON during the supply transition from 0 to V_{DD} leading to the charging of capacitor which means logic 1 at the output. During the supply transition from V_{DD} to 0, M3 will be OFF somewhere preventing the discharge.

D. 4th Case: Logic 1 input with load capacitor charged

In this case, M2 will initially be ON and during the supply transition from V_{DD} to 0, M4 also becomes ON to create a path of current pumping back to the supply which represents discharging of capacitor or logic 0.

III. MATHEMATICAL MODELING OF ENERGY

This new approach tells that when both NMOS (M3) and PMOS (M1) are ON which are connected in series acts as a resistance of value 'R' and the process of charging and discharging have same resistive paths. It's true that during charging and discharging phase all MOS are not completely ON but here we are considering same resistance value of 'R'.

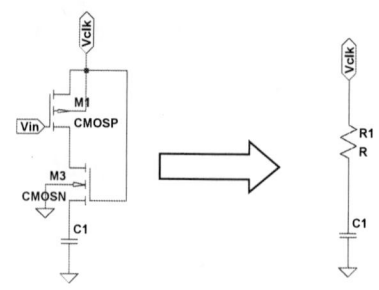

Fig.3: Charging path resembles a RC circuit

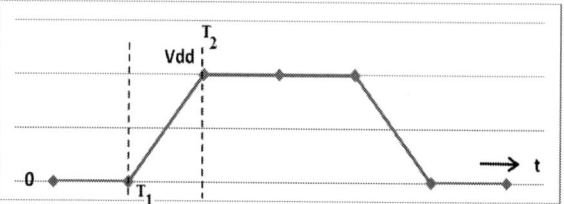

Fig. 4: Trapezoidal Power Supply (V_{CLK})

A. Charging Phase

As shown in Fig. 1 M1 is ON when input is low and as V_{CLK} increases from 0 to V_{DD} the capacitor gets charge through M3. Let's consider, T is the rise time and fall time of V_{CLK} and the equation of the V_{clk} can be written as:

$$V_{clk}(t) = \frac{V_{dd} \times t}{T} \tag{1}$$

Let M3 is ON after the voltage becomes V_{th} and capacitor (C_L) has an initial voltage V_C due to leakage. So, the voltage V_C across the load capacitance after T_{th} (the time V_{clk} reaches V_{th}) -

$$\frac{dV_c}{dt} + \frac{1}{RC_L} V_c = \frac{1}{RC_L} \times \frac{V_{dd} \times t}{T} \tag{2}$$

The solution of the above equation is given by:

$$V_C = \frac{V_{dd}}{T}(t - RC_L) + V_o \times e^{\frac{-t}{RC_L}} \tag{3}$$

The capacitor current can be derived from by differentiating equation (3). So, we have

$$i_c = C_L \frac{V_{dd}}{T} - \frac{V_0}{R} e^{\frac{-t}{RC_L}}. \tag{4}$$

B. Energy Dissipation During Charging

The magnitude of energy can be calculated as:

$$E_{Charge} = \frac{1}{R}[E' - E''] \tag{5}$$

Where, $E' = (\frac{V_{dd}}{T})^2 \times \frac{t^3}{3}$ (6)

And

978-1-5090-0037-1/16 $31.00 © 2016 IEEE   198

$$E'' = (\frac{V_{dd}}{T})^2 \times \frac{t^3}{3} - RC_L(\frac{V_{dd}}{T})^2 \times \frac{t^2}{2}$$

$$- \frac{V_{dd} \cdot V_o}{T} RC_L e^{\frac{-t}{RC_L}}(t + RC_L) \tag{7}$$

Substituting the values of equation (6) & equation (7) in equation (5), we have

$$E_{Charge} = \frac{1}{2}C_L(\frac{V_{dd}}{T})^2 t^2 + \frac{V_{dd}V_oC_L}{T}e^{\frac{-t}{RC_L}}(t + RC_L) \tag{8}$$

Equation (8) is valid between the time T_1 and T_2 when charging path is ON. Hence,

$$E_{Charge} = [\frac{1}{2}C_L(\frac{V_{dd}}{T})^2 t^2 + \frac{V_{dd}.V_oC_L}{T}e^{\frac{-t}{RC_L}}(t + RC_L)$$
$$]t=T_2$$

$$- [\frac{1}{2}C_L(\frac{V_{dd}}{T})^2 t^2 + \frac{V_{dd}V_oC_L}{T}e^{\frac{-t}{RC_L}}(t + RC_L)]t=T_1 \tag{9}$$

C. Discharging Phase

Transistor M2 is ON when input is high (logic 1) and when power clock is falling from V_{DD} to 0, the load capacitor discharges through M4 and M2. The capacitor voltage is given by

$$\frac{dV_c}{dt} + \frac{1}{RC_L}V_c = \frac{1}{RC_L}V_{dd}(1 - \frac{t}{T}) \tag{10}$$

The equation (11) is linear differential equation which gives

$$V_c = V_{dd} - \frac{V_{dd}}{T}(t - RC_L) + V_1 e^{\frac{-t}{RC_L}} \tag{11}$$

The current across capacitor can be calculated as;

$$i_c = -C_L[\frac{V_{dd}}{T} + \frac{V_1}{RC_L}e^{\frac{-t}{RC_L}}] \tag{12}$$

D. Energy Consumed During Discharging

The energy during discharge is given by

$$E_{Discharge} = \frac{1}{R}[E_1 - E_2] \tag{13}$$

Where, $E_1 = V_{dd}^2(t - \frac{t^2}{T} + \frac{t^3}{3T^2}) \tag{14}$

And $E_2 =$

$$V_{dd}^2 t - \frac{V_{dd}^2}{T}t^2 + (\frac{V_{dd}}{T})^2\frac{t^3}{3} + \frac{V_{dd}^2}{T}RC_L t - \frac{RC_L}{2}(\frac{V_{dd}}{2})^2 t^2 -$$

$$RC_LV_1V_{dd}e^{\frac{-t}{RC_L}} + RC_L\frac{V_1.V_{dd}}{T}e^{\frac{-t}{RC_L}}(t + RC_L) \tag{15}$$

Substituting the values of E_1 and E_2 in the equation (13), we have

$E_{Discharge} =$

$$V_{dd}^2 t - \frac{V_{dd}^2}{T}t^2 + (\frac{V_{dd}}{T})^2\frac{t^3}{3} + \frac{V_{dd}^2}{T}RC_L t - \frac{RC_L}{2}(\frac{V_{dd}}{2})^2 t^2 -$$

$$RC_LV_1V_{dd}e^{\frac{-t}{RC_L}} + RC_L\frac{V_1.V_{dd}}{T}e^{\frac{-t}{RC_L}}(t + RC_L) \tag{16}$$

Equation (16) is valid from T_3 and T_4 when discharge path is ON. So,

$$\therefore E_{discharge} = [-C_L\frac{V_{dd}^2}{T}.t + \frac{C_L}{2}(\frac{V_{dd}}{T})^2 t^2 + C_LV_1V_{dd}e^{\frac{-t}{RC_L}}$$

$$- \frac{C_LV_1V_{dd}}{T}e^{\frac{-t}{RC_L}}(t + RC_L)]_{at\ t=T_4} -$$

$$[-C_L\frac{V_{dd}^2}{T}.t + \frac{C_L}{2}(\frac{V_{dd}}{T})^2 t^2 + C_LV_1V_{dd}e^{\frac{-t}{RC_L}}$$

$$- \frac{C_LV_1V_{dd}}{T}e^{\frac{-t}{RC_L}}(t + RC_L)]_{at\ t=T_3} \tag{17}$$

Combining equation (9) and equation (17) we may get the total energy of the Glitch Free Adiabatic Inverter circuit. So, total energy –

$$E_{Total} = E_{Charge} - E_{Discharge} \tag{18}$$

The time T_1 at which the charging takes place can be calculated as

$$T_1 = \frac{V_{th} \times T}{V_{DD}} \tag{19}$$

Where, T is the rise time of the Power Clock. T_2 can also be calculated as

$$T_2 = \frac{(V_{DD} - V_{th}) \times T}{V_{DD}} \tag{20}$$

Equation (20) is valid if capacitance value is so small such that the charging of capacitor follows the rise time. If the the charging and discharging time constant is much larger in compared to the rise time, one has to look for the value of the output voltage at the peak of the power clock (V_{DD}). The charging will continue during the flat portion of trapezoidal

2016 International Conference on VLSI Systems, Architectures, Technology and Applications (VLSI-SATA)

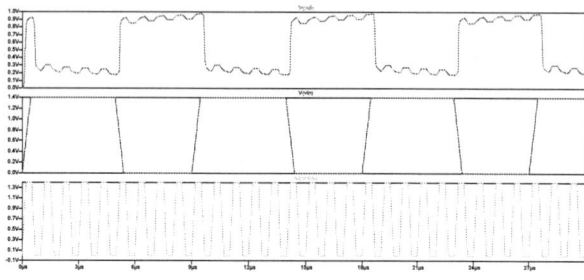

Fig. 5: Output waveform of new GFCAL inverter circuit (Top is output, middle is input and bottom is Trapezoidal power supply)

power clock till the voltage on the load capacitor is (V_{DD} - V_{th}). In the similar way, T_3 and T_4 have to be evaluated. Once T_1, T_2, T_3 and T_4 are calculated, E_{Charge} and $E_{Discharge}$ can be found out. The difference between these two will give the total energy of the circuit for one charge-discharge cycle.

IV. SIMULATION RESULTS AND ANALYSIS

In this section, the simulation results of the new GFCAL inverter circuit are analyzed. The performance of operation of the circuit is discussed in terms of how much ripple/glitches it produces and power & energy analysis of the circuit. The output of the inverter circuit is shown in Fig. 5.

We know from the literature of adiabatic that one of the main problems is the ripple or glitch, which gets generated in both logic 0 and logic 1 level of the circuit output. Our circuit has got a very little amount of glitches which can be seen from the waveform in Fig. 5.

A. Glitch Analysis

Load capacitor will take time to charge or discharge based on its capacitance value. The more the capacitance is, the less rate of charging and discharging it will have. The capacitance model to explain the ripple/glitch is shown below in Fig. 5.

In Fig. 6, 'C_{gs}' is the coupling capacitance of both the controlling transistors (M1 & M3) and 'C_load' is the load capacitance across which output is taken. As Capacitance is inversely related to reactance, the applied voltage V_{clk} is distributed among the two capacitors in an inverse ratio of their capacitance. Hence, it can be written that,

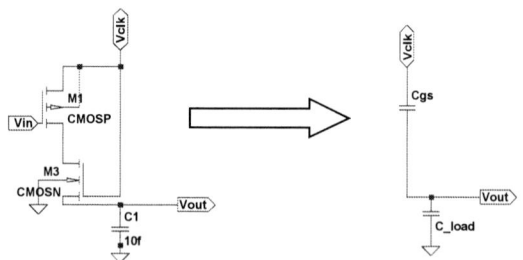

Fig. 6: Capacitance model of glitch effect

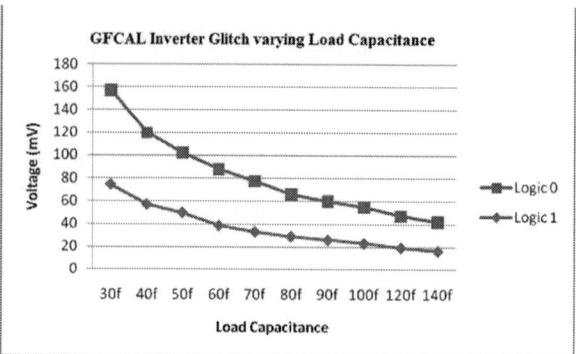

Fig. 7: Glitch analysis varying load capacitance

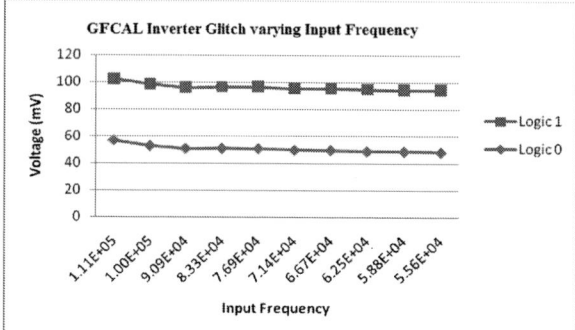

Fig. 8: Glitch analysis with Input frequency variation

$$\text{Glitch/ Ripple Voltage} = \frac{C_{gs}}{C_{Load}} \times V_{DD} \qquad (21)$$

Where, V_{DD} is the peak value of Trapezoidal Power Supply. So, a trade-off is needed in selecting coupling and load capacitance. A higher load capacitance will give almost no ripple, but it will reduce the operational speed. Thus, a n optimum value of load capacitance is to be chosen which is higher than Cgs (Device capacitance) in order to keep ripple minimum with optimal speed of circuit. Here, the glitch at logic 0 and logic 1 level has been analyzed with varying load capacitance and Input frequency as shown in Fig. 6 and Fig. 7.

From Fig. 7 it is observed that the logic circuit follows as per equation (21) as glitch is getting decreased with the increasing load capacitance. But, Glitch should be almost constant if the input frequency is varied (refer Fig. 8), as it has no effect on the generation of glitch, which mainly depends on device capacitance & load capacitance along with the peak value of V_{Clk}.

B. Power Analysis

Regarding the power of GFCAL inverter circuit, it should be almost constant with the variation of load capacitance, as because the energy is getting recycled from the load capacitance. But, there may be a small increase in the power dissipation with increasing capacitance which is due to the dissipation across switches. The power analysis has been done

978-1-5090-0037-1/16 $31.00 © 2016 IEEE   200

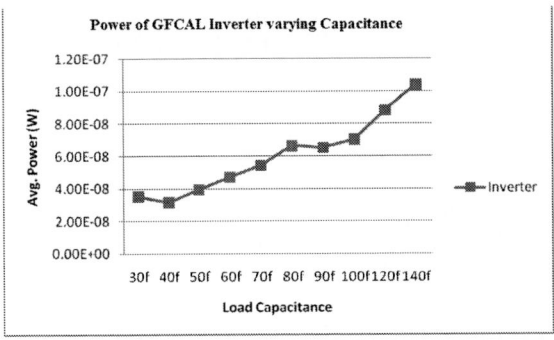

Fig. 9: Power plot of inverter varying load capacitance

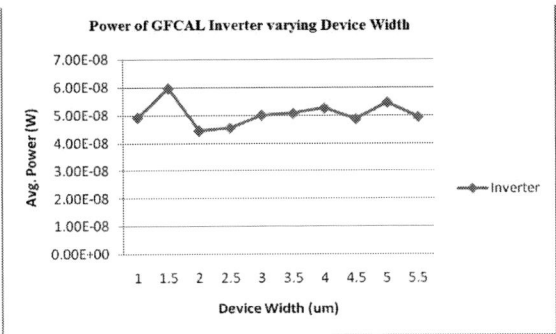

Fig. 10: Power plot of inverter varying device width

with respect to variation of load capacitance and variation in device width as shown in the following figures.

C. Energy Analysis

As the energy in getting recycled there will be less change in its value with the variation of load capacitance. The analysis of energy is plotted against variation of load capacitance and variation in device width in the following figures.

From Fig. 11 & 12, we can observe that the energy of the inverter circuit lies in femto-Joule (fJ) range which is much lesser than the conventional CMOS circuits.

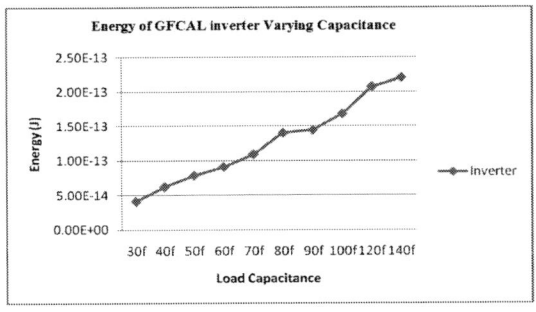

Fig. 11: Energy plot of inverter varying load capacitance

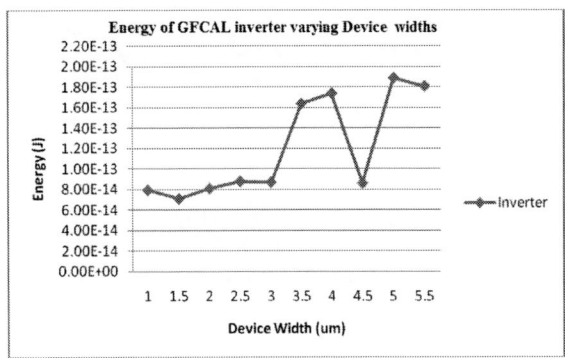

Fig. 12: Energy plot of inverter varying device width

V. CONCLUSION

In this work, a new Glitch Free Adiabatic Logic has been proposed replacing two diodes by two transistors M3 & M4 (see Fig. 1) from its old version. The circuit has been made functional using a trapezoidal power supply as it may be a good option to charge load capacitor optimally giving lesser energy in the ON path resistance and designed it in 180 nm process Technology. Mathematical modeling of the total Energy per cycle of the inverter circuit is derived and discussed. We have analyzed the glitch effect of the circuit using a capacitance model and that is verified by the simulation result which shows a huge reduction in glitch/ripple voltage. Moreover, power and energy analysis has been performed varying the load capacitance and device width separately which describes that energy has been reduced drastically as it is ranging in a few femto-Joule (fJ).

REFERENCES

[1] John P. Uyemura, "CMOS Logic Circuit Design", Kluwer Academic Publishers, 2002.

[2] P. Teichmann, "Adiabatic Logic: Future Trend and system Level Perspective", Springer series in advanced microelectronics 34, 2012.

[3] John S. Denker "A Review of adiabatic computing", IEEE symposium on Low Power electronics, 1994 .

[4] M. Sanadhya, M.V. Kumar, "Recent developments in efficient adiabatic logic circuits and power analysis with CMOS logic", Procedia Computer Science, Vol.57, pp-1299-1307, Elsevier, 2015.

[5] W.C. Athas, Lars J. Svensson, "Low power digital system based on adiabatic Principles " , IEEE transaction on very large scale intregation systems, vol.2, No.4, December 1994.

[6] Young Moon and Deog-kyoon Jeong, "An efficient charge recovery Logic Circuits", IEEE Journal of solid state circuits, vol. 31, April 1996.

[7] J. Park, S.J. Hong, J. Kim, "Energy saving design technique achieved by latched pass transistor adiabatic logic", IEEE Int. sym. on circuits & systems (ISCAS), 2005.

[8] N.S.S Reddy, M. Satyam, K. L .Kishore, " Cascadable Adiabatic logic circuits for low power applications", IET circuits device and syst.,vol.2.pp 518-526, 2008.

[9] S.Upadhay, R.K.Nagaria, R.A.Mishra, "Performance improvement of GFCAL circuits " International Journal of computer applications, vol. 78-No.5 ,September 2013.

Design and Implementation of Reconfigurable Coders for Communication Systems

Manikandan J, Shruthi S, Mangala SJ and Agrawal VK
Crucible of Research and Innovation (CORI), PES University,
100-Feet Ring Road, BSK Stage III, Bangalore 560085, Karnataka, India
{email : manikandanj@pes.edu, vk.agrawal@pes.edu}

Abstract – **In this paper, a novel attempt is made to design a reconfigurable coder system which can be reconfigured on-the-fly to work either as an encoder, or decoder, or both encoder and decoder depending on the user requirements. In order to build the proposed reconfigurable system, Convolutional encoder, Viterbi decoder, Golay encoder and Golay decoder are employed in different combinations for the proposed design. The proposed system is implemented on a Virtex-5 FPGA and the performance of the system with and without reconfigurable architecture are reported. It is observed that 56.36% of hardware resources and 72.21% of power are saved on using reconfigurable architecture over non-reconfigurable architecture. The proposed system can be easily extended to include various other encoding and decoding schemes.**

Keywords – Convolutional; Decoders; Encoders; FPGA; Golay; Reconfigurable; Viterbi.

I. INTRODUCTION

Digital Communication is extensively used for a large number of applications which includes deep space communication, satellite communication, mobile and wireless communication, ultra wideband applications and many more. The main aim of digital communication is to provide error-free data transmission and the main advantage of using digital signal is that the errors introduced by noise during transmission can be detected and possibly corrected. Encoders and decoders are considered as important blocks of any digital communication system. Encoders are used on the transmitter side of the system for the purpose of standardization, speed, secrecy, security and compression. The digitized data on the transmitter side is processed or encoded, followed by modulation and transmission over the medium. Decoders are used on the receiver side of the system wherein the data is down converted and decoded for further processing to retrieve the actual message signal. Every encoder has its corresponding decoder and usually an encoder-decoder pair is used for designing a communication system.

Convolutional coding is one of the most commonly used encoder and Viterbi decoder is used for decoding. Convolutional coding and Viterbi decoder have been successfully used for various applications including satellite communication, mobile communication, deep-space optical communication, digital broadcasting systems, Code Division Multiple Access (CDMA), Ultrawideband (UWB) signalling and many more. This has motivated the hardware implementation of convolutional coding-Viterbi decoder pair.

Similarly block codes such as Golay encoder and decoder are considered as efficient coders for forward error correction with an ability to detect four errors and correct upto three errors. Golay encoders and decoders are simple to implement and are hence used for various applications such as multiple error correcting codes in [1], for Synthetic Aperture Radar (SAR) imaging systems in [2], and for wireless communication of Software Defined Radio (SDR) in [3]. The hardware implementation of Golay encoder-decoder pair are reported in [4,5].

It is observed that FPGAs have come a long way from mere glue-logic applications to design of on-board satellite systems in [6] and design of reconfigurable systems in [7] due to their higher level of flexibility. Reconfigurable systems increase the redundancy and reliability of a system, as they are capable of reconfiguring the system to perform the task of a faulty system on-board to successfully complete missions that are criticial as required for satellite systems, aircraft system and other defense related systems [8]. Design of reconfigurable encoders and decoders for High-Data-Rate Satellite Communications is reported in [9]. Design of reconfigurable Viterbi decoder for wireless satandards is reported in [10].

In this paper, design and implementation of reconfigurable coders using Xilinx make SRAM based Virtex-5 FPGA is proposed, wherein the functionality of system as an encoder or decoder, or both are reconfigured on-the-fly based on the requirements of user. Implementation and evaluation of proposed system using five different architectures are reported in the paper. The system is also capable of reconfiguring itself to two different encoding and decoding schemes as illustrated in Fig. 1, wherein it may be observed that the proposed reconfigurable system can be easily reconfigured for usage with any application that demands these encoding and decoding schemes.

The organization of the paper is as follows: Section II gives an overview of the coders used for proposed work. Section III explains the implementation of different reconfigurable architectures for the proposed design. Experimental results and observations of the proposed design are reported in Section IV followed by conclusion and references.

978-1-5090-0037-1/16 $31.00 © 2016 IEEE

2016 International Conference on VLSI Systems, Architectures, Technology and Applications (VLSI-SATA)

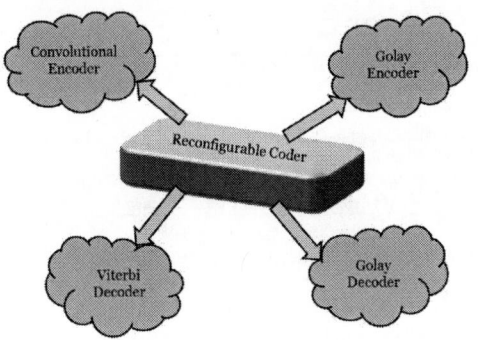

Fig. 1. Functionality of proposed Reconfigurable Coding System

II. OVERVIEW OF CODERS FOR PROPOSED WORK

In this section, a brief overview of different coders used for proposed work is reported, which includes Convolutional encoder, Viterbi decoder, Golay encoder and Golay decoder.

A. Convolutional Encoder

The ability to perform economical soft decision decoding on convolutional codes, as well as the flexibility to change block length and code rate has made convolutional coding very popular for digital communications. Convolutional encoder is mainly characterized using three parameters i.e., k - the number of input bits, n - the number of output bits and K - the constraint length. The code rate for transmission is defined as $r = k/n$. A generalized $[n,k,K]$ convolutional encoder has K-1 shift registers. The convolutional encoder used for proposed work is shown in Fig. 2, where $k=1$, $K=3$ and $n=2$. Two generator polynomials are used which are given as $g^0 = 101$ and $g^1 = 111$. Two outputs V_0 and V_1 are generated using the generator polynomials as given in (1) and (2).

$$V_0 = u_1 \oplus u_3 \qquad (1)$$
$$V_1 = u_1 \oplus u_2 \oplus u_3 \qquad (2)$$

where u_1 represents the input to the encoder, u_2 and u_3 represents the output from first flip flop and second flip flop respectively.

Convolutional encoder can be applied to either continuous stream of data or block of data. The message bits are shifted into shift registers k bits at a time. Since $K=3$, there are two shift registers and the encoder has $2^{K-1} = 4$ states. The state diagram for the convolutional encoder used in the proposed work is shown in Fig. 3. More details about convolutional encoder can be had from [11].

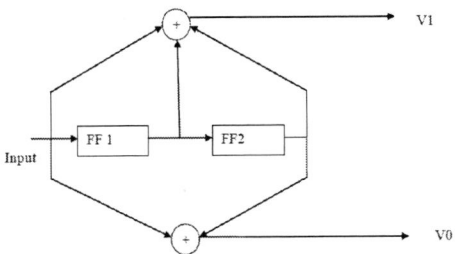

Fig. 2. Convolutional Encoder used for proposed work

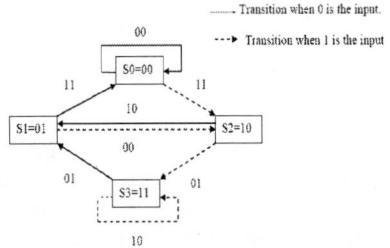

Fig. 3. State diagram of Convolutional Encoder designed

B. Viterbi Decoder

The Viterbi decoder is an optimum maximum likelihood algorithm proposed by A.J.Viterbi to decode the convolutional codes for constraint length $K \leq 15$. The Viterbi decoder comprises of three basic units – Branch metric unit, Path metric unit and Trace back unit. Branch metric is the hamming distance from current state to next state due to change in input. Path metric unit summarizes branch metrics to get metrics for 2^{K-1} paths, one of which can eventually be chosen as optimal. Fig. 4 illustrates the branch metric and path metric calculation for the proposed work. The results from these decisions are written to the memory of a trace back unit and the trace back unit restores an almost maximum-likelihood path from the decisions made by the Path metric unit. Viterbi decoding is best envisaged using a trellis diagram as shown in Fig. 5 which contains the information of the state diagram and uses time as a horizontal axis to show the possible paths through the states. More details about Viterbi decoder can be had from [11].

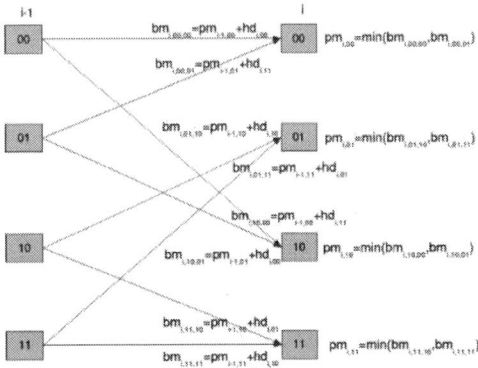

Fig. 4. Branch Metric and Path Metric Calculation for Viterbi Decoder

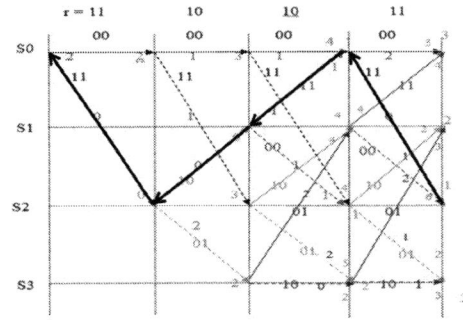

Fig. 5. Trellis diagram for Viterbi Decoder

978-1-5090-0037-1/16 $31.00 © 2016 IEEE   203

C. Golay Encoder

Golay encoder is one among the most widely used block code where a continuous stream of data is divided into blocks of 12 bits and each block is processed, one at a time. The two closely related Golay codes are extended binary Golay code and perfect binary Golay code. The extended binary Golay code, represented as G_{24} encodes 12 bits of data in a 24-bit word in such a way that any 3-bit errors can be corrected or any 7-bit errors can be detected, whereas the perfect binary Golay code, represented as G_{23} has codewords of length 23 and is obtained from the extended binary Golay code by deleting one coordinate position. The standard code notation for G_{24} and G_{23} have parameters [24,12,8] and [23,12,7] corresponding to the length of the codewords, the dimension of the code, and the minimum Hamming distance between two codewords respectively. The extended binary Golay code G_{24} is used for the proposed work, which has a generator matrix, G given as

$$G = [P \quad I] \qquad (3)$$

where I is the 12x12 identity matrix and P is the Parity matrix.

The encoded output, V of Golay encoder is given as

$$V = U * G \qquad (4)$$

where U is the 1x12 input block and the size of encoded output is 1x24. More details about Golay Encoder can be had from [11,12].

D. Golay Decoder

Golay decoder is a decoding algorithm for Golay codes which can detect upto 7 errors and correct upto 3 errors. Golay decoder uses the same parity matrix as in Golay encoder to obtain a parity check matrix, H given as

$$H = [I \quad P] \qquad (5)$$

The decoding algorithm consists of determining the error pattern $\mathbf{u} = \mathbf{v} + \mathbf{w}$, where $\mathbf{w}$ denotes the vector received and $\mathbf{v}$ is the nearest to $\mathbf{w}$ code vector. Let $\mathbf{wt(x)}$ denote the weight of the vector $\mathbf{x}$ i.e., the number of "ones" in $\mathbf{x}$, $\mathbf{p_i}$ denote the i-th row of the parity matrix P, $\mathbf{e_i}$ denote the word of length 12 with 1 in the i-th position and 0 elsewhere. Once $\mathbf{u}$ is determined, the corrected received vector will be $\mathbf{v} = \mathbf{u} + \mathbf{w}$, and the last 12 elements of v will be the input block fed to Golay encoder. The steps for Golay decoding algorithm [12] is given as a pseudocode in Fig. 6. More details about the decoding algorithm can be had from [11].

III. RECONFIGURABLE ARCHITECTURES DESIGNED FOR PROPOSED WORK

In this section, five different architectures employed for proposed reconfigurable coder systems are discussed. Reconfigurable systems are defined as systems which are capable of reconfiguring itself on-the-fly to perform different functionalities. Reconfiguration trigger is defined as the signal used to initiate reconfiguration of the system and the different approaches for generating these reconfiguration triggers include RS232 based reconfiguration, discrete input based reconfiguration, event based reconfiguration, timer based auto reconfiguration [7]. Fig. 7 shows Architecture I of reconfigurable coder which can reconfigure itself to function as either Convolutional encoder or Viterbi decoder. Architecture

Fig. 6. Pseudocode for Golay Decoder

Step 1: Compute the syndrome $\mathbf{s} = \mathbf{wH^T}$
Step 2: If $\mathbf{wt(s)} \leq 3$ then $\mathbf{u} = [\mathbf{s}, 000000000000]$.
Step 3: If $\mathbf{wt(s+p_i)} \leq 2$ for some $\mathbf{p_i}$ of P, then $\mathbf{u} = [\mathbf{s+ p_i}, \mathbf{e_i}]$.
Step 4: Compute the second syndrome $\mathbf{sP}$.
Step 5: If $\mathbf{wt(sP)} \leq 3$ then $\mathbf{u} = [000000000000, \mathbf{sP}]$.
Step 6: If $\mathbf{wt(sP+p_i)} \leq 2$ for some $\mathbf{p_i}$ of P, then $\mathbf{u} = [\mathbf{e_i}, \mathbf{sP+ p_i}]$.
Step 7: If $\mathbf{u}$ is not yet determined then request retransmission.

II is shown in Fig. 8 which is identical to Architecture I with the only difference that this system can reconfigure itself to function as either Golay encoder or Golay decoder. Fig. 9 shows Architecture III where the reconfigurable system is capable of reconfiguring itself to perform as one among the four schemes. Architecure IV shown in Fig. 10 is capable of providing two functionalities at a time and can reconfigure itself to select different combinations of encoder and decoder. Architecture V shown in Fig. 11 is identical to Fig. 10 wherein the encoder output is given to decoder input and at any instance the system can reconfigure itself to function as an encoder-decoder pair.

It may be observed from Fig. 7 to Fig. 9 that, only one reconfigurable block is selected for the system and hence only one functionality can be achieved at any instance. This approach reduces the hardware resources required which inturn reduces the power consumption and is employed for applications where only one scheme is required at any instance.

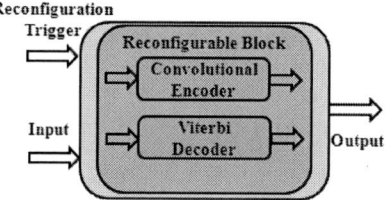

Fig. 7. Architecture I of Proposed Reconfigurable Coder System

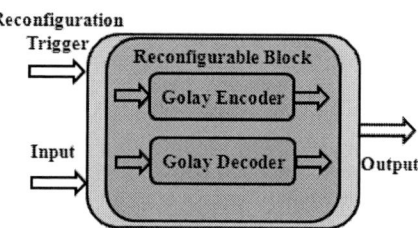

Fig. 8. Architecture II of Proposed Reconfigurable Coder System

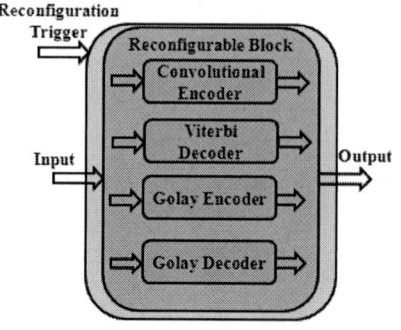

Fig. 9. Architecture III of Proposed Reconfigurable Coder System

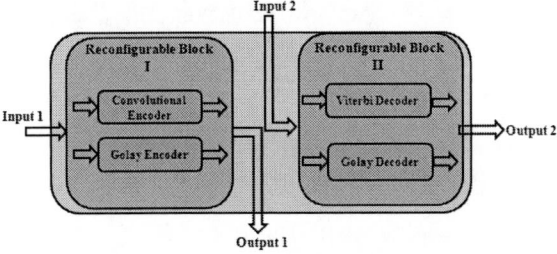

Fig. 10. Architecture IV of Proposed Reconfigurable Coder System

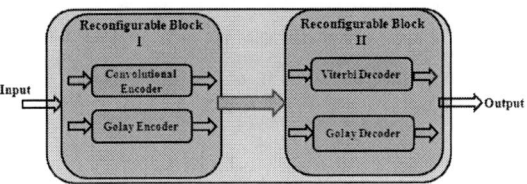

Fig. 11. Architecture V of Proposed Reconfigurable Coder System

On the other hand, it may be observed from Fig. 10 and Fig. 11 that two reconfigurable blocks are employed which enables two schemes to function simultaneously and thus the hardware resources required and power consumption will be greater than that required for system designed using single reconfigurable block. The number of reconfigurable modules in each reconfigurable block can be increased depending on the user requirements.

IV. EXPERIMENTAL RESULTS

The proposed reconfigurable system is designed and implemented using Xilinx make Virtex-5 XC5VFX70T FPGA based ML507 evaluation board. The hardware resources required for the implementation of four different schemes mentioned in Section II are reported in Table I. Details about the hardware resources saved on using reconfigurable architecture over non reconfigurable architecture for different types of architectures reported in Section III are given in Table II to Table V. It is observed from Table II, III and IV that as the number of reconfigurable modules in a reconfigurable block increases, the amount of hardware resources saved also increases. It is observed from Table V that a maximum of 24.05% of hardware resources are saved on using two reconfigurable blocks, which comes with a benefit that the system can reconfigure itself to perform two functionalities at any instance.

TABLE I. HARDWARE RESOURCES REQUIRED BY DIFFERENT SCHEMES

Resources	Conv Enc	Viter Dec	Gol Enc	Gol Dec
Slices	36	35	109	145
LUTs	40	42	211	439
Used LUT-FF pairs	35	35	97	124

TABLE II. HARDWARE RESOURCES SAVED ON USING ARCHITECTURE I

Resources	Non Reconfig. Arch	Arch I	% Saved
Slices	71	36	49.30%
LUTs	82	42	48.78%
Used LUT-FF pairs	70	35	50.00%

TABLE III. HARDWARE RESOURCES SAVED ON USING ARCHITECTURE II

Resources	Non Reconfig. Arch	Arch II	% Saved
Slices	254	145	42.91%
LUTs	650	439	32.46%
Used LUT-FF pairs	221	124	43.89%

TABLE IV. HARDWARE RESOURCES SAVED ON USING ARCHITECTURE III

Resources	Non Reconfig. Arch	Arch III	% Saved
Slices	325	145	55.38%
LUTs	732	439	40.02%
Used LUT-FF pairs	291	127	56.36%

TABLE V. HARDWARE RESOURCES SAVED ON USING ARCHITECTURE IV & V

Resources	Non Reconfig. Arch	Arch IV and Arch V			% Saved
		Block1	Block2	Total	
Slices	325	109	145	254	21.85%
LUTs	732	211	439	650	11.20%
Used LUT-FF pairs	291	97	124	221	24.05%

It should also be noted that the reconfigurable based design comes at an expense of implementing soft-processor called Microblaze onto FPGA which again consumes hardware resources as given in Table VI. This becomes negligibly small when the hardware resources required by the modules to be reconfigured is large or when the number of modules to be reconfigured increases. Details about the power consumption for proposed system using different architectures is obtained from XPower Analyzer and are reported in Table VII. It is observed that on using reconfigurable architecture, a maximum of 72.21% of power is saved.

Details about the size of bitstream file and reconfiguration time for proposed work is reported in Table VIII. The reconfiguration time is defined as the time taken by FPGA to reconfigure and is given as

$$T_{reconfig} = \frac{B_{size}}{Bw_{Max}} \qquad (6)$$

TABLE VI. HARDWARE RESOURCES UTILIZED BY THE SOFT PROCESSOR

Resources	Available	MicroBlaze
#Slice Registers	44800	2850
#Slice LUTs	44800	2690
#Block Memory	148	17
#DSP48Es	128	3
#Clock Manager	18	0
#Bonded IOBs	640	0
#BUFG/BUFGCTRLs	32	2

TABLE VII. POWER ANALYSIS REPORT FOR PROPOSED WORK

Module/Architecture	Power	Power Saved
Convolutional Encoder	1.423 W	--
Viterbi Decoder	1.423 W	--
Golay Encoder	1.423 W	--
Golay Decoder	1.423 W	--
Architecture I	1.607 W	43.53%
Architecture II	1.591 W	44.10%
Architecture III	1.603 W	43.68%
Architecture IV	1.597 W	71.94%
Architecture V	1.582 W	72.21%

TABLE VIII. DETAILS OF CONFIGURATION FILE FOR PROPOSED WORK

Module/Architecture	Size	$T_{reconfig}$
Convolutional Encoder	3300 kB	--
Viterbi Decoder	3300 kB	--
Golay Encoder	3300 kB	--
Golay Decoder	3300 kB	--
Architecture I	777kB	242 μs
Architecture II	1046kB	326 μs
Architecture III	1247kB	389 μs
Architecture IV	575kB	179 μs
	644kB	201 μs
Architecture V	621kB	194 μs
	432kB	135 μs

where B_{size} denotes the size of bitstream or configuration file for reconfiguration modules and Bw_{Max} denotes the maximum bandwidth of the configuration mode which is given as 3.2Gbps at 100MHZ clock rate for the ICAP configuration mode, which is used in the proposed work.

The FPGA floorplan for the proposed reconfigurable system using single reconfigurable block is shown in Fig. 12(a) and using two reconfigurable blocks in Fig.12(b). The partitioning of FPGA into reconfigurable area and non-reconfigurable area is illustrated in Fig. 12(a) and 12(b), wherein the pink rectangle denotes the reconfigurable p-block allocated for reconfigurable modules as described in Section III. It may be observed from Fig. 12(a) that there is only one reconfigurable block, whereas Fig. 12(b) has two reconfigurable blocks. The remaining part of FPGA is used by the non-reconfigurable part of the system. Details about the resources used and the routing inside FPGA for both the approaches are also shown in Fig. 12.

V. CONCLUSION

In this paper, design and implementation of reconfigurable coder system on a SRAM based Xilinx Virtex-5 FPGA is proposed using Convolutional encoder, Viterbi decoder, Golay encoder and Golay decoder schemes as reconfigurable modules. Different architectures of implementing the reconfigurable system is discussed and results are reported. Performance of the system designed is compared with and without reconfigurable architecture. The proposed system can be easily adapted for various communication systems and has facility to enhance the system by adding additional encoding and decoding schemes as reconfigurable modules.

ACKNOWLEDGEMENT

This work is supported by the Naval Research Board (NRB) vide project grant No. 315. and funded by the Ministry of Defence, Defence Research and Development Organization (DRDO), India. The authors thank the technical team of CoreEL Technologies, Bangalore for their support.

REFERENCES

[1] Tadao Kasami, "A Decoding Procedure for Multiple-Error-Correcting Cyclic Codes", IEEE Transactions on Information Theory, Vol.10, Issue 2, April 1964, pp 134-138

[2] David Romero-Laorden, Oscar Martinez-Graullera, Carlos Julián Martín-Arguedas and Montserrat Parrilla-Romero, "Application of Golay codes to improve SNR in coarray based synthetic aperture imaging systems", Seventh IEEE Sensor Array and Multichannel Signal Processing Workshop 2012, Hoboken, Nw Jersey, pp. 325-328.

[3] Yi Hua Chen, Jue Hsuan Hsiao, Pang-Fu Liu and Jheng-Shyuan He, "Golay (20, 8) C Code Simulation and Implementation in DSP Chip", International Conference on Consumer Electronics, Communications and Networks (CECNet), Xianning, March 2011, pp. 4990-4994.

[4] Satyabrata Sarangi and Swapna Banerjee, "Efficient Hardware Implementation of Encoder and Decoder for Golay Code", IEEE Transactions on Very Large Scale Integration (VLSI) Systems, Vol. 23, Issue 9, Sept 2015, pp. 1965-1968.

[5] Arunkumar Balasundaram, Angelo Pereira, Jun Cheol Park and Vincent Mooney, "Golay and Wavelet Error Control Codes in VLSI", Technical Report, Georgia Institute of Technology, Atlanta, Georgia, U.S.A December 2003.

[6] Manikandan.J, Jayaraman M and Jayachandran M., "Design of an FPGA-based electronics flow regulator for spacecraft propulsion system", Advances in Space Research, Vol. 47, Issue 3, Feb 2011, pp.488-495.

[7] Mangala J and Manikandan J, "FPGA Implementation of Reconfigurable Modulation System", in Proc. IEEE Int. Conf. on Advances in Computing, Communications and Informatics, Aug 2015, Kochi, pp. 493-500.

[8] Jayachandran M and Manikandan J., "Software Reconfigurable State-of-the-art Communication Suite for Fighter Aircraft", in Proc. IEEE Int. Conf. on Communication systems and Network Technologies, June 2011, Katra, Jammu, pp. 729-733.

[9] Samuel MacMullan, "Reconfigurable Encoder and Decoder for High-Data-Rate Satellite Communications", Department of Defense, 2011. https://www.sbir.gov.

[10] Vennila C, Alok Kunar Patel, Lakshminarayanan G and Seok Bum Ko, "Dynamic partial reconfigurable Viterbi decoder for wireless standards", Computers and Electrical Engineering, Vol. 39, Issue 2, Feb 2013, pp.164-174.

[11] Shu Lin, Daniel J Costello, "Error Control Coding", Pearson Education Inc., Second Edition, 2011.

[12] Czeslaw Koscielny, "Extended(24,12) Binary Golay Code: Encoding and Decoding Procedures", Application Note Maplesoft Inc., 2006.

IEEE
445 Hoes Lane
Piscataway, NJ 08854-4141

ISBN 978-1-5090-0037-1